A way to understanding an alternative future. – **Juliet Schor**, Sociology, Boston College

A book of dazzling breadth, provocative and persuasive scholarship.
– **Sylvia Marcos**, Mexican feminist activist and scholar

For too long the North has imposed its one-size-fits-all agenda on the South.
– **Dan O'Neill**, economist, University of Leeds

This *Dictionary* charts pathways for transition to an ecologically sane, politically more egalitarian, and socially more inclusive world.
– **Erik Swyngedouw**, geographer, University of Manchester

A real breakthrough in post-development thinking.
– **Gilbert Rist**, Graduate Institute of International and Development Studies, Geneva

It is about a future that died long ago . . . and about the urgency of nurturing the manifold worlds that breathe seditiously.
– **Bayo Akomolafe**, author of *These Wilds Beyond Our Fences*

A delight: stimulating, important.
– **John Holloway**, author of *Change the World Without Taking Power*

May the *Pluriverse* open our minds to what we could not see. . . .
– **Frances Moore Lappé**, founder of the Small Planet Institute

This *Post-Development Dictionary* addresses the systemic crisis we are living in by honouring cultural visions from all over the world.
– **Pablo Solon**, co-author of *Systemic Alternatives*

A wild generosity of ideas marks this book. It is a gift to celebrate and gossip about.
– **Shiv Visvanathan**, Jindal Global University

Calls out the free-market economic delusion that the imperative for survival demands.
– **Mogobe Ramose**, author of *African Philosophy Through Ubuntu*

Pluriverse helps us to re-think our societies and the meaning of being human.
– **Jingzhong Ye**, Humanities, China Agricultural University

A valuable contribution towards building a counter-epistemic community.
– **Debal Deb**, author of *Beyond Developmentality*

Development as a solution to global crises has long been criticized but a plethora of alternatives exist. – **Saral Sarkar**, author of *Eco-Socialism or Eco-Capitalism?*

A menu of narratives that supply meaning and nurture hope.
– **Marina Fischer-Kowalski**, University of Natural Resources and Life Sciences, Vienna

Whether you agree with the wisdom of plurality or not, this book will leave you thinking about radical social transformations.
– **Lourdes Beneria**, Regional Planning, Cornell University

This book's magnificent content puts forth real possibilities for building a future where we can live in peace with each other and the planet.
– **Medea Benjamin**, Co-Director, CODEPINK: Women for Peace

This strategic move towards a pluriverse destabilizes the claim to one universal knowledge as disseminated by modernist development. – **Susan Paulson**, University of Florida

There are many alternatives to the domineering, profiteering, globalizing, disempowering 'progress' of the West. – **Richard Norgaard**, author of *Development Betrayed*

Contributions from a multiplicity of thoughtful and creative minds define the path forward. – **David Korten**, author of *When Corporations Rule the World*

In this essential compendium, diverse visionaries offer both answers and inspirations. – **Paul Raskin**, founding president of the Tellus Institute

In a critical time for humanity, this volume fills a need in our knowledge. – **John Foran**, climate activist

An experimental vocabulary in movement about what comes after and beyond the trap of 'development'. – **Verónica Gago**, Universidad Nacional de San Martin, Buenos Aires

This compilation of ideas and practices helps us to rethink development. – **Diana Gómez**, anthropologist, Universidad de los Andes, Bogotá

A much welcome contribution to debates on development. – **Raquel Gutiérrez Aguilar**, author of *The Rhythm of the Pachakuti*

A look at the remarkable spectrum of experiences, proposals and radical knowledge that challenge the contemporary crisis of civilization. – **Edgardo Lander**, Venezuelan Central University, Caracas

Homogenization is the name of our civilizational malaise. – **Antonio Elizalde**, Director, *Polis: Latin American Journal of Sustainability*, Chile

The time is overdue to unsettle the cognitive supremacy of the West. – **Nina Pacari**, Kechwa indigenous leader

The spatial-temporal weaving of the *Pluriverse* inscribes each of our bodies with unique synchronicities as we participate in the existential cycles of living matter. – **Raúl Prada Alcoreza**, Bolivian writer, demographer, member of Comuna

This search enables us to bring many 'dispersed strengths' into a single ray of light illuminating the analysis and processes of change. – **Gioconda Belli**, Nicaragua

A verse is needed to express a wish, to push for change, to eradicate injustices. All of these verses and more are gathered in this book. – **Gustavo Duch**, Catalonian writer, food sovereignty activist, horticulturist apprentice

Through 'the cracks', people been able to build life-alternatives to extractive models and construct new worlds characterized by non-capitalist forms of life. – **Raúl Zibechi**, Uruguayan writer, popular educator and journalist

Absolutely thrilling. Despite the limit of 1000 words, each entry in this *Dictionary* manages to exhibit an amazing capacity for synthesis and creativity on the authors' part. – **Jûrgen Schuldt**, economist, Universidad del Pacífico, Lima

---

* View detailed comments at www.thepluriverse.org.

PLURIVERSE

A POST-DEVELOPMENT DICTIONARY

Edited by
ASHISH KOTHARI, ARIEL SALLEH, ARTURO ESCOBAR,
FEDERICO DEMARIA, ALBERTO ACOSTA

Tulika Books

First published in India in 2019 by
Tulika Books
44 (first floor), Shahpur Jat, New Delhi 110 049, India
www.tulikabooks.in

and

Authorsupfront publishing services private limited
authorsupfront.com

ISBN: 978-81-937329-8-4

eBook available on Amazon Kindle

Cover design: Neena Gupta
Cover illustration: Ashish Kothari

Many words are walked in the world. Many worlds are made. Many worlds make us. There are words and worlds that are lies and injustices. There are words and worlds that are truthful and true. In the world of the powerful there is room only for the big and their helpers. In the world we want, everybody fits. The world we want is a world in which many worlds fit. [...] Softly and gently we speak the words which find the unity which will embrace us in history and which will discard the abandonment which confronts and destroys us. Our word, our song and our cry, is so that the dead will no longer die. We fight so that they may live. We sing so that they may live.

– 'Fourth Declaration of the Lacandon Jungle' (1996)
Zapatista National Liberation Army
(Ejército Zapatista de Liberación Nacional, EZLN)

Dedicated to all those who struggle for the pluriverse, resisting injustice and seeking paths for living in harmony, with nature.

CONTENTS

FOREWORD
The Development Dictionary Revisited

Wolfgang Sachs

'The idea of development stands like a ruin in the intellectual landscape.' That's what we wrote about twenty-five years ago in 1993 in the Introduction to the *Development Dictionary*. Happy and a bit naïve, sitting on the porch of Barbara Duden's house near Pennsylvania State University in the fall of 1988, we proclaimed the end of 'the development era'. Between pasta, red wine, and onion rings, among sleeping bags, one or two personal computers and many columns of books, we began to draw up the outline of a handbook that was to expose the idea of development.

Let's remember: in the second half of the twentieth century the notion of development stood like a powerful ruler over nations. It was the geopolitical programme of the post-colonial era. Since all seventeen of us authors, coming from four continents, had grown up with the concept of development, we wanted to rid ourselves of the deeply embedded convictions of our post-war fathers. We understood that the concept had prepared the path for Western imperial power over the world. Moreover, we felt – more than we knew – that development led into a cul-de-sac, the consequences of which would hit us in the form of injustice, cultural turmoil, and ecological decline. In sum, we had realized that the idea of development had taken a direction not uncommon in the history of ideas: what once was a historical innovation became a convention over time, one that would end in general frustration. Our *spiritus mentor*, Ivan Illich, who was amongst us, commented that this idea would fit wonderfully into an archeology of modernity, which he planned to write. Even then he was of the opinion that one should talk about development with the gesture of an obituary.

Flashback

When did the era of development start? In our *Development Dictionary* we identified President Harry S. Truman as the villain. Indeed, on 20 January 1949, in his inaugural address he referred to more than half of the world's population as coming from 'underdeveloped areas'. It was the first time that the term 'underdevelopment', which would later become a key category for the justification of power, both international and national, was proclaimed from a prominent political stage. This speech opened the era of development – a period of world history, which followed the colonial era, only to be replaced some forty years later by the epoch of globalization. And today there are clear indicators that globalization might be replaced by an age of populist nationalisms.

What constitutes the idea of development? We should consider four aspects. Chrono-politically, all nations seem to advance in the same direction. Imagined time is linear, moving only forwards or backwards; but the aim of technical and economic progress is fleeting. Geopolitically, the leaders of this path, the developed nations, show the straggling countries which way to go. The bewildering variety of peoples in the world is now ranked simplistically as rich and poor nations. Socio-politically, the development of a nation is measured through its economic performance, according to gross domestic product. Societies that have just emerged from colonial rule are required to place themselves in the custody of 'the economy'. And finally, the actors who push for development are mainly experts of governments, multinational banks, and corporations. Previously, in Marx's or Schumpeter's time, development was used for an intransitive subject, like a flower that seeks maturity. Now the term is used transitively as an active reordering of society that needs to be completed within decades, if not years.

While we were ready to sing goodbye to the era of development, world history did not follow suit. On the contrary, the idea received a further boost. Just as the first drafts of our *Dictionary* were ready, in November 1989, the Berlin Wall came down. The Cold War was over and the epoch of globalization began. The gates for transnational market forces reaching into the Earth's furthest corners were thrown wide open. The nation-state became porous; the economy as well as culture was increasingly determined by global forces. Development, erstwhile a task of the state, was now de-territorialized. Transnational corporations spread out and on every continent lifestyles aligned with one another: SUVs replaced rickshaws; cell phones superseded community gatherings; air-conditioning supplanted siestas. Globalization can be understood as development without nation-states. The global middle classes – white or black, yellow, or brown – have profited the most. They shop in similar malls, buy high-tech electronic goods, watch the same movies and TV series. As tourists, they freely dispose of the decisive medium of alignment: money. Roughly speaking, already by the year 2010, half of the global middle class lived in the North and the other half in the South. Without doubt, this has been the great success of 'development thinking', yet it is a failure waiting to happen.

Downfall

'Development' is a plastic word, an empty term with positive signification. Nevertheless, it has maintained its status as a perspective, because it is inscribed in an international network of institutions from the United Nations to NGOs. After all, billions of people have made use of the 'right to development', as it is stated in the resolution of the 1986 UN plenary assembly. We, the authors of the *Development Dictionary*, were impatient to proclaim the end of the era of development; we did not anticipate that the political coma would last for decades. Yet we were right – even if we imagined it unfolding differently.

The downfall of the development idea is now obvious in the UN Agenda 2030 programme for Sustainable Development Goals (SDGs). Long gone is the time when development meant 'promise'. Back then, the talk was of young, aspiring nations moving along a path of progress. Indeed, the discourse of development held a monumental historical promise: that in the end, all societies would close the gap with the rich and partake in the fruits of industrial civilization. That era is over: everyday life is more often about survival now, not progress. While the politics of fighting poverty has been successful in some places, it has been bought at the price of even larger inequalities elsewhere; and at the price of irreparable environmental damage. Not least, global warming and the erosion of biodiversity have cast doubt on the faith that developed nations are the pinnacle of social evolution. On the contrary, progress has turned out to be regress, as the capitalist logic of the global North cannot but exploit nature. From the 'Limits to Growth' in 1972 to 'Planetary Boundaries' in 2009, the analysis is clear: development-as-growth leads to unsustainability of the planet Earth for humans. The SDGs – carrying development in their title – are a semantic deception. The Sustainable Development Goals should really be called SSGs – Sustainable Survival Goals.

The geopolitics of development has also imploded. What happened to the imperative of 'catching up', so fundamental to the idea of development? Here it is worthwhile quoting a passage from the document that announced the SDGs: 'This is an Agenda of unprecedented scope and significance . . . These are universal goals and targets which involve the entire world, developed and developing countries alike'. You cannot express the mind shift more clearly: 'the geopolitics of development', according to which industrial nations would be the shining example for poorer countries, has been disposed of. Just as the Cold War era faded in 1989, the myth of catching up evaporated in 2015. Rarely has a myth been buried so quietly. What point is there in development, if there is no country that can be called 'sustainably developed'? Apart from that, the economic geography of the world has changed. Geopolitically speaking, the rapid ascension of China as the largest economic power on Earth has been spectacular. The seven most important newly industrialized countries are now economically stronger than the traditional industrial states, although the G-7 still pretends to be the hegemon. Globalization has almost dissolved the established North–South scheme.

Furthermore, development has always been a statistical construct. Without the magic number, Gross Domestic Product (GDP), it was impossible to come up with a ranking for nations of the world. Comparing income was the point of development thinking. Only in this way could relative poverty or wealth of a country be determined. Since the 1970s, however, a dichotomy emerged in the discourse of development, juxtaposing the idea of development-as-growth to the idea of development-as-social policy. Institutions such as the World Bank, the International Monetary Fund (IMF), and the World Trade Organization (WTO) continued to bow to the idea of development-as-growth, while the United Nations Development Programme (UNDP), United Nations Environment Programme (UNEP), and most NGOs emphasized the idea of development-as-social policy. Thus the term 'development' became a catch-all phrase. The SDGs come out of this tradition. Economic growth is no longer the aim, but the reductionism of development thinking does not disappear so easily. Instead of GDP numbers we now have social indicators – nutrition, health, education, environment – in order to map a country's performance. The data allow comparison, and comparison constructs deficits along a timeline, just as between groups and nations. Reducing deficits in the world has been the aim of development for the last 70 years. In that sense the Human Development Index, not unlike the GDP, is a deficit index; it classifies countries hierarchically and thereby makes the assumption that there is only one kind of social evolution. This is how development thinking reveals its secret: it lives by the dictatorship of quantitative comparison.

Outlook

In the same year that our *Development Dictionary* was published, another book was all the rage: Francis Fukuyama's *The End of History*. This marked the atmosphere of the time: the triumph of the West with its democracy and industrialized living conditions. Twenty-five years later, in 2018, none of its promises have materialized. On the contrary, disarray, even chaos reign, fear and anger are widespread contrasting sharply with the triumphalism of the 1990s. If one had to find a word for the current atmosphere in the northern, as well as parts of the southern hemisphere, it would be: fear of the future, a fear that life prospects are shrinking and that children and grandchildren will be less well off. A suspicion spreads among the global middle class that the expectations kindled by development are not going to be fulfilled. Alienated from their traditions, aware of Western living styles through their smart phones, yet excluded from the modern world, this is the fate of too many people, and not only in poor countries. Thus cultural confusion and ecological crises fuel fear of the future.

In this way, the expansive modern age has got stuck and it is time to exit. At a glance, three narratives can be identified that respond to fear of the future: the narratives of 'the fortress', 'globalism', and 'solidarity'. 'Fortress thinking' expressed through neo-nationalism revives the glorious past of an imagined people. Authoritarian leaders bring back pride; while others are scapegoated – from Moslems to the UN. This leads to hatred of foreigners, sometimes coupled with religious fundamentalism. A kind of 'affluence chauvinism' is widespread, particularly in the new middle classes whose material goods need to be defended against the poor. By contrast, in 'globalism' we find the image of the planet as an archetypical symbol. Instead of the fortress mercantilism of 'America first', globalists promote an ideally deregulated, free-trade world, which is meant to bring wealth and well-being to corporations and consumers everywhere. The globalized liberal elite may also feel a fear of the future, but such difficulties can seemingly be overcome with 'green and inclusive growth' and smart technologies.

The third narrative – 'solidarity' – is different. Fear of the future calls for resistance against the powerful, the guarantors of an everyone-for-himself society and capitalist pursuit of profit. Instead, human rights – collective and individual – and ecological principles are valued; market forces are not an end in themselves, but means to an end. As expressed in the slogan 'think globally, act locally', a cosmopolitan localism is nurtured whereby local politics must also take into account wider needs. This means phasing out the imperial way of life that industrial civilization demands, and redefining forms of frugal prosperity. In the words of Pope Francis, currently one of the most important heralds of solidarity with his encyclical *Laudato Si'*:

> We know how unsustainable is the behaviour of those who constantly consume and destroy, while others are not yet able to live in a way worthy of their human

dignity. That is why the time has come to accept degrowth in some parts of the world, in order to provide resources for other places to experience healthy growth (§193 of *Laudato Si'*).

I feel that this *Post-Development Dictionary* is squarely rooted in the narrative of solidarity. The one hundred entries elucidate many paths to a social transformation that places empathy with humans and non-human beings first. These visions stand firmly in opposition to both xenophobic nationalism and technocratic globalism. It is deeply encouraging that the theory and practice of solidarity, as already witnessed in the geographic diversity of these *Post-Development Dictionary* authors, appears to have reached all corners of the world.

Translated from the German by Ulrich Oslender

Further Resources

His Holiness, Pope Francis (2015), *Laudato Si': On Care for Our Common Home*,
    http://w2.vatican.va/content/francesco/en/encyclicals/documents/
    papafrancesco_20150524_enciclica-laudato-si.html.
Illich, Ivan (1993), *Tools for Conviviality*. New York: Harper & Row.
Mishra, Pankaj (2017), *Age of Anger: A History of the Present*. London: Allen Lane.
Raskin, Paul (2016), *Journey to Earthland: The Great Transition to Planetary
    Civilization*. Boston: Tellus.
Sachs, Wolfgang (ed.) (2010 [1992]), *The Development Dictionary: A Guide to
    Knowledge as Power*. London: Zed Books.
Speich-Chassè, Daniel (2013), *Die Erfindung des Bruttosozialprodukts: Globale
    Ungleichheit in der Wissensgeschichte der Ökonomie*. Göttingen: Vandenhoeck &
    Ruprecht.

PREFACE

This book invites readers to join in a deep process of intellectual, emotional, ethical, and spiritual decolonization. Our shared conviction is that the idea of 'development as progress' needs to be deconstructed to open a way for cultural alternatives that nurture and respect life on Earth. The dominant Western development model is a homogenizing construct, one that has usually been adopted by people across the world under material duress. The counter-term 'post-development' implies a myriad of systemic critiques and ways of living. This *Dictionary* is intended to re-politicize the ongoing debate over socio-ecological transformation by emphasizing its multi-dimensionality. It can be used for teaching and research; to inspire movement activists; to initiate the curious, and even those in power who no longer feel at ease with their world.

The book is by no means a first on the theme of post-development. *The Development Dictionary* edited by Wolfgang Sachs, now celebrating twenty-five years of publication, set the trend. Others include *Encountering Development* by Arturo Escobar, *The History of Development* by Gilbert Rist, and *The Post-Development Reader* edited by Majid Rahnema and Victoria Bawtree. Feminist contributions include Vandana Shiva's *Staying Alive: Women, Ecology and Development* and *The Subsistence Perspective* authored by Veronika Bennholdt-Thomsen and Maria Mies. In addition to these, the work of activist-scholars such as Ashis Nandy, Manfred Max-Neef, Serge Latouche, Gustavo Esteva, Rajni Kothari, and Joan Martinez Alier, has gone a long way in drawing the contours of a post-development future.

What has been missing is a broad transcultural compilation of concrete concepts, worldviews, and practices from around the world, challenging the modernist ontology of universalism in favour of a multiplicity of possible worlds. This is what it means to call for a pluriverse. The idea to put together a compilation such as this was first discussed by three of us – Alberto Acosta, Federico Demaria, and Ashish Kothari – at the Fourth International Degrowth Conference in Leipzig, 2014. A year later, Ariel Salleh and Arturo Escobar joined the project, planning started in earnest and there are now over a hundred entries. We are conscious of thematic and geographical gaps, but offer the book as an invitation to explore what we see as relational 'ways of being'. This means remaking politics in a way that is deeply felt. Just so, in editing this book – as in any act of care – we ourselves have encountered the limits of our own cultural reflexivity, even vulnerabilities, and in turn, discovered new understandings and acceptance. The 'personal is political', as feminists say.

The book speaks to a worldwide confluence of economic, socio-political, cultural, and ecological visions. Each essay is written by someone who is deeply engaged with the world-view or practice described – from indigenous resisters to middle-class rebels. We would like to acknowledge the passion and commitment of these authors, most of whom immediately agreed to contribute. They had a rather short time to deliver. They were patient with our editorial comments, the back-and-forth that is an inevitable part of trying to achieve accessibility and consistency. A number of essays were written in languages other than English. We appreciate the role of our translators Aida Sofia Rivero S., Arturo Escobar, Eloisa Berman, Iván D. Vargas Roncancio, Kristi Onzik, Laura Gutierrez, Louise Durkin and Melanie M. Keeling in generating faithful English versions of these essays. We are also thankful to Frank Adloff for his support.

Warm thanks are due to the team at Kalpavriksh in Pune, India, as well, especially Shrishtee Bajpai and Radhika Mulay, for helping keep track of the essays, and even providing critical inputs to some essays. Special thanks to our friends and colleagues Dianne Rocheleau and Susan Paulson for their insightful comments upon an earlier version of the manuscript. We are especially grateful to Joan Martinez Alier and Marta Viana at the Institute for Environmental Sciences and Technology (ICTA). These colleagues from the Autonomous University of Barcelona enabled our mid-2017 editorial meeting in their vibrant city with financial backing from the EnvJustice project (ERC 695446). Finally, we acknowledge the enthusiastic support of our publishers and house editors at AuthorsUpFront and Tulika Books in New Delhi, who helped steer the work to completion. Since the beginning of this project we had no doubt that the book should be published from the global South, and launched in Creative Commons.

The *Dictionary* is unconventional for its genre in having three parts. These reflect the historical transition that twenty-first century scholars and activists must work in.

I **Development and Its Crises: Global Experiences.** The 'development' concept, already a few decades old, needs to be reassessed as a matter of political urgency. In this first section, a leading scholar-activist from each continent reflects on the idea and its relation to the multiple crises of modernity.

II **Universalizing the Earth: Reformist Solutions.** Here we present a range of innovations devised mostly in the global North, and often promoted as progressive 'crisis solutions'. A critical review of their rhetoric and practice exposes internal incoherencies, and suggests that they are likely to become ecologically wasteful profit-making distractions.

III **A People's Pluriverse: Transformative Initiatives.** This main section of the book is a compendium of worldviews and practices, old and new, local and global, emerging from indigenous, peasant and pastoral communities, urban neighbourhoods, environmental, feminist, and spiritual movements. They reach for justice and sustainability in a multiplicity of ways.

The visions and practices contained in this *Dictionary* are not about applying a set of policies, instruments and indicators to exit 'maldevelopment'. Rather, they are about recognizing the diversity of people's views on planetary well-being and their skills in protecting it. They seek to ground human activities in the rhythms and frames of nature, respecting the interconnected materiality of all that lives. This indispensible knowledge needs to be held safe in the commons, not privatized or commodified for sale. The visions and practices offered here put *buen vivir* before material accumulation. They honour cooperation rather than competitiveness as the norm. They see work in pleasurable livelihoods, not 'deadlihoods' to escape from on weekends or ecotouristic vacations. Again, too often in the name of 'development', human creativity is destroyed by dull, homogenizing education systems.

The entries in this book are tested against criteria such as: Are the means of economic production and social reproduction justly controlled? Are humans relating to non-humans in mutually enhancing ways? Do all people have access to meaningful livelihoods? Is there a just intergenerational distribution of bads and goods? Are traditional or modern discriminations of gender, class, ethnicity, race, caste, and sexuality being erased? Are peace and non-violence infused throughout community life? We assemble this *Dictionary* to help in the collective search for an ecologically wise and socially just world.

We envisage the book as contributing to a journey towards a Global Tapestry of Alternatives[1], strengthening hope and inspiration by learning from each other; strategizing advocacy and action; and building collaborative initiatives. In doing so, we do not underestimate the epistemological, political, and emotional challenges of remaking our own histories. As Mustapha Khayati wrote in *Captive Words* (1966):

> . . . [E]very critique of the old world has been made in the language of that world, yet directed against it . . . revolutionary theory has had to invent its own

terms, to destroy the dominant sense of other terms and establish new meanings
. . . corresponding to the new embryonic reality needing to be liberated. . . .
Every revolutionary praxis has felt the need for a new semantic field and for
expressing a new truth; . . . because *language is the house of power.*[2]

We are with you in struggle!

Ashish Kothari (Pune), Ariel Salleh (Sydney),
Arturo Escobar (North Carolina), Federico Demaria (Barcelona) and
Alberto Acosta (Quito)
March 2019
thepluriverse.org

Notes

[1] For   updates   on   the   Global   Tapestry   of   Alternatives,   see   www.
radicalecologicaldemocracy.org. This emerges from the Indian experience of Vikalp
Sangam (Alternatives Confluence), see www.vikalpsangam.org.

[2] Khayati, Mustapha (1966), 'Captive Words: Preface to a *Situationist Dictionary*',
International Situationists, 10, https://theanarchistlibrary.org/library/mustapha-
khayati-captive-words-preface-to-a-situationist-dictionary.

INTRODUCTION
Finding Pluriversal Paths

Ashish Kothari, Ariel Salleh, Arturo Escobar,
Federico Demaria, Alberto Acosta

There is no doubt that after decades of what has been called 'development', the world is in crisis – systemic, multiple, and asymmetrical; long in the making, it now extends across all continents. Never before did so many crucial aspects of life fail simultaneously, and people's expectations for their own and children's futures look so uncertain. Crisis manifestations are felt across all domains: environmental, economic, social, political, ethical, cultural, spiritual, and embodied. So this book is an act of renewal and re-politicization, where 'the political' means a collaboration among dissenting voices over the kinds of alternative worlds we want to create.

A *Post-Development Dictionary* should deepen and widen an agenda for research, dialogue, and action, for scholars, policymakers, and activists. It should offer a variety of world-views and practices relating to our collective search for an ecologically wise and socially just world. This agenda should investigate the What, How, Who, for Whom, and Why of all that is transformative, as distinct from that which is not.[1] In the transition to 'a post-development world', there will be companions with strategic vision, as well as others with good short-term tactical proposals. Democracy – as a process in permanent radicalization of itself – should speak to all areas of life, starting from the body and moving on to affirm its place in a living Earth Democracy.[2]

The seductive nature of development rhetoric – sometimes known as developmentality or developmentalism[3] – has been internalized across virtually all countries. Even some people who suffer the consequences of industrial growth in the global North accept a unilinear path of progress. Many nations of the global South have resisted attempts at environmental regulation with the charge that the North is preventing the South from reaching its own level of development. The international debate then moves on to 'monetary and technology transfers' from the global North to the South, which, conveniently for the former, does not challenge basic premises of the development paradigm. These terms, 'global North and South', are not geographic designations but have economic and geopolitical implications. 'Global North' therefore may describe both historically dominant nations as well as colonized but wealthy ruling elites in the South. Similarly, for new alter-globalization alliances,[4] 'South' can be a metaphor for exploited ethnic minorities or women in affluent countries, as much as the historically colonized or 'poorer' countries as a whole.[5]

Decades after the notion of development spread around the world, only a handful of countries called 'underdeveloped', or 'developing', or Third World – to use deprecating Cold War terms – really qualify as 'developed'. Others struggle to emulate the North's economic template, and all at enormous ecological and social cost. The problem lies not in lack of implementation, but in the conception of development as linear, unidirectional, material, and financial growth, driven by commodification and capitalist markets. Despite numerous attempts to re-signify development, it continues to be something that 'experts' manage in pursuit of economic growth, and measure by Gross Domestic Product (GDP), a poor and misleading indicator of progress in the sense of well-being. In truth, the world at large experiences 'maldevelopment', even in the very industrialized countries whose lifestyle was meant to serve as a beacon for the 'backward' ones.

A critical part of our problems lies in the conception of 'modernity' itself – not to suggest that everything modern is destructive or iniquitous, nor that all tradition is positive. Indeed, modern elements such as human rights and feminist principles are proving liberatory for many people. We refer to modernity as the dominant worldview emerging in Europe since the Renaissance transition from the Middle Ages to the early modern period, and consolidating towards the end of the eighteenth century. Not least among these cultural practices and institutions has been a belief in the individual as independent of the collective, and in private property, free markets, political liberalism, secularism,[6] and representative democracy. Another key feature of modernity is 'universalism' – the idea that we all live in a single, now globalized world, and critically, the idea of science as the only reliable truth and harbinger of 'progress'.

Among the early causes of this multiple crisis is the ancient monotheistic premise that a father 'God' made the Earth for the benefit of 'his' human children. This attitude is known as anthropocentrism.[7] At least in the West, it evolved into a philosophic habit of pitting humanity against nature, and gave rise to related dualisms such as the divide between subject versus object, mind versus body, masculine versus feminine, civilized versus barbarian. These classic ideological categories legitimize devastation of the natural world, as well as the exploitation of sex-gender, racial, and civilizational differences. Feminists emphasize the 'masculinist culture of domination' carried by these artificial pairs; intellectuals in the global South emphasize their 'coloniality'. The modern colonial capitalist patriarchal world system[8] thus marginalizes and demeans forms of knowing, such as caregiving and non-Western law, science, or economics. This is the prevailing political pattern globally, yet there have been alternative forms in Europe, as well as 'modernities' in Latin America, China, and so forth.

This book encompasses a variety of visions, reaching from the current globalizing development model to non-modern and self-defined alternatives. Many of these radical worldviews would fit into the second or third categories. In giving voice to diversity, we share a conviction that the global crisis is not manageable within existing institutional frameworks. It is historical and structural, demanding a deep cultural awakening and re-organization of relations both within and between societies across the world, as also between humans and the rest of so-called 'nature'. As humans, our most important lesson is to make peace with the Earth and with each other. Everywhere, people are experimenting with how to meet their needs in ways that assert the rights and dignity of Earth and its threatened inhabitants. The search is a response to ecological collapse, land grabs, oil wars, and forms of extractivism, such as agroindustry and plantations of genetically engineered species. In human terms, this theft brings loss of rural livelihoods and urban poverty. Sometimes Western 'progress' gives way or leads to diseases of affluence, alienation, and rootlessness. But people's resistance movements are now taking place across every continent. *The Environmental Justice Atlas* documents and catalogues over 2000 conflicts,[9] proving the existence of an active global environmental justice movement, even if it is not yet a united one.

There is no guarantee that 'development' will resolve traditional discrimination and violence against women, youth, children and intersex minorities, landless and unemployed classes, races, castes, and ethnicities.[10] As globalizing capital destabilizes regional economies, turning communities into wasted lives and refugee populations, some people cope by identifying with the macho power of the political Right. This prioritizes national identity and promises to 'take the jobs back' from the migrant scapegoats. At times, an insecure working-class Left can adopt this stance too, not recognizing the culpability of banks and corporations for their predicament. A drift towards authoritarianism is taking place all over the world, from India to the USA and Europe. The illusion of representative democracy is kept alive by a privileged technocratic class with its neoliberal trajectory of innovation for green growth. There is a fuzzy line between the Right and the orthodox Left when it comes to productivism, modernization, and progress. Moreover, each such ideology builds on eurocentric and masculinist values, so reinforcing the status quo.

Karl Marx reminded us that when a new society is born from inside the old, it drags many defects of the old system along with it. Later Antonio Gramsci would observe of his time that 'the crisis consists precisely in the fact that the old is dying and the new cannot be born; in this interregnum a great variety of morbid symptoms appear'.[11] What these European Left intellectuals did not anticipate was how, today, alternatives are also emerging from the political margins – from both the colonial periphery and the domestic periphery of capitalism. The Marxist analysis remains necessary but it is not sufficient; it needs to be complemented by perspectives such as feminism and ecology, as well as imaginations emanating from the global South, including Gandhian ideals. In a time of transition such as this, critique and action call for new narratives combined with hands-on material solutions. Doing more of the same, but better, or less of the same is not enough. The way forward is not simply to make corporations more accountable or to set up regulative bureaucracies; it is not even a matter of recognizing full citizenship for the 'coloured', 'elderly', 'disabled', 'women', or 'queer' through liberal pluralist policy. Likewise, the conservation of a few 'pristine' patches of nature at the margins of urban capitalism will have little effect on the collapse of biodiversity.

This *Post-Development Dictionary* speaks to a time when the great twentieth-century political models – liberal representative democracy and state socialism – have become incoherent and dysfunctional forms of governance, even if achieving welfare and rights for a few. Accordingly, the book opens with some reflections on the development idea, drawing on the experience of a scholar-activist from each continent – Antarctica aside. These are the voices of Nnimmo Bassey (Africa), Vandana Shiva (Asia), Jose Maria Tortosa (Europe), Phil McMichael (North America), Kirk Huffman (Oceania), and Maristella Svampa (South America).

Following these critics, the *Dictionary* turns to examine the limits of developmentalism as it shapes reformist solutions to global crises. Here we see the ghost of modernity reincarnated in infinite ways, as short-sighted crisis remedies of those in power keep the North–South status quo in place. This section covers, amongst other topics, market-mechanisms, geo-engineering, and climate-smart agriculture, the population question, green economics, reproductive engineering, and transhumanism. An over-arching theme is the much-celebrated political gesture of 'sustainable development'. Of course, even well-intentioned people may inadvertently promote superficial or false solutions to global problems. Then again, it is not so easy to distinguish mainstream or superficial initiatives from 'radical, transformative' ones, in these days when the military-industrial-media complex and greenwashing industry promotions are at their seductive best.

Criticism of industrialization is not new. Mary Shelley (1797–1851), Karl Marx (1818–83) and Mohandas Gandhi (1869–1948), each in their own way, expressed misgivings about it, as have many people's movements throughout the last two centuries. The early twentieth-century debate on sustainability was strongly influenced by the Club of Rome's *Limits to Growth* argument,[12] and official circles have expressed concern about mass production technologies and consumption patterns since the 1972 United Nations Conference on Environment and Development in Stockholm. Regular conferences at a global level would reiterate the mismatch between 'development and environment', with the 1987 report *Our Common Future* bringing it sharply into focus. However, the United Nations (UN) and most nation-state analyses have never included a critique of social structural forces underlying ecological breakdown. The framing has always been on making economic growth and development 'sustainable and inclusive' through appropriate technologies, markets, and institutional policy reform. The problem is that this mantra of sustainability was swallowed up by capitalism early on, and then emptied of ecological content.

In the period from the 1980s onwards, neoliberal globalization advanced aggressively across the globe. The UN now shifted focus to a programme of 'poverty alleviation' in developing countries, without questioning the sources of poverty in the accumulation-driven economy of the affluent global North. In fact, it was argued that countries needed to achieve a high standard of living before they could deploy resources in protecting the environment.[13] So, economic 'growth' was redefined as a necessary step.[14] This watering down of earlier debates on limits opened the way for the ecological modernist 'green economy' concept. The new millennium has seen a plethora of such Keynesian proposals – bio-economy, the Green Revolution for Africa, the Chinese and European promotion of the circular economy, and the 2030 Agenda for Sustainable Development.[15]

At the UN Conference for Sustainable Development in 2012, this hollow sustainability ideology was the guiding framework for multilateral discussions. For some time, the United Nations Environmental Programme (UNEP), together with the corporate sector, some even on the political Left,[16] had been talking enthusiastically about the need for a 'green new deal'. In preparation for Rio+20, UNEP published a report on the Green Economy, defining it 'as one that results in improved human well-being and social equity, while significantly reducing environmental risks and ecological scarcities'.[17] In line with the pro-growth policy of sustainable development advocates, the report conceptualized all living natural forms across the planet as 'natural capital' and 'critical economic assets', so intensifying the marketable commodification of life on Earth. However, opposition from alter-globalization activists was fierce.

The official Rio+20 final declaration advocates economic growth in more than twenty of its articles. This approach is based on a supposed greening of neoclassical economic theory called 'environmental economics', a belief that growth can de-link or decouple itself from nature through dematerialization and de-pollution by what is called 'eco-efficiency'. However, empirical cradle-to-grave and social metabolism studies from ecological economics show that such production has 'dematerialized' in relative terms – using less energy and materials per unit of GDP – but it has not reduced overall or absolute amounts of materials and energy, which is what matters for sustainability. Historically, the only periods of absolute dematerialization coincide with economic recession.[18] The popular idea of 'economic efficiency' falls far short of respecting biophysical limits – in nature and natural resources, in ecosystem assimilative capacity or in planetary boundaries.

The international model of green capitalism carried forward in the declaration, *Transforming Our World: The 2030 Agenda for Sustainable Development*[19] reveals the following flaws in what have become known as the Sustainable Development Goals (SDGs):[20]

- No analysis of how the structural roots of poverty, unsustainability, and multidimensional violence are historically grounded in state power, corporate monopolies, neo-colonialism, and patriarchal institutions
- Inadequate focus on direct democratic governance with accountable decision-making by citizens and self-aware communities in face-to-face settings
- Continued emphasis on economic growth as the driver of development, contradicting biophysical limits, with arbitrary adoption of GDP as the indicator of progress
- Continued reliance on economic globalization as the key economic strategy, undermining people's attempts at self-reliance and autonomy

- Continued subservience to private capital, and unwillingness to democratize the market through worker–producer and community control
- Modern science and technology held up as social panaceas, ignoring their limits and impacts, and marginalizing 'other' knowledge
- Culture, ethics, and spirituality sidelined and made subservient to economic forces
- Unregulated consumerism without strategies to reverse the global North's disproportionate contamination of the globe through waste, toxicity, and climate emissions
- Neoliberal architectures of global governance becoming increasingly reliant on technocratic managerial values by state and multi-lateral bureaucracies.

This 2015 framework of SDGs, now global in its reach, is a false consensus.[21] For instance, it calls for 'sustained economic growth', which enters into contradictions with the majority of the SDGs. If 'development' is seen as a toxic term to be rejected,[22] then 'sustainable development' becomes an oxymoron. More specifically, the degrowth theorist Giorgios Kallis has commented: 'Sustainable development and its more recent reincarnation "green growth" depoliticize genuine political antagonisms between alternative visions for the future. They render environmental problems technical, promising win-win solutions and the impossible goal of perpetuating economic growth without harming the environment.'[23] This is what happens with reformist solutions.

We do not mean to belittle the work of people who are finding new technological solutions to reduce problems, for instance, in renewable energy, nor do we mean to diminish the many positive elements contained in the SDG framework.[24] Rather, our aim is to stress that in the absence of fundamental socio-cultural transformation, technological and managerial innovation will not lead us out of the crises.[25] As nation states and civil societies gear up for the SDGs, it is imperative to lay out criteria to help people make such a distinction.

In counterpoint to the blinders of conventional political reason, the main section of the *Dictionary* gathers together a range of complementary notions and practices that form radical and systemic initiatives.[26] Some of these revive or creatively re-interpret long-standing indigenous worldviews; others come from recent social movements; yet others revisit older philosophies and religious traditions. All of them ask: What is so badly wrong with everyday life today? Who is responsible for it? What would a better life look like, and how do we get there? As feminists for the '*sostenibilidad de la vida*' ask: 'What is a life worth living? And, how can conditions that allow it to happen be met?'[27]

Together, these perspectives compose a 'pluriverse': a world where many worlds fit, as the Zapatistas of Chiapas put it. All people's worlds should co-exist with dignity and peace without being subjected to diminishment, exploitation and misery. A pluriversal world overcomes patriarchal attitudes, racism, casteism, and other forms of discrimination. Here, people re-learn what it means to be a humble part of 'nature', leaving behind narrow anthropocentric notions of progress based on economic growth. While many pluriversal articulations synergize with each other, unlike the universalizing ideology of sustainable development, they cannot be reduced to an overarching policy for administration either by the UN or some other global governance regime, or by regional or state regimes. We envision a world confluence of alternatives, provoking strategies for transition, including small everyday actions, towards a great transformation.

Our project of deconstructing development opens into a matrix of alternatives, from universe to pluriverse. Some visions and practices are already well-known in activist and academic circles. For instance, *buen vivir*, 'a culture of life' with various names throughout South America; *ubuntu*, emphasizing the southern African value of human mutuality; *swaraj* from India, centred on self-reliance and self-governance.[28] This book rests on the hypothesis that there are thousands of such transformative initiatives around the world. Others, less well-known but equally relevant would be *kyosei, minobimaatisiiwin, nayakrishi*, as well as critically reflective versions of major religions including Islam, Christianity, Hinduism, Buddhism, and Judaism. So, too, political visions such as eco-socialism and deep ecology share points of convergence with earlier communal ideals. While many terms have a long history, they reappear in the narrative of movements for well-being, and again, co-exist comfortably with contemporary concepts such as degrowth and ecofeminism.[29]

From North, South, East, or West, each strand in the post-development rainbow symbolizes human emancipation 'within nature'.[30] It is the latter bond that distinguishes our pluriversal project from cultural relativism. As Aldo Leopold would say, 'A thing is right when it tends to preserve the integrity, stability, and beauty of the "biotic community". It is wrong when it tends otherwise.'[31] While making peace with the Earth, another peacemaking goal is linking ancestral and contemporary knowledge together in a process that will demand horizontal and respectful dialogue. That said, there are no blueprints valid for all times and places, just as no theory is immune to questioning. Indeed, this kind of historical reflexivity is only now becoming recognized as a terrain of politics. The response to macro-power structures like capital and empire is a well-travelled landscape; what is still largely unexplored is the field of micro or capillary power that feeds everyday violence. Honourable rhetorics of abstract justice, even spiritual paeans to Mother Earth, will not suffice to bring about the changes we want. Building a pluriversal house means digging a new foundation.

Transformative initiatives differ from mainstream or reformist solutions in a number of ways. Ideally, they will go to the roots of a problem. They will question what we have already identified as core features of the development discourse – economic growth, productivism, the rhetoric of progress, instrumental rationality, markets, universality, anthropocentrism, and sexism. These transformative alternatives will encompass an ethic that is radically different from the one underpinning the current system. The entries in this section of the book reflect values grounded in a relational logic; a world where everything is connected to everything else.

There are many paths towards a bio-civilization, but we envisage societies that encompass the following values and more:

- diversity and pluriversality
- autonomy and self-reliance
- solidarity and reciprocity
- commons and collective ethics
- oneness with and rights of nature
- interdependence
- simplicity and enoughness
- inclusiveness and dignity
- justice and equity
- non-hierarchy
- dignity of labour
- rights and responsibilities
- ecological sustainability
- non-violence and peace.[32]

Political agency will belong to the marginalized, exploited, and oppressed. And transformations will integrate and mobilize multiple dimensions, though not necessarily all at one go. One exemplar of that vision might be the set of confluences called Vikalp Sangam taking place in India since 2014.[33] The values advanced by this movement are:

Ecological wisdom, integrity and resilience. Where primacy is given to maintaining eco-regenerative processes that conserve ecosystems, species, functions, and cycles; respecting ecological limits from local to global; and where there is an infusion of ecological ethics into all human activities.

Social well-being and justice. Where fulfilment is physical, social, cultural, and spiritual; there is equity in socio-economic and political entitlements and responsibilities; non-discriminatory relations and communal harmony replace hierarchies based on faith, gender, caste, class, ethnicity, ability, and age; and where collective and individual human rights are ensured.

Direct and delegated democracy. Where consensus decision-making occurs at the smallest unit of settlement, in which every human has the right, capacity, and opportunity to take part; establishing democratic governance by directly accountable delegates in ways that are consensual, respectful, and supportive of the needs and rights of those currently marginalized, for example, young people or religious minorities.

Economic democratization. Where private property gives way to the commons, removing the distinction between owner and worker; where communities and individuals – ideally 'prosumers' – have autonomy over local production, distribution, and markets; where localization is a key principle, with trade built on the principle of equal exchange.

Cultural diversity and knowledge democracy. Where a plurality of ways of living, ideas and ideologies is respected; creativity and innovation are encouraged; and the generation, transmission, and use of knowledge – traditional or modern, including science and technology – is accessible to all.

So where are women – 'the other half' of humanity – in all this? How do we ensure that a post-development pluriverse does not dissolve coloniality while keeping women 'in their place' as the material bearers of everyday life activities? A first step in the anticipation of deep systemic change is to question how both traditional and modern practices and knowledge privilege 'masculinity' and the opportunities that go with it. Originally, the two words – 'economy' and 'ecology' shared the one Greek root – *oikos*, meaning 'our home'. But soon enough, this unity was broken apart as men's self-appointed domination over nature came to include the exploitation of women's energies. Whole civilizations have been built on the sex-gendered control of women's fertility – the quintessential resource for continuance of any political regime. This turned women into 'means' not 'ends', mere chattels, so taking away their standing as full human individuals in their own right.

Ironically, the economy or productive sector, as it is known in the global North, now destroys its very own social and ecological foundations in the reproductive sector. The book contains several entries on aspects of women's challenge to this irrational development ethos – Latin American and Pacific feminisms; PeaceWomen; matriarchies; wages for housework; body politics; gift economies, and ecofeminism. Most of these initiatives are grounded in women's struggles for survival. They link political emancipation with environmental justice, local problems with global structures, usually arguing for sustainable subsistence against linear progress and 'catch up development'.[34] Conversely, mainstream Western feminism tends to be anthropocentric, such that liberal and even socialist feminists may be pacified with the goal of 'equality'. In this way, their politics unwittingly band-aids existing masculinist institutions.

Official UN and government analyses have never included a thorough critique of structural forces underlying ecological breakdown. Similarly, the deep structure of ancient patriarchal values carried forward by global developments remains unexamined. Known as 'the longest revolution', the liberation of women from social domination by men will be no easy matter. Even policy experts, too often conflate the well-being of household or community with the well-being of the breadwinner, ignoring a domestic power hierarchy. In academia, the postmodern tendency to reduce embodied sexual identity to the construct of 'gender' is another unhelpful convention. In the same way, treating 'class, race, and gender' as abstract 'intersectional structures' can deflect attention from the raw materiality of lived experience. Formal democratic gestures – the vote or wage equality for women – barely scratch the surface of centuries-old habits of sex-gender oppression.[35] An adherence to spiritual virtues, or strong secular principles such as diversity and solidarity, can help, but does not guarantee an end to the biophysical impacts of sex-gender violence.

Activists seeking just and sustainable alternatives need to acknowledge this unspoken level of political materiality. To varying degrees, women both in the North and the South face silencing and harassment; they lack not only resources, but often freedom of movement. They live with issues such as culturally sanctioned indignities over menstruation, clitoral excision, polygamy, dowry murder, honour killing, suttee, pinching, groping, and now digitized revenge porn. They endure enforced child bearing, domestic violence, marital rape, youth gang rape, genocidal rape as a weapon of war, stigmatiszation as widows and persecution as 'witches' in old age. In the twenty-first century, a combination of female foeticide, privatized violence, and militarized collateral damage on civilian populations is resulting in a falling demographic ratio of women to men globally. In Asia alone, one and a half million women have lost their lives in the last decade due to such factors.

The abuse of children and cruelty to animals are further aspects of the ancient yet widespread patriarchal prerogative over 'lesser' life forms. These activities are a form of extractivism; a gratification through energies drawn from other kinds of bodies, those deemed 'closer to nature'. Following Elizabeth Dodson Gray's pioneering analysis, ecofeminist scholars have offered a deep historical critique of the global capitalist patriarchal order – its religions, economics, and science. In deconstructing the continuing potency of ancient ideological dualisms – humanity over nature, man over woman, boss over worker, white over black, they have shown different forms of social domination to be interrelated.[36] Thus, a 'politics of care' enacted by women from the global North and South converges with the mores of *buen vivir*, *ubuntu* and *swaraj*, because across the hemispheres women's everyday labours teach 'another epistemology', not based on instrumental logic, but 'relational' – like the rationality of ecological processes.[37] In its deepest articulation, these pluriversal voices contest both modernity and traditionalism by locating the material embodiment of class, race, sex-gender and species inside an eco-centric frame. There can be no pluriverse until the historical underpinnings of masculine entitlement are part of the political conversation.

Readers will rightly question the confidence that we, and numerous authors in the *Dictionary*, invest in the idea of 'community'. True, it is a contested term, one that can readily hide oppressions based on sex-gender, age, class, caste, ethnicity, race, or ability. We recognize, too, that 'localized' governance or economies are often xenophobic; a parochialism seen currently in the nationalist opposition to refugees in many parts of the world. Beset by intolerance on the Right, and a defensive 'identity politics' on the Left, our compendium of alternatives reaches for integrative and inclusive practices. Hopefully, life-affirming elements can be discovered even in some of the world's patriarchal religions, and we hope to cultivate that potential.

The ideal of communality envisaged here carries the paradigmatic sense of today's movements towards 'commoning' or *la comunalidad*. As in the case of the initiatives networked by Vikalp Sangam, these collectives are based on autonomous decision-making via face-to-face relations and economic exchange directed at meeting basic needs through self-reliance.[38] Our understanding of community is a critical one: 'in process' and always questioning the modern capitalist patriarchal hegemony of the 'individual' as kernel of society. We hope this book inspires counter movements to that globally colonizing pressure, just as we are inspired, in turn, by cultural groups across the world that still enjoy a collective existence.[39] In this context, Mexican sociologist Raquel Gutiérrez Aguilar proposes a concept of *entramados comunitarios* or communitarian entanglements:

> [T]he multiplicity of human worlds that populate and engender the world under diverse norms of respect, collaboration, dignity, love, and reciprocity, that are not completely subjected to the logic of capital accumulation even if often under attack and overwhelmed by it . . . such community entanglements . . . are found under diverse formats and designs. . . . They include the diverse and immensely varied collective human configurations, some long-standing, others younger, that confer meaning and 'furnish' what in classical political philosophy is known as 'socio-natural space'.[40]

Many radical worldviews and practices in this book make the pluriverse visible. In speaking about them, we enhance their existence and viability. In fact, the very proliferation of assertions coming from these 'other' worlds makes this book possible. Conversely, it is in this sense that the mainstream or reformist development solutions can said to be proven false. In responding to the ecological crisis, 'experts' in the global North take the categories of One World responsible for devastation of the planet as the very point of departure for their alleged solutions! However, their commitment to *la dolce vita* cannot enlighten us on the fundamental task of making the pluriverse sustainable. To repeat: the notion of the pluriverse questions the very concept of universality that is central to Eurocentric modernity. With their phrase 'A world where many worlds fit', the Zapatistas give us the most succinct and apt definition of the pluriverse.

Whereas the West managed to sell its own idea of One World – known only by modern science and ruled by its own cosmovision – the alter-globalization movements propose pluriversality as a shared project based on the multiplicity of 'ways of worlding'. Under conditions of asymmetric power, indigenous peoples have had to alienate their own common sense experience of the world and learn how to live with the Eurocentric masculinist dualism between humans and non-humans, which led them to treat indigenous peoples as non-human and 'natural resources'. They resist this separation when they mobilize on behalf of mountains, lakes or rivers, arguing that these are sentient beings with 'rights', not mere objects or resources. Conversely, many thinking people in the industrialized world are demanding rights for the rest of nature to be expressed in law and policy. In so doing they are taking a step towards incorporating something that indigenous peoples have always integrated into their worldview, but they are doing so in the formal ways that they are familiar with.[41] There is a long way to go for the multiplicity of worlds to become fully complementary with each other, but movements for justice and ecology are finding increasing common ground. So, too, women's political struggles converge upon this same point.

In both the global North and South, it is most often ordinary caregiving mothers and grandmothers who join this entanglement – defending and reconstituting communal ways of being and place-based forms of autonomy. In doing so, they, like the indigenous others described earlier, draw on non-patriarchal ways of doing, being and knowing.[42] They invite participation, collaboration, respect and mutual acceptance, and horizontality; they honour sacredness in the cyclic renewal of life. Their tacitly matriarchal cultures resist ontologies, founded on domination, hierarchy, control, power, the negation of others, violence, and war. From the worldwide movement of Peace Women to African anti-extractivist networks, women are defending nature and humanity with the clear message that there can be no decolonization without de-patriarchalization.

Such initiatives resonate powerfully with the post-development concepts profiled here.[43] For the pluriverse is not just a fashionable concept, it is a practice. Societal imaginaries based on human rights and the rights of nature are impossible to arrive at through top-down intervention. Initiatives such as the Transition Movement or ecovillages can contain a mix of reformist and broader systemic changes. Emancipatory projects will rely on solidarity across continents and they can work hand in hand with resistance movements. This is exemplified by the Yasuní – ITT initiative in Ecuador which urges: 'leave the oil in the soil, the coal in the hole, and the tar sands in the land'.[44] To live according to the insights of multiple partially connected, if radically different, worlds may mean holding traditional and modern certainties and universals at bay in our personal and collective lives. As editors of a *Post-Development Dictionary* we strive to provide some concept-tools and practices for honouring a pluriverse; fostering a bio-civilization that is eco-centric, diverse, and multidimensional, and one that is able to find a balance between individual and communal needs. This living, pre-figurative politics is based on the principle of creating right now the foundations of the worlds we want to see come to fruition in the future; it implies a contiguity of means and ends.

How do we get from here to there? We are, after all, talking about profound shifts in the spheres of economy, politics, society, culture, and lived sexuality! Transitioning implies accepting an ensemble of measures and changes in different domains of life and at different geographic scales. Transitions can be messy and not fully radical, but can be considered 'alternative' if they at least hold a potential for living change. Given the diversity of imaginative visions across the globe, the question of how to build synergies among them remains open. There will be setbacks; strategies will fade along the way and others will emerge. Differences, tensions, even contradictions, will exist, but these can become a basis for constructive exchange. The ways towards a pluriverse are multiple, open, and in continuous evolution.

Notes

[1] For initial thoughts on the *Post-Development Dictionary* agenda, see Demaria and Kothari (2017). For an early attempt to articulate different alternatives to development, see Kothari *et al.* (2015) and Beling *et al.* (2017). The latter discusses discursive synergies for a 'great transformation' towards sustainability among Human Development, Degrowth, and Buen Vivir advocates.

[2] See Shiva, http://www.navdanya.org/earth-democracy.

[3] Nandy (2003: 164–75); Mies (1986); Deb (2009); Shrivastava and Kothari (2012).

[4] For a fundamentally different form of globalization than the currently dominant one, see the essay under alter-globalization in this volume.

[5] Salleh (2006).

[6] Used here in the sense of an anti- or non-spiritual and religious orientation, not in the sense of an orientation that equally respects all faiths and non-faith belief systems.

[7] Or, as Dobson (1995) argues, 'human instrumentalism', since we may all inevitably be a bit human-centred in a neutral way. However, the analysis of ideological dualism as such, is owed to ecofeminist thinker Elizabeth Dodson-Gray (1979).

[8] Grosfoguel and Mielants (2006).

[9] The *Environmental Justice Atlas (EJ Atlas)* collects the stories of struggling communities and is the largest worldwide inventory of such conflicts. It aims to make these mobilizations more visible, highlight claims and testimonies and make a

case for true corporate and state accountability for injustices inflicted through their activities (Martinez-Alier *et al.* 2016; Scheidel *et al.* 2018), see https://ejatlas.org/.

[10] Navas *et al.* (2018).

[11] Gramsci (1971[1930]), pp. 275–76.

[12] Meadows *et al.* (1972).

[13] See, for instance, a presentation by former Indian Prime Minister Manmohan Singh (1991), and a critique of this in Shrivastava and Kothari (2012), pp. 121–22.

[14] Gómez-Baggethun and Naredo (2015).

[15] Salleh (2016).

[16] For example, the New Economics Foundation, London, and Rosa Luxemburg Stiftung, Berlin.

[17] UNEP (2011); Salleh (2012).

[18] Ecological economists have provided significant empirical evidence with their socio-metabolic analyses, which measure the energy and material flows of the economy. For an example, see Krausmann *et al.* (2009) and Jorgenson and Clark (2012). For a discussion of method, see Gerber and Scheidel (2018).

[19] SDSN (2013); UNEP (2011); United Nations Secretary General Panel (2012); United Nations (2013); United Nations (2015).

[20] Adapted from Kothari (2013).

[21] This phenomenon was anticipated in pioneering work by Shiva (1989) and Hornborg (2009).

[22] Dearden (2014).

[23] Kallis (2015).

[24] For a critical but appreciative view of the potential of the SDG framework, see Club de Madrid (2017).

[25] See also http://www.lowtechmagazine.com/about.html.

[26] For earlier contributions: Salleh (2017 [1997]); Kothari *et al.* (2015); Escobar (2015); Beling *et al.* (2018).

[27] The phrase is Spanish for 'sustainability of life': Peréz Orozco (2014).

[28] Gudynas (2011); Metz (2011); Kothari (2014).

[29] Demaria *et al.* (2013); D'Alisa *et al.* (2014); Bennholdt-Thomsen and Mies (1999); Salleh (2017 [1997]).

[30] Salleh (2017 [1997]); Sousa Santos (2009).

[31] Leopold (1949), p. 224.

[32] For an extensive and intensive process of visioning the elements and values of radical alternatives, see the Vikalp Sangam (Alternatives Confluences) process in India, ongoing since 2014: http://kalpavriksh.org/our-work/alternatives/vikalp-sangam/; and the vision note emerging from this at http://www.vikalpsangam.org/about/the-search-for-alternatives-key-aspects-and-principles/.

[33] Adapted from the Vikalp Sangam vision note, at http://www.vikalpsangam.org/about/the-search-for-alternatives-key-aspects-and-principles/.

[34] Bennholdt-Thomsen and Mies (1999).

[35] Wages for women in developed economies stand at approximately 70 percent of the male wage for equivalent work. Men in developed economies spend less than 20 minutes a day with their children. In modern India, only 15 percent of women are in the paid workforce.

[36] Dodson Gray (1979), Merchant (1980), Waring (1987).

[37] Salleh (1997 [2017], 2011, 2012).

[38] For a detailed narrative on the legitimacy of using 'community' and its various derivatives, one acknowledging the contestations, see Escobar (2010, 2014).

[39] See http://www.congresocomunalidad2015.org/ for details of the First International Congress on Comunalidad, 2015, convened at Puebla, Mexico, where these issues were discussed at length.

[40] Aguilar (2013), p. 33.

[41] See, for instance, Kauffman and Sheehan (2018); and https://therightsofnature.org.

[42] This ethic should not be read through the lens of liberal ideology; that is, as women's 'essential nature'. It is a learned outcome of the experience of care-giving labours, historically assigned to women across most cultures.

[43] Acosta and Brand (2017).

[44] Acosta (2014).

Further Resources

Acosta, Alberto (2014), 'Iniciativa Yasuní-ITT: La difícil construcción de la utopía'. *Rebelión.* http://www.rebelion.org/noticia.php?id=180285.

Acosta, Alberto and Ulrich Brand (2017), *Salidas del laberinto capitalista. decrecimiento y postextractivismo.* Barcelona: Icaria.

Beling, Adrian, Julien Vanhulst, Federico Demaria, Violeta Rabi, Ana Carballo and Jérôme Pelenc (2018), 'Discursive Synergies for a "Great Transformation" towards Sustainability: Pragmatic Contributions to a Necessary Dialogue between Human Development, Degrowth, and Buen Vivir', *Ecological Economics.* 144: 304–13.

Bennholdt-Thomsen, Veronika and Maria Mies (1999), *The Subsistence Perspective.* London: Zed Books.

Club de Madrid (2017), *A New Paradigm for Sustainable Development?* Summary of the deliberations of the Club de Madrid Working Group on Environmental Sustainability and Shared Societies: http://www.clubmadrid.org/en/publicacion/a_new_paradigm_for_sustainable_development.

D'Alisa, Giacomo, Federico Demaria and Giorgios Kallis (2014), *Degrowth: A Vocabulary for a New Era.* London: Routledge.

Dearden, Nick (2015), 'Is Development Becoming a Toxic Term?', *The Guardian.* 22 January 2015, http://www.theguardian.com/global-development-professionals-network/2015/jan/22/development-toxic-term?CMP=share_btn_tw.

Deb, Debal (2009), *Beyond Developmentality.* Delhi: Daanish Books.

Demaria, Federico, Francois Schneider, Filka Sekulova and Joan Martinez-Alier (2013), 'What Is Degrowth? From an Activist Slogan to a Social Movement', *Environmental Values.* 22 (2): 191–215.

Demaria, Federico and Ashish Kothari (2017), 'The Post-Development Dictionary Agenda: Paths to the Pluriverse', *Third World Quarterly*, 38 (12): 2588–99.

Dobson, Andrew (1995), *Green Political Thought.* London: Routledge.

Dodson-Gray, Elizabeth (1979), *Green Paradise Lost.* Wellesley, MA Roundtable Press.

Escobar, Arturo (1995), *Encountering Development.* Princeton: Princeton University Press.

———— (2010), 'Latin America at a Crossroads: Alternative Modernizations, Postliberalism, or Postdevelopment?', *Cultural Studies*. 24 (1): 1–65.

———— (2011), 'Sustainability: Design for the Pluriverse', *Development*. 54 (2): 137–40.

———— (2014), *Sentipensar con la tierra: Nuevas lecturas sobre sobre desarrollo, territorio y diferencia*. Medellín: Universidad Autonoma Latinoamericana (UNAULA).

———— (2015), 'Degrowth, Post-development, and Transitions: A Preliminary Conversation', *Sustainability Science*. 10.(3): 451–62.

———— (2018), *Designs for the Pluriverse: Radical Interdependence, Autonomy, and the Making of Worlds*. Durham: NC: Duke University Press.

EZLN, Ejército Zapatista de Liberación Nacional (1997), 'Fourth Declaration of the Lacandon Jungle', http://www.struggle.ws/mexico/ezln/jung4.html.

Gandhi, Mohandas Karamchand (1997 [1909]), 'Hind Swaraj', in Anthony Parel (ed.), *M. K. Gandhi: Hind Swaraj and Other Writings*. Cambridge: Cambridge University Press.

Gerber, Julien-Francois and Arnim Scheidel (2018), 'In Search of Substantive Economics: Comparing Today's Two Major Socio-Metabolic Approaches to the Economy – MEFA and MuSIASEM', *Ecological Economics*. 144: 186–94, https://doi.org/10.1016/j.ecolecon.2017.08.012.

Gómez-Baggethun, Erik and Jose Manuel Naredo (2015), 'In Search of Lost Time: the Rise and Fall of Limits to Growth in International Sustainability Policy', *Sustainability Science*. 10 (3): 385–95.

Gramsci, Antonio (1971 [1930]), *Selections from the Prison Notebooks*. New York: International Publishers.

Grosfoguel, Ramón and Eric Mielants (2006), 'The Long-Durée Entanglement between Islamophobia and Racism in the Modern/Colonial Capitalist/Patriarchal World-System: An Introduction', *Human Architecture: Journal of the Sociology of Self-Knowledge*. 5 (1), http://scholarworks.umb.edu/humanarchitecture/vol5/iss1/2.

Gudynas, Eduardo (2011), 'Buen Vivir: Today's Tomorrow', *Development*. 54 (4): 441–47.

Gupte, Manisha (2017), 'Envisioning India Without Gender and Patriarchy: Why Not?' in Ashish Kothari and K. J. Joy (eds), *Alternative Futures: India Unshackled*, edited by. New Delhi: AuthorsUpFront.

Gutiérrez Aguilar, Raquel (2013), 'Pistas reflexivas para orientarnos en una turbulenta época de peligro', in Raquel Gutiérrez, Natalia Sierra, Pablo Davalos, Oscar Olivera, Hector Mongragon, Vilma Almendra, Raul Zibechi, Emmanuel Rozental and Pablo Mamani (eds), *Palabras para tejernos, resistir y transformar en la época que estamos viviendo*. Oaxaca: Pez en el árbol.

Hornborg, Alf (2009), 'Zero-Sum World', *International Journal of Comparative Sociology*. 50 (3/4): 237–62.

Jorgenson, Andrew and Brett Clark (2012), 'Are the Economy and the Environment Decoupling? A Comparative International Study: 1960–2005', *American Journal of Sociology*. 118 (1): 1–44.

Kallis, Giorgios (2015), 'The Degrowth Alternative', Great Transition Initiative, http://www.greattransition.org/publication/the-degrowth-alternative.

Kauffman, Craig and Linda Sheehan (2018), 'The Rights of Nature: Guiding Our Responsibilities through Standards', in Stephen Turner, Dinah Shelton, Jona Razaqque, Owen Mcintyre, and James May (eds), *Environmental Rights: The Development of Standards*. Cambridge: Cambridge University Press.

Kothari, Ashish (2013), 'Missed Opportunity? Comments on Two Global Reports for the Post-2015 Goals Process', Pune, Kalpavriksh and ICCA Consortium, http://www.un-ngls.org/IMG/pdf/Kalpavriksh_and_ICCA_Consortium_-_Post-2015_reports_critique_-_Ashish_Kothari_July_2013.pdf.

———— (2014), 'Radical Ecological Democracy: A Way for India and Beyond'. *Development*. 57: 36–45, http://doi.10.1057/dev.2014.43.

———— (2016), 'Why Do We Wait So Restlessly for the Workday to End and for the Weekend to Come?', *Scroll*. 17 June, https://scroll.in/article/809940/.

Kothari, Ashish, Federico Demaria and Alberto Acosta (2015), 'Buenvivir, Degrowth and Ecological Swaraj: Alternatives to Development and the Green Economy', *Development*. 57 (3): 362–75.

Krausmann, Fridolin, Simone Gingrich, Nina Eisenmenger, Karl-Heinz Erb, Helmut Haberl and Marina Fischer-Kowalski (2009), 'Growth in Global Materials Use, GDP and Population during the 20th Century'. *Ecological Economics*. 68: 2696–705, https://doi.org/10.1016/j.ecolecon.2009.05.007.

Latouche, Serge (2009), *Farewell to Growth*. London: Polity.

Leopold, Aldo (1994), *A Sand County Almanac and Sketches Here and There*. New York: Oxford University Press.

Martinez-Alier, Joan, Leah Temper, Daniela Del Bene and Arnim Scheidel (2016), 'Is There a Global Environmental Justice Movement?', *Journal of Peasant Studies*. 43: 731–55.

Marx, Karl and Frederick Engels (1959 [1872]), *The Manifesto of the Communist Party*. Moscow: Foreign Languages Publishing House.

Meadows, Donella, Dennis Meadows, Jorge Randers and William Behrens (1972), *The Limits to Growth*. New York: Universe Books.

Merchant, Carolyn (1980), *The Death of Nature: Women and the Scientific Revolution*. New York: Harper.

Metz, Thaddeus (2011), 'Ubuntu as a Moral Theory and Human Rights in South Africa', *African Human Rights Law Journal*. 11 (2): 532–59.

Mies, Maria (1986), *Patriarchy and Accumulation on a World Scale*. London: Zed Books.

Nandy, Ashis (2003), *The Romance of the State and the Fate of Dissent in the Tropics*. New Delhi: Oxford University Press.

Navas, Grettel, Sara Mingorria and Bernardo Aguilar (2018), 'Violence in Environmental Conflicts: The Need for a Multidimensional Approach'. *Sustainability Science*. 13 (3): 649–60.

Pérez Orozco, Amaia (2014), *Subversiónfeminista de la economía: Aportesparaun debate sobre el conflicto capital-vida*. Madrid: Traficantes de Sueños.

Rahnema, Majid and Victoria Bawtree (1997), *The Post-Development Reader*. London: Zed Books.

Rist, Gilbert (2003), *The History of Development: From Western Origins to Global Faith*. London: Zed Books.

Sachs, Wolfgang (ed.) (2010 [1992]), *The Development Dictionary: A Guide to Knowledge as Power*. London: Zed Books.

Salleh, Ariel (2006), 'We in the North are the Biggest Problem for the South: A Conversation with Hilkka Pietila', *Capitalism Nature Socialism*. 17 (1): 44–61.

————— (2011), 'Climate Strategy: Making the Choice between Ecological Modernisation or "Living Well"', *Journal of Australian Political Economy*. 66: 124–49.

————— (2012), 'Green Economy or Green Utopia? Rio+20 and the Reproductive Labor Class', *Journal of World Systems Research*. 18 (2): 141–145.

————— (2016), 'Climate, Water, and Livelihood Skills: A post-development reading of the SDGs', *Globalizations*. 13 (6): 952–59.

————— (2017 [1997]), *Ecofeminism as Politics: nature, Marx, and the postmodern*. London: Zed Books.

Scheidel, Arnim, Leah Temper, Federico Demaria and Joan Martínez-Alier (2018), 'Ecological distribution conflicts as forces for sustainability: an overview and conceptual framework', *Sustainability Science*. 13 (3): 585–98.

SDSN, Sustainable Development Solutions Network (2013), *An Action Agenda for Sustainable Development*, Report for the UN Secretary General, Sustainable Development Solutions Network, United Nations, http://unsdsn.org/wp-content/uploads/2013/06/140505-An-Action-Agenda-for-Sustainable-Development.pdf.

Shelley, Mary (2009 [1818]). *Frankenstein*. London: Penguin.

Shiva, Vandana (1989), *Staying Alive: Women, Ecology and Development*. London: Zed Books.

Shrivastava, Aseem and Ashish Kothari (2012), *Churning the Earth: The Making of Global India*. Delhi: Viking/Penguin.

Singh, Manmohan (1991), 'Environment and the New Economic Policies', Foundation Day Lecture, Society for Promotion of Wastelands Development, Delhi, 17 June.

Sousa Santos, Boaventura de (2009), 'A Non-Occidentalist West? Learned Ignorance and Ecology of Knowledge', *Theory, Culture and Society*. 26 (7–8): 103–25.

UNEP (2011), *Towards a Green Economy: Pathways to Sustainable Development and Poverty Eradication: A Synthesis for Policy Makers*. Nairobi: United Nations Environment Programme, www.unep.org/greeneconomy.

United Nations (2013), *A New Global Partnership: Eradicate Poverty and Transform Economies Through Sustainable Development*. The Report of the High-Level Panel of Eminent Persons on the Post-2015 Development Agenda. New York: United Nations.

————— (2015), *Transforming Our World: The 2030 Agenda for Sustainable Development*. New York: United Nations, https://sustainabledevelopment.un.org/post2015/transformingourworld/publication.

United Nations Secretary-General's High-level Panel on Global Sustainability. (2012), *Resilient People, Resilient Planet: A Future Worth Choosing*. New York: United Nations.

Waring, Marilyn (1988), *Counting for Nothing*. Sydney: Allen &Unwin.

Ziai, Aram (2015), 'Post-development: Pre-mature Burials and Haunting Ghosts'. *Development and Change*. 46 (4): 833–54.

ABBREVIATIONS

AIMES	African Initiative on Mining Environment and Society
AMV	African Mining Vision
BECCS	BioEnergy with Carbon Capture and Storage
BUKO	Bundeskongress Entwicklungspolitischer Aktionsgruppen
CADA	Andean Centre for Agriculture and Livestock Development
CASA	Council of Sustainable Settlements of the Americas
CDR	Carbon Dioxide Removal
CITES	Convention on International Trade in Endangered Species
CRISPR	Clustered regularly interspaced short palindromic repeats
CSW	Community Seed Wealth
EFF	European Feminist Forum
EITI	Extractive Industries Transparency Initiative
EJOs	Environmental Justice Organizations
ESG	Earth System Governance
EZLN	Ejército Zapatista de Liberación Nacional
FAO	Food and Agriculture Organization
FINRRAGE	Feminist International Network of Resistance for Reproductive and Genetic Engineering
GACSA	Global Alliance for Climate-Smart Agriculture
GATT	General Agreement on Tariffs and Trade
GDP	Gross Domestic Product
GEC	Green Economy Coalition
GGR	Greenhouse Gas Removal
GHG	Greenhouse Gas
GNH	Gross National Happiness
ICC	International Chamber of Commerce
ICCAs	Territories and Areas Conserved by Indigenous Peoples and Local Communities
ICEA	International Confucian Ecological Alliance

ICSU	International Council for Science
IHDP	International Human Dimensions Programme on Global Environmental Change
ILO	International Labour Organization
IMF	International Monetary Fund
IPCC	Inter-Governmental Panel of Experts on Climate Change
ITUC	International Trade Union Confederation
LETS	Local Exchange Trading Schemes
MAB	Movement of Dam Affected People
MAREZ	Municipios Autónomos Rebeldes Zapatistas
MEAs	Multilateral Environment Agreements
NDB	New Development Bank
NEF	New Economics Foundation
NIPTs	Noninvasive Prenatal Tests
NSN	Nayakrishi Seed Network
ODA	Official Development Assistance
OECD	Organisation for Economic Co-operation and Development
OPEC	Organisation of Petroleum Exporting Countries
PES	Payments for Ecosystem Services
PGD	Pre-implantation Genetic Diagnosis
PPPs	Public–Private Partnerships
PRATEC	Andean Project of Peasant Technologies
RED	Radical Ecological Democracy
RIPESS	Intercontinental Network for the Promotion of Social Solidarity Economy
SMEs	Small and Medium-Sized Enterprises
SRM	Solar Radiation Management
TISS	Tata Institute of Social Sciences
UDHR	Universal Declaration of Human Rights
UNEP	United Nations Environment Programme
WTO	World Trade Organization

DEVELOPMENT AND ITS CRISES: GLOBAL EXPERIENCES

BREAKING THE CHAINS OF DEVELOPMENT

Nnimmo Bassey

Keywords: Africa, development, colonialism, climate change

The pursuit of 'development' has promoted butchery on the African continent. The notion that the path to development taken by others is what we must follow is essentially imperialist, used to justify colonialism, neo-colonialism, and neo-liberalism. The fact that this still holds sway is a testament to the resilience of capitalist primitive accumulation. The forces behind these phenomena now promote the enslavement of nature and the enthronement of warfare with ultramodern weapons.

Development, in the linear pattern etched by the global North, is a rigged idea that stacks nations into developed and underdeveloped categories. Development suggests growth, expansion, enlargement, and spread, none of which captures the sense of justice or equity, or considers the ecological limits of a finite planet.

Most African governments have not interrogated the concept of development itself. Political leaders are yet to dissect the fact that the industrialized world got to where it is now through unsustainable exploitation of nature and unjust exploitation of territories and peoples. Thinkers such as Walter Rodney (1973), Chinweizu Ibekwe (1975), and Frantz Fanon (1963) have produced excellent exposés that should have ignited critical introspection. Or, perhaps our leaders are not bold enough to reject an unacceptable pathway, having witnessed how agents of imperial powers assassinated Thomas Sankara of Burkina Faso, Amílcar Cabral of Guinea Bissau, or Patrice Lumumba of Congo – three leaders with alternative notions. Does not the continuous payment of the so-called colonial debt to France by its former colonies in Africa tell of a continent still colonized?

What are the indicators of development in Africa? The first is Gross Domestic Product (GDP), renamed Gross Domestic Problem by Lorenzo Fioramonti (2013). This is measured by the extent of physical infrastructure and the amount of currency reserves. A higher degree of both indicates over-extraction of both nature's resources and human resources. The accumulation of foreign reserves bears testimony to the fact that such resources are stored away to support foreign industries and to pay for imports. When nations are liquid, according to the parameters of the World Bank and the International Monetary Fund (IMF), they are quickly encouraged to obtain foreign loans from the holders of their 'reserves'; and, once they are in dire straits, they are offered extreme conditionalities that they must meet before the noose can be slackened.

The pillage that went with colonialism is often overlooked. Some even see colonialism as a form of aid that helped to shine light on an allegedly 'dark' continent. As a commentator noted, 'The reparations debate is threatening because it completely upends the usual narrative of development. It suggests that poverty in the global South is not a natural phenomenon, but has been actively created. And it casts western countries in the role not of benefactors, but of plunderers' (Hickel 2015). As the writer went on to state, there is not enough money in the world to compensate for the evils of colonialism. Today, besides aid from bilateral relations, there are philanthropist foundations that have taken upon themselves the messianic roles of determining Africa's development path and pattern, which, ironically, is not substantially different from what has already been happening in the name of 'development'.

Today, the climate regime is one arena in which the poor, in both the global North and South, take all the mitigating actions while the rich and powerful compound the problems. Droughts, famines, and water stress grow as governments everywhere neglect sociocultural and ecological realities in pursuit of foreign exchange. To grab more cash, governments buy the carbon offset and market environmentalism baits and loosen environmental and financial controls for transnational corporations. These manifest in land grabs and in the displacement of forest communities, either to make way for export-oriented monoculture plantations, or to lock up forests for carbon stock.

The South's new love of foreign exchange and readiness to be a dumping ground for 'goods', leaves it oblivious to how this harms local production, A proliferation of unjust trade rules, trade-free zones, violent conflicts, and even wars follow. It is instructive to note that all the blood flowing in mineral-rich nations such as the Congo has not stopped mineral extraction. Call them 'blood diamonds' or 'blood crude oil', neither the exploitation of the resources, nor their export has stopped.

The way out is to realize that we are in a rigged game and to become agents of social change. But what is the change we want? At the risk of sounding romantic, one may say that our future may well be found in our past. Africa has to recover her history and affirm that great steps were recorded in the immediate post-colonial years, until the 1980s when the World Bank forced its so-called Structural Adjustment Programmes (SAPs) on her. These were the programmes that disrupted social investments and damaged manufacturing and agricultural productivity. Then, the artificial boundaries that separate our peoples into various nationalities and foist divisive foreign languages on our peoples require interrogations and resolutions. We will also do well to recapture the interconnectivity of our humanity – *ubuntu* – that made the collective the basis of communal organizing.

The 'Africa Rising' narrative may well be another means of blunting critical reflections on the nature of economic relations among our nations (Bond 2013) as well as with the nations of the global North and China. If the 'rise' is based on conventional GDP figures, these do not reflect the objective realities of the citizens because the figures are driven mostly by the export of raw materials from the extractive sector. Yet, Africa must rise! For us to rise, we must hold firmly onto the Earth, which opens our eyes to our contexts and realities, to the forces that have dominated our culture, belief, and thought patterns. That is when we can question the notion of 'development'. That is when we can see the chains around our ankles, and break them. And then we may sound the call for an African awakening.

Further Resources
Bond, Patrick (2013), 'Africa "Rising", South Africa Lifting? Or the Reverse?',
    http://www.dailymaverick.co.za/opinionista/2013-02-06-africa-rising-south-africa-lifting-or-the-reverse/#.U0wu3sdn_R0.
Fanon, Frantz (1963), *The Wretched of the Earth*. New York: Grove Press.
Fioramonti, Lorenzo (2013), *Gross Domestic Problem: The Politics behind the World's Most Powerful Number*. London/New York: Zed Books.
Hickel, Jason (2015), 'Enough of Aid: Let's Talk Reparations',
    https://www.theguardian.com/global-development-professionals-network/2015/nov/27/enough-of-aid-lets-talk-reparations.
Ibekwe, Chinweizu (1975), *The West and the Rest of Us: White Predators, Black Slaves and the African Elite*. New York: Random House.
Rodney, Walter (1973), *How Europe Underdeveloped Africa*. Dar Es Salaam: Tanzanina Publishing House.

Nnimmo Bassey is Director of the ecological think-tank Health of Mother Earth Foundation (HOMEF) based in Nigeria. He chaired Friends of the Earth International from 2008 to 2012. His books include *To Cook a Continent: Destructive Extraction and the Climate Crisis in Africa* (Pambazuka Press, 2012) and *Oil Politics: Echoes of Ecological War* (Daraja Press, 2016).

DEVELOPMENT — FOR THE 1 PER CENT

Vandana Shiva

Keywords: global crises, capitalist patriarchal logic, economic violence, poverty, *oikos*

We need to move beyond the discourse of 'development' and Gross Domestic Product (GDP) as shaped by capitalist patriarchal thinking, and reclaim our true humanity as members of the Earth family. As Ronnie Lessem and Alexander Schieffer write:

> [I]f the fathers of capitalist theory had chosen a mother rather than a single bourgeois male as the smallest economic unit for their theoretical constructions, they would not have been able to formulate the axiom of the selfish nature of human beings in the way they did. (2010: 124)

Capitalist patriarchal economies are shaped through war and violence – wars against nature and diverse cultures, and violence against women. And while the objective is to own and control the real wealth that nature and people produce, there is an increasing replacement of material processes with economic fictions such as 'the logic' of competitive markets.

Separation is the key characteristic of paradigms emerging from the convergence of patriarchal values and capitalism. First, nature is separated from humans; then, humans are separated on the basis of gender, religion, caste, and class. This separation of what is interrelated and interconnected is the root of violence – first in the mind, then in everyday actions. It is not an accident that social inequalities of the past have taken a new and brutal form with the rise of corporate globalization. It is often observed today that by current trends, 1 per cent of the global population will soon control as much wealth as the remaining 99 per cent.

Today, corporations claim legal personhood over the rights of real people. But the distancing of fictitious constructs from real sources of wealth creation has gone even further. Finance is now substituted for capital, with tools and technologies that allow the rich to accumulate wealth as 'rentiers' while doing nothing. Money making in the financial economy is based on speculation. And financial deregulation lets the rich speculate using other people's hard-earned wages. The idea of 'growth' has emerged as the measure of success among individuals and governments. It speaks of a paradigm designed by capitalist patriarchal Big Money, just for Big Money to grow bigger.

What the paradigm of economic growth fails to take cognizance of is the destruction of life in nature and society. Both ecology and economics are derived from the Greek word *oikos* which means 'home', and both words imply a form of household management. When economics works against the science of ecology, it results in mismanagement of the Earth, our home. The climate crisis, the water crisis, the biodiversity crisis, the food crisis are each different symptoms of mismanagement of the Earth and her resources. People mismanage the Earth and destroy her ecological processes by not recognizing nature as the 'real capital' and 'source' of everything else derived from it. Without nature and her ecological processes to sustain life on Earth, the grandest economies collapse and civilizations disappear.

Under the contemporary neoliberal development model, the poor are poor because the 1 per cent has grabbed their livelihood resources and wealth. We see this today in the displacement of both the Rojava communities in the Middle East and Rohingya peoples of Myanmar. Peasants are getting poorer because the 1 per cent promotes an industrial agriculture based on purchase of costly seeds and chemical inputs. This traps them in debt and destroys their soil, water, biodiversity, and their freedom. My book, *Earth Democracy* (2005), describes how Monsanto corporation monopolized the cotton seed supply through hyped-up marketing of engineered Bt cotton. Often forced into debt by purchase of these expensive GMO (Genetically Modified Organism) seeds and other so-called Green Revolution technologies, some 300,000 Indian farmers have committed suicide over the past two decades, with most suicides concentrated in the cotton belt. I have started a rural research farm called Navdanya to resist these violent monopolies. We save the farmers' own traditional varieties of organic cotton to distribute in a movement for Seed Freedom.

If farmers are getting poorer, it is because the Poison Cartel – now reduced to three players: Monsanto Bayer, Dow Dupont, and Syngenta Chem China – makes them dependent on buying costly seeds and chemicals. Vertically integrated corporations, linking seed to chemicals to international trade to processing of junk food, are stealing 99 per cent of the value that farmers produce. They are getting poorer because 'free trade' promotes dumping, destruction of livelihoods and depression of farm prices. Beyond this, small farmers are actually more productive than large industrial corporate farms, without using environmentally damaging commercial additives such as fertilizers, pesticides, and genetically engineered seeds. By contrast, the global peasant union, Via Campesina, points out that traditional ways of provisioning not only allow more autonomy for farmers, but can even mitigate the effects of global warming.

It goes without saying that the 'growth economy' of the 1 per cent is deeply anti-life, and many of these effects are felt by working people in the global North as well. The Filipino peoples' NGO, IBON International, affirms that if masculine violence was used traditionally to keep women exploitable both as productive workers and reproductive bodies, now masculine violence works in the service of capitalist profit making. People everywhere are getting poorer because governments captured by the 1 per cent impose profit-making privatization policies for health and education, transport and energy, reinforced by World Bank and IMF mandates. Workers, farmers, housewives, and nature at large are made into 'colonies' by the dominant capitalist patriarchal economic paradigm. The capitalist model of development by globalization expresses a convergence of two forms of violence – the power of ancient patriarchal cultures combined with the modern neoliberal rule of money.

Further Resources
Lessem, Ronnie and Alexander Schieffer (2010), *Integral Economies*. Farnham, UK: Ashgate/Gower.
Navdanya, www.navdanya.org.
Resurgence Magazine (2007), 'How Wealth Creates Poverty',
    http://www.resurgence.org/magazine/article250-how-wealth-creates-poverty.html.
Shiva, Vandana (2005), *Earth Democracy*. Boston: South End.
———— (2009), *Soil not Oil*. London: Zed Books.
Via Campesina (2009), *Small Scale Farmers Are Cooling Down the Earth*. Jakarta: Via Campesina.

Vandana Shiva is Director, Research Foundation for Science, Technology, and Ecology, New Delhi. A former quantum physicist, she has become an influential global environmental activist and author of several books including *Staying Alive: Women Ecology and Development* (1989), *Monocultures of the Mind* (1993), and *Stolen Harvest* (2001). She is a recipient of the Alternative Nobel Prize and of the Sydney Peace Prize.

MALDEVELOPMENT

José María Tortosa

Keywords: basic needs, maldevelopment, underdevelopment, well-being, freedom, identity

The word 'maldevelopment' is partially a reaction to the weaknesses and harmful secondary effects of 'development' as a programme. The term 'underdevelopment' became a part of public language after President Truman's 1949 inaugural speech in which a clear anti-communist slant, typical of that time, prevailed. In the use of the word 'development', there is an underlying metaphor, taken from biology, that living beings develop, growing according to their genetic code. It is a process that is natural, gradual, without irregularities, and beneficial. In case that this development does not occur, the doctor might intervene, re-directing it, as was stated in the fourth point of that speech, towards the agreement amongst companies and governments of developed countries to transfer technology and produce growth in underdeveloped ones. But with an exception – the old imperialism-exploitation for foreign profit has no place in our plans. What we envisage is a programme of development based on the concepts of democratic fair-dealing, claimed Truman.

As in other metaphors, development has the risk of hidden ideology. This is even more so if it is concentrated on the objective of growth, expressed as GDP, without any type of reference to its limits. GDP-based economics implies limitless growth, leaving out the second part of the biological metaphor, aging, as well as systematically forgetting the relationship between this process and its environment.

The metaphor of maldevelopment is different. Living beings suffer maldevelopment when their organs do not follow their code. They become unbalanced and deformed. Its use in social science appears to have begun with an article by Sugata Dasgupta in 1967. The classic work was written by Samir Amin in 1990 and was also cited in the book edited by Jan Danecki in 1993 with intercontinental participation reflecting discussions of the project, 'Goals, Processes and Indicators of Development' (GPID Project) of the United Nations, 1978–82.

It is a metaphor, but unlike development, maldevelopment attempts to refer to verification – first, of the failure on a global scale of the development programme and, second, to the verification of mal-living that can be observed in the structure and workings of the world system and its components. If development implies a normative element (desirable), maldevelopment contains an empirical component (observable), or even a critical element (undesirable).

Extending the metaphor, consider a medical clinic that starts out from a diagnosis, carries out a prognosis and establishes a therapy, in terms of an ideal health, not always well defined, but whose absence tends to be clearly defined and classified as disease. In this sense, maldevelopment can be understood as part of an illness whose components can be numbered according to the following table in which, on the one hand, basic necessities (well-being, freedom, identity, security) according to those specified by Galtung are found, and on the other, there are three or four levels (local as differentiated from the state, the ecosystem, and the world) in which a diagnosis can be carried out.

The table can be read horizontally, establishing the cases of lack of fulfilment of basic human necessities. However, a vertical reading might prove more fruitful. Such a reading can be initiated in the third column which makes reference to the relationship amongst different actors of the world system, characterized by the asymmetry of their capacity and power of decision-making and influence. It is not about developed and underdeveloped characterized by growth and technology, but about what is central and peripheral, characterized by their power.

The second column refers to subject matter that has been present, at least on a rhetorical level, in some of the approaches towards development,

	State/Local	*Ecosystem*	*World system*
Well-being	Poverty Inequity, inequality Stagnation	Global warming Resource exhaustion Pollution	Polarization Peripheralization Exploitation
Freedom	Limited democracy Repression Marginalization	Dependence on Nature, without partnership	Dependence Repression Marginalization
Identity	Internal colonization Nationalism Fundamentalism	Alienation from Nature, loss of roots	Colonization Homogenization Identitary reactions
Security	Violence Civil war Terrorism	Human-made catastrophes	War between countries Transnational terrorism Nuclearization

such as eco-development. These are placed here to draw attention to a dual reality: on the one hand, that causes of ecosystem degradation tend to be more in countries where power is centralized, but also more recently in emerging ones. On the other hand, we find ourselves faced with certain problems, which in some cases can be more dramatic in their effects within peripheral countries compared to others, in the form of human-made catastrophes. However, the effects of these problems on the survival of the species and the maintenance of the current system could be generalized.

Finally, the first column indicates the points in which current maldevelopment is best recognized or verified. Its incidence is high in peripheral countries, even higher in emerging countries and, finally, the highest in currently hegemonic countries. However, poverty, repression, fundamentalism, or criminal violence are not the exclusive patrimony of the periphery, but are found, sometimes with greater intensity, in countries where power is centralized.

The word 'maldevelopment' does not bring with it a classification, more or less contrived, of developed and underdeveloped countries, as was outlined in Truman's speech along with the proposal that the developed come to the aid of the underdeveloped. The perspective that it offers is different: all countries are, in one way or another, maldeveloped and the ultimate reason is their immersion in the world system that produces, in a word, capitalism, which is where the problem seems to reside.

Further Resources

Amin, Samir (1990), *Maldevelopment: Anatomy of a Global Failure*. Tokyo: United Nations University Press; London: Zed Books.

Danecki, Jan (ed.) (1993), *Insights into Maldevelopment: Reconsidering the Idea of Progress*. Warsaw: University of Warsaw, Institute of Social Policy.

Dasgupta, Sugata (1968), 'Peacelessness and Maldevelopment: A New Theme for Peace Research in Developing Nations', in *Proceedings of the International Peace Research Association*, Second Conference.

Galtung, Johan (1980), 'The Basic Needs Approach', in Katrin Lederer (ed.), *Human Needs: A Contribution to the Current Debate*. Cambridge, Mass: Oelgeschlager, Gunn & Hain.

Tortosa, José María (2011), *Maldesarrollo y Mal Vivir. Pobreza y violencia a escala mundial*. Quito: Abya Yala.

Unceta, Koldo, 'Desarrollo, subdesarrollo, maldesarrollo y postdesarrollo: una mirada transdisciplinar sobre el debate y sus implicaciones', *Carta Latinoamericana – Contribuciones en Desarrollo y Sociedad en América Latina*. 7, Montevideo.

José María Tortosa earned his doctorate in Social Science (Roma, 1973) and Sociology (Madrid, 1982). He was professor in the Department of Sociology II (1991 to 2009),

Universidad de Alicante; Director (2006 to 2007), and honorific collaborator (2009 to present) in the Instituto Interuniversitario de Desarrollo Social y Paz, as well as in Development Projects with the United Nations University, 1978–82. He has authored thirty books.

THE DEVELOPMENT PROJECT

Philip McMichael

Keywords: development, climate change, food sovereignty

The mid-twentieth century 'development project' materialized in the wake of economic depression, world war, and decolonization as a blueprint for nation-building and a strategy for world order in the context of the Cold War. The United States led post-war reconstruction to stabilize world capitalism and extend its economic empire to the post-colonial world. By describing so-called Third World cultures as undeveloped, the role of colonial exploitation in the rise of the West was erased. Further, idealized Western-style industrial development was imposed as the universal standard, as represented in Gross National Product metrics. As First World surrogates, the Bretton Woods institutions (World Bank, International Monetary Fund) served as key financial agencies complementing US aid programmes in targeting Third World states as commercial clients and gaining access to strategic resources in an anti-communist crusade. Here, even as development was identified as national economic growth, it evolved as market rule on a world scale.

Represented as a civilizational improvement on the past and participation in a future of boundless consumption, development is normalized in modern discourse as a seductive inevitability. But development is imagined and measured only on the positive side of the material ledger. However, it has revealed itself to be a paradox, betraying its initial promise. The prospect of endless material prosperity for all via an expanding commodified global market is denied by global inequality with an affluent minority on the one hand, and overexploited labour population and ecosystems on the other. Instead of mass consumption, selective enrichment amidst a precarious employment situation, rising debt, and footloose labour register the paradox.

Moreover, increasing entropy is now evident in deteriorating ecological conditions and fragile social institutions, with political and economic elites practising self-preservation, discounting public needs, and a looming climate emergency. A world awash with commodities has no environmental sensors – as the UK's *Stern Review on the Economics of Climate Change* (2006) observed, the world's greatest market failure now is climate change.

Current neoliberal development amplifies capitalism's focus on short-term gain via increasing the velocity of circulation of capital and commodities, juxtaposing short-term 'economic time' with long-term 'geochemical-biological time controlled by the rhythms of Nature' (Martinez-Alier 2002: 215). For example, when shrimp aquaculture destroys coastal mangroves, the 'all the shrimp you can eat' development undermines not only fish breeding grounds and local livelihoods, but also repositories of biodiversity, carbon sinks, and coastal defence against rising seas (ibid.: 80). Furthermore, while economic time claims linear improvement on the past, its past is ever-present – in climate change:

> For every year global warming continues and temperatures soar higher, living conditions on earth will be determined more intensely by the emissions of yore, so that the grip of yesteryear on today intensifies – or, put differently, *the causal power of the past inexorably rises*, all the way up to the point when it is indeed "too late". The significance of that terrible destiny, so often warned of in climate change discourse, is the final *falling in of history on the present* (Malm 2016: 9; emphasis in the original).

Unfortunately, those humans least responsible for climate change are its most vulnerable – having been marginalized by development: from extant rural cultures and climate refugees to slum dwellers (a third of the global urban population). From a biophysical perspective, the UN *Millennium Ecosystem Assessment* stated that recent economic development 'has resulted in a substantial and largely irreversible loss in the diversity of life on Earth. These problems, unless addressed, will substantially diminish the benefits that future generations obtain from ecosystems' (United Nations 2005: 1).

In response, the High Level Panel for the UN's Sustainable Development Goals initiative observed that 'environment and development were never properly brought together', adding the somewhat questionable proposal that 'because we "treasure what we measure" an important part of properly valuing earth's natural abundance is to incorporate it into accounting systems' (UNDESA 2013). Such a vision extends a particular kind of control over nature in the name of private interest, foreclosing other meanings and uses of land and privileging investor rights over land-user rights. Further, it deepens the externalization of nature by separating out elements of interactive biophysical processes as 'ecosystem services', commodified in the name of environmental governance and 'sustainable development', possibly under an illusion that there remains sufficient natural world to sustain. Sustaining development in a compromised environment is a poor substitute for rehabilitating degraded ecosystems and promoting biodiverse practices.

Post-development addresses these multiple contradictions by embracing principles of natural repair and regeneration, beginning with local responsibility. A variety of rural cultures with low-input farming systems producing more than half of the world's food, hold this potential (Hilmi 2012). This is the genuine centuries-old conventional agriculture. The development narrative, concentrating capital, and centralizing control, has appropriated the term 'conventional' for industrial agriculture, designating low-input practices as obsolete. But this, too, is the development paradox – as the 200-million-strong international peasant coalition, La Via Campesina, claims: with support, local farming systems can 'feed the world and cool the planet', adapting farming practices to natural cycles via restorative agro-ecological methods across the global South and North, anchoring a broad food sovereignty movement for democratizing food systems. The urban variant includes nested food markets and solidarity economies close relatives of over 300 Transition Towns that have sprouted over the last decade in the UK, North America, South Africa, Europe, and Australia, managing energy descent via permaculture, urban commoning, and cross-community alliance-building.

Such initiatives promote natural regeneration to restore rather than ruin biodiversity, sequester rather than release emissions, and convert rather than consume energy. These are post-development principles, already in existence, orthogonal to market discounting or the act of reducing nature to a commodity. For the survival of humanity, and indeed all life on Earth, the protection of these materially and socially significant principles is imperative to the sustenance of both rural and urban environments. In this sense, post-development means ending the fetishism of the market.

Further Resources

Hilmi, Angela (2012), *Agricultural Transition: A Different Logic*. The More and Better Network, http://ag-transition.org/pdf/Agricultural_Transition_en.pdf.

Malm, Andreas (2016), *Fossil Capital: The Rise of Steam Power and the Roots of Global Warming*. London and New York: Verso Books.

Martinez-Alier, Joan (2002), *The Environmentalism of the Poor: A Study of Ecological Conflicts and Valuation*. Cheltenham: Edward Elgar.

Stern, Nicholas (2006), *Stern Review: The Economics of Climate Change.* Cambridge: Cambridge University Press.

United Nations (2005), *Ecosystems and Human Well-being: Synthesis.* Washington, DC: Island Press.

United Nations (UNDESA) (2013), 'A New Global Partnership: Eradicate Poverty and Transform Economies through Sustainable Development', https://sustainabledevelopment.un.org/index.php?page=view&type=400&nr=893 &menu=1561.

Philip McMichael is a Professor of Development Sociology, Cornell University, New York. He has worked with FAO, UNRISD, La Vía Campesina and the International Planning Committee for Food Sovereignty. He is the author of *Settlers and the Agrarian Question: Foundations of Capitalism in Colonial Australia; Development and Social Change: A Global Perspective;* and *Food Regimes and Agrarian Questions.* His current research is on land grabs and land rights, food regimes, and food sovereignty.

———————

OCEANIA'S KASTOM EKONOMI

Kirk Huffman

Keywords: Pacific Ocean, Australia, Melanesia, Regenvanu, *Kastom Ekonomi*

In conversations with Pacific village chiefs, you will typically hear:

> Foreigners used to tell us we needed to "Change"; then they told us we needed "Progress", and now they tell us we need "Development". It usually means they are after something we have – either our forests or our land or what is under our land, or our souls or language or culture, or our feeling of contentment with our way of life . . . (*Huffman* 2008–15)

The said foreign representatives may be from the World Bank, Asian Development Bank (ADB), foreign governments, logging or mining companies, or NGOs. Some newer Pentecostalist churches promote the same developmentalist views among their converts. It is all part of a new 'perpetual growth' world where the real God is money accumulation. So many projects across the Pacific have either not brought promised results, or have failed so completely that the word 'development' is now often jokingly used to refer to something that goes wrong.

Most of the destructive development is concentrated in the larger Melanesian islands of the western Pacific. In Polynesia, that is central and eastern Pacific, and in Micronesia to the north, islands are smaller and lack resources. Notable exceptions are Nauru and Banaba, almost destroyed by phosphate extraction. In Melanesia, which comprises the islands of New Guinea, the Solomons, Vanuatu, New Caledonia and the border nation of Fiji, horrific development proceeds apace. West Papua is logged to clear land for palm oil plantations and the world's largest open-cut copper and gold mine. Meanwhile, local Melanesians there are now outnumbered by poor Javanese families, brought in by an Indonesia government-sponsored *Transmigrasi* programme. Originally funded by the World Bank, this development model has resulted in severe repression of West Papuan indigenous peoples by the Indonesian military and paramilitary forces.

In neighbouring Papua New Guinea (PNG), Exxon Mobil's vast Liquid Natural Gas project antagonizes tribal populations, and a series of Special Agricultural Business License scandals is unfolding. The latest chapter of this saga pits landowners on the island of New Britain against Malaysian logging giant Rimbunan Hijau. The PNG government sides with the company, which also owns a local newspaper and transport company, as well as a massive new hotel complex in the capital, Port Moresby.

Another worst case development scenario is the island of Bougainville, subject to PNG military operations for nearly a decade from 1989 after islanders closed down Rio Tinto's Panguna copper mine. Jubilee Australia (2014:11) estimates islander deaths in the order of 10,000 to 15,000 during this period of conflict. Australia still lobbies behind the scenes to get the mine reopened and provides large consultancy fees for academics, to get the now-autonomous government of Bougainville to agree. Local press reports show that women strongly resist these moves. The tragic Bougainville situation inspired the 2009 film *Avatar*, and the 2013 film, *Mr Pip* is based on these events.

Logging and some mining operations create ongoing problems in the Solomon Islands. Further south and east, the nation of Vanuatu has few mineral resources, and it closed two major Asian logging companies in the 1980s and 1990s on the island of Mala Kula after their operations had antagonized local people. However, Vanuatu became prey from the early 2000s to foreign land alienators' misuse of a new Strata Title Act to lease and subdivide indigenous land and sell it to foreigners. This threat galvanized the ni-Vanuatu people, and a National Land Summit was held in 2006. Ralph Regenvanu, former Director of the Vanuatu Cultural Centre, then founded his Land and Justice Party (GJP). In 2014, as Minister for Lands, he introduced stricter legislation to protect indigenous holdings from alienation by commercial interests. As a result, the Opposition party, supported by investors and real estate agents, regularly mounts motions of no confidence in the government.

Known in recent history as the New Hebrides, Vanuatu had the unique experience of two colonial powers ruling it at the same time – Britain and France. Thus, since independence in 1980, many ni-Vanuatu have been wisely suspicious of outside influences. Their traditional way, known as *Kastom* in Pidgin English, is seen by economists as 'blocking development'. However, smart Melanesians tend to see *Kastom* as protecting them from bad development and the disease that comes with it – *Sik blong Mane* or money addiction. In 2005, the Vanuatu Cultural Centre began promoting the traditional Melanesian lifestyle and economy, and in 2007, the government declared a *Yiablong Kastom Ekonomi*.

Regenvanu has described the traditional economy as the source of Melanesian resilience. His address at the Lowy Institute's 2009 conference on 'The Pacific Islands and the World: The Global Economic Crisis' was undoubtedly the most significant paper presented at this Brisbane event. Even then, World Bank economists and regional politicians ignored its wisdom. The ni-Vanuatu wish to promote *Kastom Economi* and protect land rights, agriculture, and self-sufficiency is a much more sustainable trajectory than artificial growth models, such as construction proposals, over-reliance on tourism, or 'retirement lifestyles for Australian baby boomers'.

Australia is part of Oceania, but often has difficulties realizing this. It also has issues with its original First Peoples, who as yet have no Constitutional standing and whose Native Title Law is constantly threatened by development proposals. Australia's First Peoples also have the highest rate of indigenous incarceration anywhere in the world. From 1863 to the end of that century, Queensland sugar plantations relied on an imported Pacific island labour force known as Kanakas. Pacific remittance workers still provide seasonal labour on Australian farms, but islanders have mixed views about this 'big brother country'. These would improve if the country lifted the 2008 legal ban on *kava* drinking, and toned down rhetoric on economic development in favour of more culturally and climatically aware political visions. Failing this, China and Indonesia are waiting in the wings with development models that can only spell doom for Pacific ways of life.

Further Resources

Ginzburg, Oren (2006), *There you go!* London: Survival International, www.survivalinternational.org/thereyougo.

Huffman, Kirk (2005), *Traditional Money Banks in Vanuatu*. Port Vila: Vanuatu National Cultural Council/UNESCO.

———— (2010), 'Review and Reflections on Tim Anderson and Gary Lee (eds) "In Defence of Melanesian Customary Land"', *Pacific Currents*. 1 (2) and 2 (1), http://intersections.anu.edu.au/pacificcurrents/huffman_review.htm.

Jubilee Australia (2014), *Voices of Bougainville Report*, Sydney.

Regenvanu, Ralph (2010), 'The Traditional Economy as Source of Resilience in Vanuatu' in Tim Anderson and Gary Lee (eds), *In Defence of Melanesian Customary Land*. Sydney: AID/WATCH.

Robie, David (2014), *Don't Spoil My Beautiful Face: Media, Mayhem & Human Rights in the Pacific*. Auckland: Little Island Press.

Kirk Huffman is a Sydney-based anthropologist/ethnologist, with 18 years of field experience in Vanuatu, and also in the Solomons, the Maghreb, and parts of the Sahara, northern Colombia, and the western Mediterranean. He is Honorary Curator at the National Museum, Vanuatu Cultural Centre.

THE LATIN AMERICAN CRITIQUE OF DEVELOPMENT

Maristella Svampa

Keywords: consumerism, underdevelopment, colonial matrix of power, extractivism

Critical approaches to the hegemonic notion of development have existed in Latin America since the early discussions of the Club of Rome's *Limits to Growth*. Critiques have ranged from debates on sustainable development to contemporary critique of the expansion of the commodity frontier. I would like to highlight three key moments in Latin American thought: the critique of consumer society (1970s–80s); the post-development critique (1990s–2000s); and critical perspectives on extractivism (early 2000s–present).

The first phase is best illustrated by Brazilian economist, Celso Furtado, who, in gaining distance from the classic perspectives of the Economic Commission for Latin America and the Caribbean, ECLAC, argued that one of the indirect conclusions of the 'limits' argument, was that the lifestyle promoted by capitalism would only be viable for industrialized countries and elite minorities within underdeveloped ones. Any attempt to generalize the consumerist way of life would lead to the collapse of the system. In the same vein, the Argentina-based interdisciplinary group Fundación Bariloche, coordinated by Amilcar Herrera, maintained that behind the report there lay the neo-Malthusian logic characteristic of hegemonic development discourses. In 1975, this group created an alternative model titled *Catástrofe o Nueva Sociedad? Modelo Mundial Latinoamericano* [*Catastrophe or New Society? A Latin American World Model*], which argued that environmental degradation and the devastation of natural resources were not due to population growth but due to the high consumption rates in rich countries, de facto enacting a division between 'developed' and 'underdeveloped' countries. The corollary of this perspective was that the privileged populations of the planet would have to lower their excessive consumption patterns and diminish their rates of economic growth in order to reduce the pressure on natural resources and the environment. Although these critiques did not escape the dominant logic of productivism, which saw limitless economic growth as a value in and of itself, they did have the virtue of questioning the dominant episteme.

Other concepts in the 1980s similarly emphasized critiques of consumption. Among them were the notion of 'human-scale development' and the 'theory of human needs' developed by Chilean economist, Manfred Max Neef. A further poignant cultural critique of post-industrial society emphasizing its instrumental rationality and crass materialism, came with Ivan Illich's highly influential notion of 'conviviality'. Hence, in this first moment, the thrust of the development critique involved re-thinking consumption and cultural patterns in favour of the common good, and egalitarian societies founded on more austere life styles and longer-lasting productive systems.

The second moment, associated with the post-developmentalist perspective, centred on development as a discourse of power. Here, we highlight the contribution of Gustavo Esteva in the *Development Dictionary*, coordinated by Wolfgang Sachs (1992), who elaborated a radical critique underscoring the colonial matrix of the idea as a post-war (1949) invention on the part of the United States and other Western powers. Another notable contribution within this line of thinking was Arturo Escobar's deconstruction of the modern concept of development as an instrument of domination, revealing its main mechanisms of operation: the division between development and underdevelopment; the professionalization of development 'problems' and the rise of 'development experts'; and the institutionalization of development through a network of national, regional, and international organizations. Escobar highlighted the ways in which development made invisible diverse local experiences and knowledge. He further suggested, already in the mid-1990s, a shift away from thinking about 'alternative Development' towards 'alternatives *to* Development'.

A third, and current, phase began in the early 2000s with the critique of existing neo-extractivism and the start of the Commodity Consensus. This phase spurred a critique of the productivist logic underlying development and of the expansion of extractive mega-projects (large-scale mining, oil extraction, new agrarian capitalism with its combination of genetically modified organisms and agro-chemicals, large-scale dams, mega-real estate projects, among others). Such new forms of extractivism are characterized by the intensive occupation of territories, land grabbing, and the destructive appropriation of nature for export. While extractivism refers to the over-exploitation and large-scale export of primary goods from Latin America to core and emerging economies, the notion of Commodity Consensus suggests that, similar to the Washington Consensus, there is an agreement – increasingly explicit every year – around the irreversible or irresistible nature of the current extractivist model. This inevitability forecloses the possibility of considering alternatives to current development models. Beyond alleged comparative advantages, such as high international prices, these trends have deepened the historic role of the region as a provider of raw materials. It has also intensified asymmetries between the global economic centre and its peripheries, as reflected in the trend towards re-primarization of national economies and the uneven distribution of socio-environmental conflicts.

Unlike the previous two analytic phases, the current one has seen explicit re-signification of the environmental question, this time in relation to territories, politics, and civilization. This 'environmentalization of struggles' as Enrique Leff would say, is reflected in diverse eco-social-territorial movements directed against private sector transnational corporations and the state. Such movements have broadened and radicalized their discursive positions, incorporating other issues such as the critique of mono-cultural development models. This politics reveals a crisis of the instrumental and anthropocentric view of nature with its dualist and hierarchical ontology.

Given this epistemic–political landscape, we are witnessing the consolidation of a radical new environmental rationality and post-developmentalist vision. Horizon-concepts such as *buen vivir, bienes communes* or common goods, ethics of care, food sovereignty, autonomy, rights of nature, and relational ontologies are key elements of this recent dialectical turn in Latin American critical thought. This turn synthesizes contributions of previous periods, integrates the critique of consumption models and dominant cultural patterns and recasts the post-development perspective.

Further Resources

Escobar, Arturo (2014), *Sentipensar con la tierra: Nuevas lecturas sobre desarrollo, territorio y diferencia*. Medellín, Colombia: Ediciones Unaula, https://mundoroto. files.wordpress.com/2015/03/sentipensar-con-la-tierra.pdf.

Esteva, Gustavo (1992), 'Development', in Wolfgang Sachs (ed.), *The Development Dictionary: A Guide to Knowledge as Power*. London and New York: Zed Books.

Grupo Permanente de Trabajosobre Alternativas al Desarrollo, http://www.rosalux. org.ec/grupo/.

Gudynas, Eduardo (2015), *Extractivismos: Ecología, economía y política de un modo de entender el desarrollo y la Naturaleza*. Cochabamba: Cedib/Claes.

Illich, Ivan (1973), *Tools for Conviviality*. London: Boyars.

Svampa, Maristella (2016), *Debates Latinoamericanos. Indianismo, Desarrollo, Dependencia y Populismo*. Buenos Aires: Edhasa.

Maristella Svampa is an Argentinian sociologist, writer, and researcher within the Argentinean National Scientific and Technical Research Council (CONICET). She is Professor at the Universidad Nacional de La Plata, Argentina, and the author of several books on political sociology and social movements, as well as several fiction books. She is a Member of the Permanent Group on Alternatives to Development established by the Rosa Luxemburg Foundation.

UNIVERSALIZING THE EARTH: REFORMIST SOLUTIONS

BRICS

Ana Garcia and Patrick Bond

Keywords: BRICS, sub-imperialism, development banks, multilateral reform

The BRICS states – Brazil, Russia, India, China, and South Africa – were not initially considered a potential geopolitical bloc when Jim O'Neill of Goldman Sachs Asset Management conceived the acronym BRIC in 2001 as the next group of high-growth economies. The world financial crisis in 2008–09 had consolidated the G20, including all the BRICS countries, for the purpose of generating a world response to financial instability, based upon short-lived Keynesian principles of state deficit financing, loose monetary policy and lending coordination for bank bailouts. In 2009, the first BRIC summit took place in Yekaterinburg, Russia, creating expectations of a future challenge to the dominance of western countries in multilateral institutions (Bond and Garcia 2015). At Beijing's behest, South Africa was added in 2010 for continental balance.

Since then, a contradiction between the BRICS' economic potential and their political role in shoring up western-centric multilateralism has become obvious. For instance, the world crisis offered four BRICS states the opportunity to lobby for greater voting power in the International Monetary Fund (IMF) during the 2010–15 'quota reform', for which the BRICS contributed $75 billion in recapitalization funds. China's vote rose from 3.8 per cent to 6.1 per cent of the total, but this was at the expense of lowering the input of poorer countries such as Nigeria (whose voting share fell by 41 per cent), Venezuela (by 41 per cent), and even South Africa (by 21 per cent).

High commodity prices and low wages fuelled the BRICS' accelerated growth prior to the price peak in 2011 and the subsequent crashes in 2015. BRICS-based corporations also become major international investors. Economic modernization offers the BRICS a capitalist path to development based on exploitation of the workforce and nature. The BRICS' economic growth has been marked by extreme inequality, despite their leaders' demand for greater equality in the international system.

The BRICS' agreement to create a New Development Bank (NDB) was signed at the Fortaleza summit in 2014, the same year that Beijing created the Asian Infrastructure Investment Bank. Both banks focus on infrastructure and energy projects, ultimately serving the interests of the extractive and agribusiness industries (Garcia 2017). New logistics corridors within and near the BRICS states connect territories and natural resources to foreign markets, such as China's Belt and Road Initiative, generating a major conflict with India over passage through Pakistan-controlled Kashmir, and Mozambique's Nacala Corridor.

The rapid operationalization of these new banks is partly due to the initial absence of socio-environmental standards that require impact assessments or negotiations with communities, although the first set of credits went towards renewable energy projects. The principle of non-interference in internal affairs distinguishes the NDB from traditional multilateral financial institutions such as the World Bank. On the other hand, in 2016, the NDB began wide-ranging operational collaboration with the World Bank which involved joint participation in project preparation, co-financing and staff exchange.

Likewise, the Contingent Reserve Arrangement (CRA), which has an initial fund of $100 billion, can be activated in the event of a BRICS country's balance of payments crisis, but only in a manner complementary to the IMF. A borrowing country must maintain an IMF structural adjustment programme, otherwise it will be able to borrow only 30 per cent of its quota of CRA funds. In the case of South Africa, which will be first to borrow to service a $150 billion debt, that quota is only $10 billion. Thus, both the NDB and the CRA function in a complementary way to the Bretton Woods Institutions.

Similarly, within the United Nations Framework Convention on Climate Change, four of the BRICS countries – Brazil, South Africa, India, and China – played a central role alongside US President, Barack Obama, in ending the Kyoto Protocol provision for binding emissions cuts. This alliance began at the UN summit in 2009 in Copenhagen, and continued in 2011 in Durban. It culminated in a non-binding climate deal in Paris in 2015 with numerous fatal flaws that include prohibition on climate debt liabilities. That deal is beneficial mainly to the historic polluters as well as to the BRICS countries, all of whose economies have a high carbon intensity.

There is certainly potential for the BRICS to contest western hegemony. The best example may be in the struggle over intellectual property rights on medicines within the WTO, in which Brazil and India confronted western governments and pharmaceutical companies during the 1990s. In 2001, HIV+ activists celebrated the generic AIDS medicines that resulted from a WTO intellectual property exemption, and in South Africa alone, huge gains in life expectancy resulted (from 52 in 2004 to 64 today). Moreover, genuine geopolitical tensions between the West and at least two BRICS states – Russia and China – will continue to play out in incidents such as Edward Snowden's 2013 Moscow asylum, the 2014 Russian invasion of the Crimea, chaotic Syrian war alliances, Polish missile deployment, and Chinese economic expansion, as well as South China Sea conflicts.

Yet, these tensions have been exceptions, and the overall BRICS contribution to multilateralism is accommodating to Western hegemony. As a project of national elites and their multinational corporations, the BRICS have not formulated an ideological alternative to neoliberal globalization of which China is the world's leading proponent at present. Rather, they work within the capitalist order and occupy an increasingly important place in the expanded reproduction of global capital.

To explain this, Brazilian dependency theorist, Ruy Mauro Marini (1965), developed the concept of 'sub-imperialism' during the 1960s to identify key countries in imperialism's expansion. Today, the 'deputy sheriff' role strengthens imperialism's ongoing commodification of everything, neoliberal economic policies, mineral and oil extractivism, and repressive control of dissident populations. As a result, oppositional forces from both the BRICS and their hinterlands are mobilizing in solidarity to demand real change, local and global, as a still nascent 'BRICS from below' process (Bond and Garcia 2015).

Further Resources
Bond, Patrick and Ana Garcia (eds) (2015), *BRICS: An Anti-capitalist Critique.* Johannesburg: Jacana Media.
Garcia, Ana (2017), 'BRICS Investment Agreements in Africa: More of the Same?', *Studies in Political Economy.* 98 (1): 24–47.
Marini, Ruy Mauro (1965), 'Brazilian Interdependence and Imperialist Integration', *Monthly Review.* 17 (7): 10–29.

Ana Garcia is Professor of International Relations at the Federal Rural University of Rio de Janeiro and an associate of the Institute of Alternative Policies for the Southern Cone of Latin America. Her doctorate was from the Pontifical Catholic University of Rio de Janeiro.

Patrick Bond is Distinguished Professor of Political Economy at the University of the Witwatersr and in Johannesburg. His books cover urban environments, climate change, global financial crisis; Africa-wide, Zimbabwean, and South African political economy.

CIRCULAR ECONOMY

Giacomo D'Alisa

Keywords: societal metabolism, resource efficiency, reuse, sustainability, growth

Circular Economy (CE) is an emerging strategic vision aiming to decouple economic growth and environmental impacts. Its core strategy aims to:

1. Reduce the use of raw material in order to revert the extractive model of the current economic system;
2. Boost reuse practices, avoiding discarding patterns for matters and stuffs that still have use value for different actors in society; and
3. Increase the recyclability of goods by implementing effective market arrangement for secondary materials.

Technical and design-related as well as managerial solutions are at the forefront of CE research and applications. These solutions counteract the planned obsolescence – the engineered limited useful life of products in order to increase their long-term sales volumes – of the standard business model and extend the life usability of materials.

The appearance of the CE concept can be traced back to pioneers of ecological economics such as Kenneth Boulding. In the mid-60s, he criticized the idea of an economy in a continuous and linear expansion, a cowboy's economy based on ever-more land to colonize and expand their livestock production. He foresaw the advent of the spaceship economy, where the expansion towards new extractive frontiers is not possible anymore and the recycling of materials and energy become the main worry of a businesses. Later on, in the '80s, the material balance of the economy also became a central argument of very influential environmental economists such as David Pearce and Kerry Turner (1990), who, probably for the first time, used the term 'circular economy'. They explained that only if one ignores the environment – the closed system that sets the limits and boundaries for the extraction and discard of stuff – can the economy appear to be a linear system in expansion. In the same period, industrial ecologists and eco-designers started to develop applied research on how to increase the efficiency in material use and extend the life of products. Those studies contributed to the development of the so-called industrial metabolism, that is, the integrated assessment of labour activities as well as technological and physical processes necessary to convert raw and secondary materials and energy into finished products and waste. The urgency of reducing waste has influenced enormously the development of CE ideas and application. This also explains why CE policies originate from or are directly part of waste legislative framework and waste programmatic plans (Ghisellini *et al.* 2016). Insights for a CE are coming from many other disciplines. In the field of architecture, for example, the concept of cradle-to-cradle is pushing designers to imagine regenerative products. Natural scientists and natural resources managers are fostering the spread of the biomimicry approach, which tries to imitate the qualities of well-adapted elements and structures present in nature to solve human issues. Circular Economy also applies principles emerging out of permaculture, an integrated system of cultivation that simulated the evolution of a self-organizing biological ecosystem.

Nonetheless, the current patterns of the 'real' economy of material and energy flows suggest caution is needed about the redeeming qualities of CE. The current economy is much more efficient than the one existing a century ago, but it uses resources at a level never seen before. It extracts an unprecedented amount of raw materials and releases unsustainable quantities of solid and gaseous wastes. The material footprint of nations, an indicator that accounts for the impact associated with raw material extractions to the nation that actually consume the end-chain products, shows that no absolute decoupling is near the horizon. Becoming wealthier does not ease pressure on natural resources at all (Wiedmann *et al.* 2015). A first-of-its-kind empirical study aiming to estimate the circularity of world economy clarifies that only 6 per cent of the materials extracted are recycled and go back to feed the loop of production and consumption. The current maximum potential for recycling is actually about 30 per cent; the other 70 per cent is composed mainly of energy and to a lesser extent of waste rock that cannot be recycled (Haas *et al.* 2015). It is not difficult to conclude then that the current pattern of global economy is very far from the CE objective.

Furthermore, while there are expectations, though not always fully demonstrated, that CE will boost employment and create meaningful jobs; it is astonishing that there is no discussion on the possibility of increasing unequal distribution of and access to resources, products, and services even under CE scenarios.

The wariness expressed earlier should not be conducive to a shallow dismissal of CE principles and applications. Indeed, CE actors deserve attention. These actors would be, among others, the open source circular economy communities, that is, experts, designers, and innovators who wish to promote transparency, open access to information, products and technologies and who offer open source solutions to environmental and resource problems. These grassroots movements challenge not only the business model, but the very essential institution of capitalism, that is, the private property of knowledge and information. The hesitation to engage with these actors of the blooming commons-based digital economies could be a missed opportunity, since some of the most important innovations that may make a low-carbon 'degrowth' society technically and socially possible are taking place there. It is thus extremely important to follow and create synergy with them.

Further Resources

Ghisellini, Patrizia, Catia Cialani and Sergio Ulgiati (2016), 'A Review on Circular Economy: The Expected Transition to a Balanced Interplay of Environmental and Economic Systems', *Journal of Cleaner Production*. 114: 11–32.

Haas, Willi, Fridolin Krausmann, Dominik Wiedenhofer and Markus Heinz (2015), 'How Circular Is the Global Economy? An Assessment of Material Flows, Waste Production, and Recycling in the European Union and the World in 2005', *Journal of Industrial Ecology*. 19 (5): 765–77.

Open Source Circular Economies, https://oscedays.org.

Pearce, David William and Kerry R. Turner (1990), *Economics of Natural Resources and the Environment*. London: Harvester Wheatsheaf.

Wiedmann, Thomas O., Heinz Schandl, Manfred Lenzen, Daniel Moran, Sangwon Suh, James West and Keiichiro Kanemoto (2015), 'The Material Footprint of Nations', *PNAS*. 112 (20): 6271–76.

Giacomo D'Alisa is an ecological economist and political ecologist. His research interests range from waste management to environmental justice, from illegal waste trafficking to environmental crime. He promotes degrowth visions and is interested in exploring how a degrowth society centred on care and commons would look. He is currently a Post-Doc at the Centre for Social Studies (CES), University of Coimbra.

CLIMATE-SMART AGRICULTURE

Teresa Anderson

Keywords: agriculture, climate change, climate-smart agriculture, agribusiness, greenwashing, adaptation, mitigation

'Climate-smart agriculture' is a buzzword used to describe agriculture that supposedly mitigates or adapts to climate change. However, the absence of a clear definition or specific criteria to distinguish it, has enabled the agribusiness industry to enthusiastically adopt the term to re-brand their activities as good for the climate. Many organizations in the food movement are wary of – or even opposed to – the concept of climate-smart agriculture. They share growing concerns that the term is so vague that it is being used to greenwash practices that are, in fact, damaging for the climate and for farming. Many are worried that the promotion of climate-smart agriculture does more harm than good and, in reality, undermines the transition towards sustainability and justice urgently needed in our food systems.

The term 'climate-smart agriculture' was initially framed by the UN Food and Agriculture Organization (FAO), with the aim of promoting approaches to agriculture that could mitigate and adapt to climate change, while increasing yields.

The problem is that there are no specific definitions for what can – or cannot – be called 'climate-smart'. As a result of the vague conceptualization by FAO, corporations freely adopt the phrase to describe practices that they claim drive innovation and reduce environmental impact. However, there are no meaningful criteria or evidence required in order to use the term. Climate-smart agriculture practices are not required to follow agro-ecological or similar principles. Nor are there any social safeguards to prevent so-called climate-smart activities from undermining farmers' livelihoods, driving land grabs, or pulling farmers into debt.

Thus, while some may assume that climate-smart agriculture means that such activities are beneficial for the climate, there are no guarantees to ensure that that is the case. Unfortunately, the term is now so widely used that it is considered too late to develop meaningful definitions or criteria. For agriculture to genuinely address the multiple challenges associated with climate change, a profound systemic shift is needed.

As the effects of climate change are felt around the world, agriculture, food security, food sovereignty, and farming communities are particularly at risk from its impacts. Erratic rainfall patterns, droughts, floods, and temperature extremes are increasingly playing havoc with farmers' ability to grow food.

At the same time, agriculture – particularly the green revolution model of industrial agriculture – is a major cause of climate change. A significant proportion of global greenhouse gas (GHG) emissions are created through industrial livestock production, as well as the widespread use of synthetic nitrogen (Gilbert 2012).

Furthermore, while contributing to the climate change problem, the 'green revolution' approach to agriculture leaves food systems particularly vulnerable to its impacts. Seeds sold by agribusinesses have largely been bred to require large amounts of water and synthetic nitrogen fertilizers. But these synthetic fertilizers cause the water-retaining organic matter in the soil to break down and disappear, causing the soil to quickly dry out in times of low rainfall or drought, leading to poor yields or crop failure. The soil's reduced ability to absorb water also leaves crops more exposed to damage from heavy rainfall or flooding.

It is therefore clear that action is urgently needed, both to reduce agriculture's contribution towards climate change and to help food systems cope with and adapt to impacts now and in the future. Fortunately, one of the most effective solutions to reducing agriculture's contribution to climate change is also one of the most effective adaptation strategies available. By replacing synthetic fertilizers with natural techniques, agro-ecology reduces emissions and improves the soil's ability to absorb and retain water in times of drought and flood. Improving farmers' access to a diversity of locally adapted seed varieties is also critical to ensuring that they can deal with a range of unpredictable weather events. And by strengthening local economies, food miles can be reduced while at the same time enhancing food sovereignty and farmers' control over their food systems.

Agribusiness corporations, however, are seeking to delay these necessary transformations in our food systems. Instead of recognizing the need to transform farming practices, many of these businesses are simply using the 'climate-smart agriculture' term to rebrand their own harmful industrial agriculture practices, so that they can continue doing what they were doing – business as usual.

Corporations such as Monsanto, McDonalds, Syngenta, Walmart, and Yara (the world's largest fertilizer manufacturer) all claim that they are pioneering climate-smart agriculture practices. They argue that the biggest climate benefits will come from the biggest players taking action, and that polluting corporations must be part of the solution.

Monsanto claims that their GM 'Roundup-Ready' herbicide-tolerant seeds reduce CO_2 emissions from soil due to the application of herbicides instead of tilling for weeds. And they are hoping to develop systems that would monitor GHG emissions and would advise on weather patterns. Yara is developing fertilizer products and application techniques that they claim will have a reduced emissions impact. McDonalds claims to be leading on the development of 'sustainable beef'. Many proponents of industrial agriculture claim that 'sustainable intensification' can be a strategy for climate-smart agriculture because it increases yields while lowering emissions per unit of output. Some climate-smart agriculture projects have also been linked to controversial funding through carbon offsetting.

By calling themselves climate-smart, corporations hope to avoid scrutiny and regulation, so that they can continue their activities and expand their business even as they probably go on increasing their overall GHG emissions. But by continuing business-as-usual, they also undermine local food sovereignty and bring a host of other socio-economic and environmental problems associated with their industrial agriculture practices.

Complicating the debate, many groups promoting small-scale, agro-ecological farming practices that really do benefit the climate, also call their activities climate-smart. Confusion arises when different actors use the same term to describe very different approaches. The term 'climate-smart agriculture' continues to be used by some governments, NGOs, and corporates, but with widely differing meanings and agendas. Several governments, corporations, and NGOs have joined the Global Alliance for Climate-Smart Agriculture (GACSA). Meanwhile, hundreds of civil society organizations have expressed their opposition to climate-smart agriculture, the GACSA, and to any formal endorsement of the term in UN climate negotiations.

Further Resources

ActionAid (2014), 'Climate Resilient Sustainable Agriculture Experiences from ActionAid and its partners', http://www.actionaid.org/publications/climate-resilient-sustainable-agriculture-experiences-actionaid-and-its-partners-0

Climate Smart Agriculture Concerns (2014). 'Corporate-Smart Greenwash: Why We Reject the Global Alliance on Climate-Smart Agriculture', http://www.climatesmartagconcerns.info/rejection-letter.html.

Food and Agriculture Organization (FAO), 'Climate-Smart Agriculture', http://www.fao.org/climate-smart-agriculture/overview/en/.

Gilbert, Natasha (2012). 'One-Third of Our Greenhouse Gas Emissions Come from Agriculture', Nature, https://www.nature.com/news/one-third-of-our-greenhouse-gas-emissions-come-from-agriculture-1.11708.

Monsanto (2017). 'Driving Innovation in Modern Agriculture to Combat Climate

Change', https://monsanto.com/company/sustainability/articles/climate-smart-agriculture-practices/.
ScienceDirect (2014). 'Sustainable intensification: What is its role in climate smart agriculture?',
http://www.sciencedirect.com/science/article/pii/S1877343514000359.
Yara, 'Sustainability', http://yara.com/sustainability/climate_smart_agriculture/.

Teresa Anderson is Action Aid International's Policy and Communications Officer on Climate Change and Resilience; she is based in London. She is has authored several reports and articles including *Clever Name, Losing Game: How Climate-Smart Agriculture is Sowing Confusion in the Food Movement* and *Hotter Planet, Humanitarian Crisis: El Niño, the 'New Normal' and the Need for Climate Justice*, published by Action Aid.

DEVELOPMENT AID

Jeremy Gould

Keywords: Aidland, imperialism, colonialism, ngoization, privatization

Development aid – aka development assistance or development cooperation – is a diverse assemblage of governmental and discursive practices that first emerged in the geopolitical ferment after World War II and the concomitant unravelling of Europe's global colonial empires. Seen in historical context, development aid reflects concerted strategic efforts among the western industrial powers to sustain the economic and political privileges of empire within a radically transformed post-colonial world order. The rhetoric justifying development aid, on the other hand, portrays it as an effort to help populations inadvertently left behind in humanity's inevitable march towards modernity. The conflation of raw self-interest with a professed will to do good is a structural paradox that resonates throughout every aspect of development aid.

Different motivations were at work in the gestation of development aid. On the one hand, decolonization offered a unique opportunity for the United States to take advantage of a Europe weakened to a large extent by war and internal conflict. In line with its aspirations to global dominance, the United States deployed development aid to advance its interests within the newly liberated countries of Asia and Africa. At the same time, Europe's imperial metropoles were anxious to safeguard their interests in their former colonies. Aid provided a 'technical' means to sustain their economic (and ideological) dominance. A third factor was how an emerging agenda of multilateralism enabled the United Nations to be consolidated as an instrument of inter-governmental cooperation and authority, amidst deepening post-war competition between capitalist and socialist camps. It is notable that while the Soviet Union and its allies provided substantial economic assistance to many ex-colonies, these investments were never accounted for under the auspices of 'Official Development Assistance' (ODA). In this sense, ODA was an exclusively capitalist enterprise.

Amidst this mishmash of ideological contestation, commercial competition, and geopolitical strategizing, the discourse of development aid took on a rarified technical veneer. It is characterized by two implicit premises. One is the categorical virtue of capitalism, economic efficiency, productivity, and increased consumption as an inevitable historical goal for all peoples; the second premise is the immanent moral imperative of 'advanced' populations to provide succour to those who were not yet 'developed'. The fact that the 'underdeveloped' populations of the ex-colonies were the originators of sophisticated contextualized technologies and stewards of complex and fragile environments was ignored, as was their status as political subjects of independent sovereign powers. These silences and implicit assumptions underscore one of several constitutive paradoxes of development aid – the strategic deployment of aid to promote Western self-interest is systematically obscured via a 'technical' language, which appears at once ontologically decontextualized, morally unassailable and politically neutral.

Many scholars have noted the contradictory nature of the knowledge practices within the development aid apparatus. Despite the well-documented failure of development aid to achieve its primary aims, as well as the substantial evidence of aid's 'perverse' effects, the self-referential discourse that circulates in 'Aidland' (Apthorpe 2011) exudes unshakable faith in its potential for improvability and ultimate success. Indeed, David Mosse (2005) has argued that the goal of portraying a project as successful can take precedence over attaining substantial development outcomes.

The end of the Cold War witnessed a significant shift in the organization of development aid. From the 1950s until the 1990s, aid was overwhelmingly a public, intergovernmental enterprise; the past two decades have seen its dramatic privatization. This is reflected, above all, in the expanding role of wealthy private international aid agencies – Oxfam, ActionAid, Care

International, Save the Children, World Vision, and so on. These powerful agencies, whose base capital consists primarily of private donations, have taken on a significant role in policy formulation and implementation as subcontractors for public development aid agencies. Typically, such international private agencies work with or through weaker private organizations in 'recipient' countries, thus contributing dramatically to the 'ngo-ization' of local social movements. Ngoization refers to a process by which a local body or movement loses its political and intellectual autonomy by becoming a client contractor to the international body. The influential role of private agencies, such as the Carnegie, Ford, Gates, and Rockefeller Foundations, in the production of knowledge related to the global South, and in defining global development agendas, is another facet of the creeping privatization of development aid.

The idea of development, as natural and inevitable human progress, remains seductive. The narrative of humanity's rise from austerity to abundance, from grueling manual toil to an epoch of leisure and creativity made possible by technological innovation lies at the core of modernity's appeal. The rhetoric of Aidland draws on this imaginary of historical inevitability, and adds to it the universal moral imperative of succour. The volumes of money allocated to development aid fluctuate from year to year but show no sign of disappearing. In 2015, total volumes of public and private development aid amounted to 315 billion dollars.

That said, the hegemonic sway of western development aid is unravelling. The emergence of China as a key geopolitical player, and her dramatic incursion into the development aid market, especially in Africa, has irreparably challenged the West's pretensions to post-colonial dominance. More importantly, countless actors in so-called recipient contexts – be these politicians, government officials, community leaders, activists, journalists, artists, or academics – today have evolved sophisticated strategies to appropriate facets of the aid apparatus and its resources to their own social, economic, and political agendas. In some instances, these local strategies merely transpose corrupt practices from Aidland to other ends. And yet, it is also evident that militant justice movements, often spearheaded by indigenous peoples, are emerging across the global South to challenge the silences and tacit assumptions that naturalize the work of 'development'.

Further Resources

Apthorpe, Raymond (2011), 'With Alice in Aidland: A Seriously Satirical Allegory', in David Mosse (ed.), *Adventures in Aidland: The Anthropology of Professionals in International Development*. New York and Oxford: Berghahn Books.

Escobar, Arturo (2011), 'Introduction to the Second Edition', *Encountering Development: The Making and Unmaking of the Third World*. Princeton: Princeton University Press.

Gould, Jeremy (2005), 'Timing, Scale and Style: Capacity as Governmentality in Tanzania', in David Mosse and David Lewis (eds), *The Aid Effect: Giving and Governing in International Development*. London/Ann Arbor: Pluto Press.

Mosse, David (2005), *Cultivating Development: An Ethnography of Aid Policy and Practice*. London: Pluto Press.

Jeremy Gould is an anthropologist by training and has taught Development Studies at the Universities of Helsinki and Jyvaskyla, Finland. He has been employed as an expert consultant by several international development agencies and conducted ethnographic research on development aid in African countries including Tanzania and Zambia. He is currently writing about the interaction of law and politics in postcolonial Zambia.

DIGITAL TOOLS

George C. Caffentzis

Keywords: computers, Foxconn, exploitation, environmental damage

As development leads to the adoption of digital tools in almost every sphere of daily life, the expression 'blood computers' has been coined. The analogy is to 'blood diamonds', following increasing evidence of the trail of blood that computer production involves. More specific is the connection between some digital companies and militias responsible for the displacement and killing of millions of people in the Democratic Republic of the Congo. The term dates from a report published in 2009 by the England-based NGO Global Witness: 'Faced with a Gun What Can You Do?' This accused mining, metal, and electronics companies of being silent accomplices in the violence inflicted by armed groups, mostly operating in the mineral-rich provinces of North and South Kivu, who 'frequently force civilians to mine the minerals, extorting taxes, and refusing to pay wages' (Dias 2009).

Coltan, being a mineral necessary for the production of smartphones and laptop computers, is of particular concern. The purpose of the Global Witness report, and others in its wake, was to alert consumers and authorities to the need for stricter controls over the process by which such extracted minerals reach the international market and buyers. As a result of campaigns organized under the slogan 'No blood for my mobile' in 2010, the US Congress passed legislation, aka the Dodd-Frank Act, that required corporate disclosure of mineral sources. It is acknowledged that there are no such things as conflict-free phones or computers, because the 'channels through which coltan circulates are . . . labyrinthine, and frequently clandestine' (Brophy and de Peuter 2014: 63).

However, broader issues concerning the place of digital tools in social production and struggles are not sufficiently addressed by social justice movements. All too often digital technology is uncritically praised as a key organizational tool, connecting activists across the world, providing fast, effective means of debate and mobilization, with little or no mention of ecological and social costs of production. As theorist Saral Sarkar noted in *Eco-Socialism or Eco-Capitalism?*, digital production is an ecological disaster: 'The fact that computers and most electronic goods are becoming smaller and smaller also constitutes a disadvantage for the environment. Such products are highly complex and contain a mix of several materials. Miniaturization makes it more and more difficult, sometimes impossible to separate these materials, and this is an obstacle to recycling' (Sarkar 1999: 128). He added that Germany alone generates 120,000 tonnes of computer scrap every year, all containing highly toxic substances.

A key element of the self-congratulatory ideological aura radiating from the computer industry is its purported 'cleanliness'. Spokespeople for the computer industry continually contrast their industrial revolution of information and dematerialization to the dirty age of steam engines and internal combustion engines using coal and oil. This ideology is an unintended gift of the computer's original theorist, Alan Turing, who in the 1930s described the computer in totally abstract terms. Within that framework he was able to show that there were internal limits on the powers of computation, that is, there are numbers that are not computable by any computer. But Turing was entirely unconcerned about where the materials a computer is composed of came from; what the source of its free energy was; and what to do with the waste heat that its operation would generate. As Charles Bennett wrote, 'Computers may be thought of as engines for transforming free energy into waste heat and mathematical work' (Gleick 2011: 360). The need for free energy and its corollary waste heat – are inevitably intensifying as the use of communication and computational technology is more integral to the production and reproduction cycle of the capitalist system than ever before.

The fact that digital electronic products leave a path of destruction behind them is also confirmed by the notoriously exploitative conditions under which they are produced. Workers at mainland China's Foxconn plants – the largest producer of computers in the world – have often threatened to commit suicide in protest against long hours of work and starvation wages. Of course, digital tools have also been used to enhance political struggles globally (Brophy and de Peuter 2014: 66). But activists and scholars should refrain from celebrating digitalization without accounting for the conditions under which its technologies are produced. It is impossible, for instance, to assert un-problematically – as by some radical circles, that the internet represents a new kind of common when its material production is based upon the destruction of many natural commons, expropriation and toxic contamination of vast areas of land, and displacement or murder of those who once lived there.

These concerns are especially important as the question of technology is central to all twenty-first-century transformative projects. It is crucial to develop a comprehensive outlook on the question of computers by investigating both the conditions of their production and effects of their use. A guideline in this process should be the realization that, historically, capitalist technology has been produced to control working-class struggle and destroy existing forms of organization at the base of working-class resistance. Digitalization cannot be simply appropriated, and directed to different goals.

Further Resources

Brophy, Enda and Greig de Peuter (2014), 'Labor Mobility: Communicative Capitalism and the Smartphone Cybertariat', in Andrew Herman, Jan Hadlaw and Thomas Swiss (eds), *Theories of the Mobile Internet: Materialities and Imaginaries*. New York: Routledge.

Dias, Elizabeth (2009), 'First Blood Diamonds Now Blood Computers?', *Time Magazine*. http://www.time.com/time/world/article/08599,1912594,00.html.

Gleick, James (2011), *The Information: A History, A Theory, A Flood*. New York: Vintage Books.

Sarkar, Saral (1999), *Eco-Socialism or Eco-Capitalism? A Critical Analysis of Humanity's Fundamental Choices*. London: Zed Books.

George C. Caffentzis is Professor Emeritus in the Department of Philosophy at the University of Southern Maine, Portland, and the author of *In Letters of Blood and Fire: Work, Machines, and the Crisis of Capitalism* (2013) and *No Blood for Oil: Essays on Energy, Class Struggle and War 1998–2016* (forthcoming).

EARTH SYSTEM GOVERNANCE

Ariel Salleh

Keywords: Earth System Governance, Anthropocene, neoliberal hegemony, development, cultural autonomy, embodied knowledge

In the twenty-first century, the relation between rulers and ruled is a widening gap, although the potential for a locally grounded global democracy inheres in grassroots alter-globalization movements. With capitalism currently in overproduction and stagnation, the neoliberal accumulation drive turns towards financial speculation. State functions are captured by the corporate sector; labour regulation and welfare measures contract. Proposals for Earth System Governance (ESG) seek an international political architecture where climate and biodiversity are 'post-sovereign' issues. Earth System Governance speaks to 'political actors' other than states, namely intergovernmental bureaucracies, business, and elite scientific networks. Beyond this transnational ruling class, the creative agency of workers, indigenous peoples, and care-giving women are relegated to the background.

Earth System Governance is offered as a new 'knowledge paradigm' for an environmentally responsive global economy and polity. Its website posits the '5-As': analytical problems of architecture, agency, adaptiveness, accountability, allocation, and access. These are paired with four cross-cutting research themes: power, knowledge, norms, and scale. In addition, ESG has four case-study domains or 'flagship activities': water system, food, climate, and economy. Like the Anthropocene argument, to which ESG is linked, it deflects historical tensions between capital and labour, geographic core and periphery, production, and reproduction. By 'naturalizing' man-made problems, both the Anthopocene notion and ESG potentially displace social responsibility while defending the capitalist status quo.

Early in the '70s, US foreign policy strategist George Kennan had called for a global management body located outside the United Nations. Support came from the Mont Pelerin Society and US Heritage Foundation, right-wing promoters of individualism, private enterprise, competitiveness, and free trade. Partly in response to this call, a World Economic Forum emerged in 1987, and a World Business Council for Sustainable Development was proactive at the 1992 Rio Earth Summit. The Rio Agenda 21, Biodiversity, and Climate Change Conventions reflect this influence. A Global

Environment Facility was installed in the World Bank soon after. By the late '90s, proposals for a 'World Environment Organization' to operate alongside the neoliberal World Trade Organization came from President Chirac of France and Chancellor Kohl of Germany, backed by Brazil, Singapore, and South Africa.

While European scientists were talking about Earth System Analysis, the Potsdam Institute was reviewing the 800 or so Multilateral Environment Agreements (MEAs). The point of this was to bring the Convention on International Trade in Endangered Species (CITES), hazardous wastes (Basel Convention), ozone levels (Montreal Protocol), and biosafety (Cartagena Protocol) into line with the General Agreement on Tariffs and Trade (GATT). Other key participants in the ongoing dialogue on environmental governance are the International Chamber of Commerce (ICC), the World Bank, the Organisation for Economic Co-operation and Development (OECD), UNESCO, International Trade Union Confederation (ITUC), and the New Economics Foundation (NEF). Since the 2008 financial crisis, Green Economy Coalition (GEC) has embraced large NGOs.

With a head office at Lund university, Earth System Governance research gains standing from the International Human Dimensions Programme on Global Environmental Change (IHDP), the United Nations University, the International Council for Science (ICSU), and it actively enlists academic centres around the world. It appears extremely well funded, with sponsors including the Potsdam Institute and Volkswagen Foundation. The ESG website displays projects, conferences and publications. A central theme is the idea of a 'World Environment Organization', possibly achieved through upgrading the United Nations Environment Programme (UNEP) by endowing it with sanction powers over nation states as the World Trade Organization (WTO) has. Alternatively, some ESG advocates see the International Labour Organization (ILO) offering a model; others argue that an agency designed to mediate government, business, and labour is unsuited to resolving complex environmental conflicts among actors with diverse cultural interests. Neoliberal innovations such as public-private partnerships (PPPs) are also within the purview of ESG researchers.

When the Rio+20 Summit was held in 2012, the ESG network put forward a World Environment Organization proposal for deliberation. This coincided with a corporate-government-UN-sponsored 'green economy' agenda – *The Future We Want* – drafted with help from public relations firms. People's ecology and social justice movements met this establishment proposal with a global vision entitled *Another Future Is Possible!* In Via Campesina's words: 'We demand a complete ban on geo-engineering

projects and experiments under the guise of "green" or "clean" technology
. . . We struggle for small-scale sustainable food production for community
and local consumption'. By 2015, the mission embedded in the UN's
Millennium Development Goals was transferred to a set of Sustainable
Development Goals reflecting the corporate 'green economy' agenda.

Civil society voices rightly resist the promotion of market values as an
organizing principle of everyday life and political decision-making. While
paying lip service to the democratic 'principle of subsidiarity', capitalist
development projects and free trade colonize resources, labour, and markets
in the global periphery; so, people lose local livelihoods and cultural
autonomy. Top-down plans for 'sustainable development' are underwritten
by extractivism. Market logic such as Payments for Ecosystem Services
(PES) are simply an opportunity cost for the global South. The just principle
of 'common but differentiated responsibilities and respective capacities',
embedded in the original Kyoto Treaty, is put aside as international
negotiations drag on at meetings of the UN Framework Convention on
Climate Change.

With its '5-As' and technico-legal engineering of authority at multiple
scales, ESG constitutes a hegemonic process, talking change while walking
hand in hand with those in power. The belief that nature exists for
human convenience and 'instrumental rational' assumptions, that it can
be controlled reflect a masculinist hubris born of the European scientific
revolution. Earth System Governance and the Anthropocene thesis reinforce
this violence through the abstractions of systems theory. As a 'knowledge
paradigm', ESG bypasses critical scholarship as well as peasant, indigenous,
and ecological feminist perspectives grounded in the labour of maintaining
living processes. Today's transnational ruling class, with its objectifying,
life-dissociated ways of knowing, can only promote an illusion of 'earth
governance'. A post-development response to ecological and social crises
must be embodied, empirical, and democratic.

Further Resources

Biermann, Frank and Steffen Bauer (2005), 'The Rationale for a World Environment
    Organization' in Frank Biermann and Steffen Bauer (eds), *A World Environment
    Organization: Solution or Threat for Effective International Environmental
    Governance*. London: Ashgate.
Davos World Economic Forum Privatization Agenda,
    https://www.weforum.org/focus/davos-2018.
Earth System Governance, https://www.earthsystemgovernance.org/.
La Via Campesina, www.viacampesina.org.
Salleh, Ariel (2015), 'Neoliberalism, Scientism, and Earth Systems Governance',

in Raymond L. Bryant (ed.), *International Handbook of Political Ecology*. Cheltenham: Edward Elgar.

Steffen, Will, Jacques Grinevald, Paul Crutzen and John McNeill (2011), 'The Anthropocene: Conceptual and Historical Perspectives', *The Royal Society Publishing*. 369 (1938): 842–67.

The Global Governance Project, www.glogov.org.

UNCSD-Rio+20 (2012), *The Future We Want*, https://sustainabledevelopment. un.org/content/documents/733FutureWeWant.pdf.

Ariel Salleh is an activist, author of *Ecofeminism as Politics: nature, Marx, and the postmodern* (1997/2007) and editor of *Eco-Sufficiency and Global Justice: women write political ecology* (2009). She was a founding editor of the US journal *Capitalism Nature Socialism*; is an Honorary Associate in Political Economy at the University of Sydney; Senior Fellow, Friedrich-Schiller-Universitat Jena; and Visiting Professor, Nelson Mandela University. She is a Member of the Permanent Group on Alternatives to Development established by the Rosa Luxemburg Foundation.

<hr />

ECOMODERNISM

Sam Bliss and Giorgos Kallis

Keywords: post-environmentalism, decoupling, Anthropocene, technology

Ecomodernism foresees salvation in technology. The solution to the environmental problems created by technology, ecomodernists argue, is more technology. They call for concentrating human activity in dense cities and factory farms to leave more room for wildlife. They advocate deploying nuclear energy, genetic modification, synthetic materials, and more new technologies that 'decouple' humans from nature. Their goals are to shrink humanity's total environmental impact and to achieve economic development for all. These two objectives, they argue, can only be accomplished by breaking free from biological resources and natural cycles. Peasants, pastoralists, forest dwellers, and fishing communities who directly rely on the ecosystems they inhabit are cast as environmental villains destroying otherwise unspoiled nature. Ecomodernists acknowledge that greater energy and material efficiency makes society wealthier and leads to greater consumption. But they have faith that innovation will unleash cheap, abundant, clean, and dense energy sources that will make growth sustainable.

Ecomodernism is an idea born in the United States. Unlike the 'eco-efficiency' and 'ecological modernization', schools of environmental thought more prominent in Europe, ecomodernists do not emphasize energy conservation or renewables. They rarely talk of free markets or carbon pricing; instead, they propose that governments fund research towards the necessary technological breakthroughs.

In April 2015, a group of eighteen scientists and intellectuals published *An Ecomodernist Manifesto* with 'the conviction that knowledge and technology, applied with wisdom, might allow for a good, or even great, Anthropocene'. The Manifesto is an easy-to-read, 3000-word text of simple arguments articulating an ecomodernist future full of high-tech, artificialized cities and pristine, untouched wilderness. It was choreographed by the Breakthrough Institute, an advocacy think tank in Oakland, California, founded in 2003 by Michael Shellenberger and Ted Nordhaus, both long-time strategists for environmental groups. In 2004, this duo authored *The Death of Environmentalism*, an essay that attacked the political strategies of the environmental movement, and called for a new 'post-environmentalism'. The 'post' prefix signalled a departure from the classic environmentalist demands for limits and regulation. It also hinted at the 'post-materialist' thesis that people appreciate nature and pay to protect it once they get rich. Over the following decade, the pair built a network of pro-nuclear, anti-resource-conservation conservationists around the Breakthrough Institute[1]. Ecomodernism *is* post-environmentalism.

The Manifesto in 2015 attempted to unite distant poles of the political continuum behind its technologically optimistic 'upwinger' vision framed in positive, apolitical language. No social movement has materialized. A June 2016 'march for environmental hope' to save a nuclear power plant in California was organized by Shellenberger's new pro-nuclear group Environmental Progress[2] and was attended by just 80 people. Apparently saving the environment by accelerating the very industrial development that destroyed it is not a narrative that mobilizes the masses. So why pay any attention at all to ecomodernism?

Because powerful actors do, the mainstream media and the academic world paid the Manifesto plenty of attention. Reviews ranged from enthusiasm to scepticism to harsh criticism, but what is noteworthy is the fact that publications such as the *New York Times*, the *Guardian*, and even the editorial board of *Nature*, the world's most-cited scientific journal, read and relayed the Manifesto's message. The authors of the text include distinguished environmental scholars such as David Keith of Harvard University and Ruth De Fries of Columbia University.

There is no social movement associated with post-environmentalism because post-environmentalism does not need one. It is simply an exaggerated version of some of society's most dominant attitudes and convictions: that consumerism-as-usual can continue if we support clean technologies, that poor communities degrade their environments by directly managing and harvesting the resources they subsist on, that climate change is a technical challenge that does not require social or cultural transformation, that economic growth is a natural, inevitable process. In a sense, post-environmentalism is anti-environmentalism with the twisted belief, supposedly based on science, that doing what we thought was bad for the environment is the only way to save it. One can imagine that it might comfort people to hear that sustainability and poverty alleviation can be achieved without sacrificing the affluent lifestyles that they enjoy or strive for. There is not even any need to protest or demonstrate, because all that must be done is simply to accelerate processes that have long been underway: urbanization, agricultural intensification, economic growth, substitution of biological resources and labour with minerals and modern energy, now nuclear fission or fusion. The Manifesto provides conservative politicians with a strong discourse for claiming that they are on the side of the environment while authorizing more of the same destruction.

But all the processes celebrated by the Manifesto have historically led to greater and greater, not lesser, environmental damage. Believing that speeding them up will reverse that trend runs counter to scientific evidence. Our thorough review of the literature reveals that the foundational claims of the post-environmentalists are not based on facts. Urbanization comes with increased resource use and pollution when one accounts for the footprint of cities. Intensified agriculture does not spare land for wilderness. New sources of energy add to old ones instead of replacing them. Developed countries only appear to reduce environmental impacts by exporting them to less-developed countries. People who 'decouple' from nature care less about protecting it. Social movements change the world for the better. Technologies alone do not.

Ecomodernism suppresses passion to organize and mobilize for socio-ecological transformation by assuring people that advanced artificial technology will replace nature in the economy, so that we can just let the natural world be. The message is that we humans will never learn to inhabit our common planet in a way that is more mindful of the other species with whom we share it; instead we must separate our economies from their ecologies. It has the makings of a self-fulfilling prophecy. One that, if realized, will be a total disaster.

Notes

[1] The Breakthrough Institute is a research and advocacy organization that serves as the unofficial headquarters of ecomodernism, http://www.thebreakthrough.org.

[2] Environmental Progress is ecomodernist Michael Shellenberger's new project, a campaign group fighting to keep old nuclear power plants from being shut down, http://www.environmentalprogress.org.

Further Resources

Asafu-Adjaye, John, Christopher Foreman, Rachel Pritzker, Linus Blomqvist, David Keith, Jayashree Roy, Martin Lewis, Stewart Brand, Mark Sagoff, Barry Brook, Mark Lynas, Michael Shellenberger, Ruth Defries, Ted Nordhaus, Robert Stone, Erle Ellis, Roger Pielke, Jr and Peter Teague (2015), *An Ecomodernist Manifesto*, http://www.ecomodernism.org.

Blomqvist, Linus, Ted Nordhaus and Michael Shellenberger (2015), *Nature Unbound: Decoupling for Conservation*. Oakland: Breakthrough Institute.

Shellenberger, Michael and Ted Nordhaus (2004), *The Death of Environmentalism: Global Warming Politics in a Post-environmental World*, http://www.thebreakthrough.org/images/Death_of_Environmentalism.pdf.

Sam Bliss is a PhD student in ecological economics at the University of Vermont, US. He is the US correspondent of the academic collective Research & Degrowth and a founding member of Degrow US.

Giorgos Kallis is an environmental scientist working on ecological economics and political ecology. He is a member of the Catalan Institute for Research and Advanced Studies and Professor at the Autonomous University of Barcelona.

ECOSYSTEM SERVICE TRADING

Larry Lohmann

Keywords: ecosystem services, environment, climate change, biodiversity, neoliberalism, carbon markets, Paris Agreement

Official responses to environmental crisis increasingly revolve around trading units of environmental benefit. The 1997 Kyoto Protocol, the 2005 EU Emissions Trading Scheme and the 2015 Paris Agreement claim to be tackling climate change through exchange of pollution rights. Similar schemes license trading in biodiversity tokens, which for instance, industrialists or developers can buy to 'neutralize' the very destruction they are responsible for.

None of these 'market environmentalist' initiatives has any potential to address the climate crisis, the biodiversity crisis, or any other ecological crisis. That is not their function. They are better understood as components of the capitalist struggle to find responses to the collapse of compromises it was forced into during the twentieth century.

One compromise involved the northern welfare state, demand management, and a high-wage, high-consumption deal for a northern white male labour aristocracy coupled with 'underconsumption' in the global South cheap oil supplies. This compromise faltered from the 1970s onwards: oil producers refused to keep prices low, women refused to do unpaid reproductive work, minorities refused racism, fed-up workers looked for ways off the treadmill. To cope with falling profit rates, fresh supplies of low-cost labour were assembled in the global South by separating historically unprecedented numbers of people from the land, and in the global North by separating workers from the welfare state, unions, and existing wage contracts. To give the new labourers something to work with, sweeping new offensives were launched to extract raw materials from commons and indigenous territories worldwide. Accompanying this re-energized extractivism was a 'neo-Keynesian' response to the problem of how low-paid workers were supposed to be able to buy all the new goods on offer: a vast expansion of private credit, in effect a colonization of the future wages of the poor. Finance also helped fill the profit gap by promoting speculative bubbles, asset-strips, derivative fabrication, real-estate speculation, industrial-scale tax evasion, thefts of public goods, and other swindles.

A second compromise that crumbled during the late twentieth century was national developmentalism, which capital had viewed as a way of damping down the revolutionary energies of post-colonial nationalist movements. With its promise of independence-oriented national-level divisions of labour between agriculture and industry, developmentalism inevitably stood in the way of more globalized relations of property and value. It also fell victim to contradictions inherent in its promotion of capitalist substitutes for communal approaches. The Green Revolution, food aid, and expanded infrastructure, together with 'land reform' centred on privatized individual holdings, only increased dependency and class divisions. Luckily for capital, the need for compromise diminished as the spectre of a socialist alternative faded after the Chinese reforms of 1979 and collapse of the Soviet Union a decade later. Equally fortunately, capital was able to turn the rise of the Organisation of the Petroleum Exporting Countries (OPEC) to its advantage by deploying petrodollar debt as a post-developmentalist means of disciplining the global South into a world market. A return to a more colonial-style global order was heralded by a new wave of coercive trade treaties and intercontinental infrastructure corridors, headlined by the World Trade Organisation (WTO) slogan 'made in the world'.

A third failed compromise was conventional environmental regulation, which both expressed and contributed to the 'maxing out' of the free waste sinks that industrial capital had long relied on. Regulatory bureaucracies had claimed to be able to manage crisis by applying pastiches of commons principles such as the unconditional right to life of various species including humans. But like welfarism – marked by a half-hearted defence of the human right to subsist – this compromise could not last. As soon as the landmark US environmental legislation of the 1970s was promulgated, it came under attack for being a 'growth ban'. Luckily, neoliberal ideologues, Washington-based think tanks and environmental NGOs were on hand to offer a way out. The regulation would stay, but its commons elements would go. Limits to degradation would be set not from 'outside' by experts ignorant of the needs of capital, but through collaboration with business. Physical science would be replaced by 'econoscience'. None of the rights of humans or nonhumans would be unconditional.

The key was to construct 'a new nature' consisting of standardized ecosystem services that could be traded worldwide. To avoid the expense of reducing environmental impact at home, businesses could now comply with environmental laws by buying, from near or far, low-cost units of ecological compensation. These would be CO_2 emissions-reduction equivalents, units of bat conservation, 'internationally transferred mitigation outcomes', and so on. Nature was retooled and 'averaged out' to mass-produce tokens of cheap regulatory relief alongside cheap resources and cheap labour, helping to keep open extractive and pollution pipelines that conventional environmental legislation had threatened to pinch off. The catch was that investment would flow into the new ecosystem services only when there was enough demand from extraction, fossil-fuelled manufacture, and infrastructure development. In an ultimate Orwellian reconciliation, a 'healthy' environment had come to depend on environmental degradation.

Thus power plants in Europe, say, could 'offset' their greenhouse gas emissions by colonizing the photosynthetic capacity of tracts of land in Latin America, Africa, or Asia. Corporations could also mine a hypothetical future by purchasing units of 'avoided degradation': as long as they could claim to be preventing what they decreed to be otherwise 'unavoidable' degradation elsewhere, private firms were legally authorized to continue business as usual at home. This logic amounted to a machine for regenerating self-fulfilling colonial mythologies. The rhetoric contrasted unimaginative Third Worlders fated to trash their environment through irresponsible industrial development, or slash-and-burn farming with enlightened Northern investors who alone were capable of independent action to ensure the future of nature. Like welfarism and developmentalism, conventional environmental regulation had given way both to more globalized value relations and to new colonialisms of space and time.

Because the point of ecosystem service trading is to cheapen regulation to facilitate capital accumulation, downward pressure on prices is as great as that in primary commodity markets. No southern country will make its fortune through ecosystem service trading any more than it will through neo-extractivism. Popular movements must oppose both as part of their struggles against austerity programmes, wage cuts, new enclosures of commons, financialization, free trade agreements, and other aspects of neoliberalism.

Further Resources

Araghi, Farshad (2009), 'The Invisible Hand and the Visible Foot: Peasants, Dispossession and Globalization', in A. Haroon Akram-Lodhi and Cristobal Kay (eds), *Peasants and Globalization: Political Economy, Rural Transformation and the Agrarian Question*. New York: Routledge.

Felli, Romain (2014), 'On Climate Rent', *Historical Materialism*. 22 (3–4): 251–80.

Pena-Valderrama, Sara (2016), 'Entangling Molecules: An Ethnography of a Carbon Offset Project in Madagascar's Eastern Rainforest', PhD thesis, Durham University, http://etheses.dur.ac.uk/11475/.

Larry Lohmann Works with The Corner House, a British NGO. He has lived in Thailand and Ecuador and is on the advisory board of the World Rainforest Movement. He is the autor of many academic articles as well as books such as *Energy, Work and Finance* (with Nicholas Hildyard, Sturminster Newton: The Corner House, 2014).

EFFICIENCY

Deepak Malghan

Keywords: efficiency, norm-deviation, history, political economy, quantification

The idea of efficiency is ubiquitous in the contemporary world. Constant improvement in efficiency is the goal of automobile engineers as well as that of central bank governors. The venerable *Oxford English Dictionary* tells us that efficiency has more than a dozen current avatars just as a noun. Efficiency is a measure of efficacy or 'success in accomplishing [an intended] purpose'. It is not surprising therefore that efficiency arguments have been central to both theory and practice of post-war development in the global South. Efficiency ostensibly provides an objective yardstick to measure the effectiveness of projects undertaken in the name of development. However, an efficiency metric is no more objective than its underlying 'purpose'. Efficiency is able to maintain a cloak of objectivity because from being 'an operative agent or efficient cause', efficiency has become an end in itself – the 'final cause'.

Efficiency is so pervasive in the current social order that disciplines ranging from computer science to culture studies, and everything else in between, study its manifestations. Despite apparent diversity in the various uses of efficiency, it is possible to identify a stable generative mechanism that explains its evolutionary history and political economy from its origins in the long English eighteenth century to the present time. Any measurement of efficiency involves three stages. First, a normative benchmark is defined for the phenomenon being studied. Examples might include maximum work that can be extracted from an ideal heat engine or the maximum pleasure that can be obtained from a human pursuit. Maximum human development would be another one. Second, observations are made about the actual state of the world measuring the performance of real steam engines. Finally, any efficiency measure simply reports the deviation of the observed state of the world from the normative benchmark. This norm-deviation structure has provided, from the nineteenth century to the present day, an enduring framework for addressing policy questions of improvement, progress, modernization, and development.

Efficiency is a quantified norm-deviation device, and its history intersects significantly with the rise of quantification in nineteenth century. Efficiency's intersection with quantification connects the first cotton spinning mills in Industrial Revolution Britain to the US Ford assembly line. It was the logic of efficiency and quantification that mediated the transition of the peasant and the craftsperson into industrial proletariat. By the turn of the twentieth century, quantified efficiency measures had made their way into homes with the rise of the home economics movement. The norm deviation framework underlying efficiency measurements was indeed central to the spread of quantification and objectivity. Subjective norms could now be buried inside a norm-deviation efficiency measure. There is an irreducible normative component to any efficiency metric that we no longer concern ourselves with. A monotonic increase in efficiency is an unalloyed good even if increasing efficiency results in the large-scale plunder of people and of the planet itself.

Efficiency has been integral to development paradigms on either side of the Cold War divide. From the very beginning, the norm-deviation structure underlying efficiency was central to measuring development performance. For example, when development economists study the progress of a country over time or make cross-country comparisons, they track a country's national income as a percentage of United States' national income for the corresponding years. Contemporary metrics to measure a broader set of development outcomes beyond national income, such as the Human Development Index (HDI), were directly derived from ideological and material battles over questions of efficiency. Efficiency goals are always cited as the rationale for the liberalization-privatization-globalization prescriptions that are supposed to result in improved development performance. However, the use and abuse of efficiency is not limited to advocates of neoliberal globalization as a tool for rapid economic development. Efficiency arguments were at the heart of the communist enterprise from the time of Bolshevik ascendency in Russia through the end of the Cold War. As Stalin suggested, 'the essence of Leninism' was the coming together of the Russian revolutionary sweep and American efficiency.

'Accumulation by dispossession' being the 'new imperialism' is part of a long chain of events punctuated by the idea of efficiency. A commitment to efficiency drives a political economy of production that necessitates centralization of productive resources, in turn, making conflicts and contestations over development inevitable. Efficiency as a national and social virtue has been at the heart of modern liberal thought at least since the beginning of nineteenth century and provided the normative justification for empires built on the back of colonial conquests. Development conflicts in large parts of the South over displacement and dispossession are fuelled by the very ideas of efficiency-inspired progress that sustained the colonial enterprise. While the core and periphery of present-day development conflicts are often situated within a common state or nation-state boundary, it is still useful to study the political economy of such conflicts through the efficiency lens.

In the age of development, there is no idea more important than economic growth that sustains our continued commitment to efficiency. Beyond recognition of the 'big tradeoff' between equality and efficiency, and episodic ameliorative interventions, the 'gospel of efficiency' has largely held sway. First, in the wake of the global oil crisis in the 1970s, and even more forcefully with the recognition of the existential threat of climate change, improving energy efficiency, and more broadly, ecological efficiency has been the favourite response to the ecological predicament. Efficiency improvements alone did not solve the Victorian fears around coal and will not solve our contemporary conundrums.

The efficiency revolution has run its course. Given the historical political economy of efficiency, a post-development world can only be created – or even imagined – if we let go of the commitment to efficiency. A post-development world cannot be built with efficiency as its guiding principle.

Further Resources

Alexander, Jennifer Karns (2008), *The Mantra of Efficiency: From Waterwheel to Social Control*. Baltimore: Johns Hopkins University Press.
Chatterjee, Partha (2011), *Lineages of Political Society: Studies in Postcolonial Democracy*. New York: Columbia University Press.

Deepak Malghan is an ecological economist and historian. He teaches at the Indian Institute of Management Bangalore in India.

GEO-ENGINEERING

Silvia Ribeiro

Keywords: technofixes, climate change, fourth industrial revolution, precautionary principle

Technology could play a positive role in addressing the deepening environmental, climate, social, health, and economic crises. To do so, the technologies must be ecologically sustainable, culturally and locally appropriate, socially just, and must integrate a gender perspective. In industrial societies, however, technology has become primarily a tool used to increase the profits of large corporations and powerful economic groups. This is certainly the case with the technologies fuelling the so-called fourth industrial revolution – biotechnology, genomics, nanotechnology, informatics, artificial intelligence, and robotics.[1] The convergence of these technologies has far-reaching implications and impact on societies.

When technology is presented as the solution to all crises, this serves those who control the technologies. The myth of technology as a silver bullet is based on the false assumption that questioning the root causes of crises is not necessary because every problem has a technological solution. Faced with the food crisis, for instance, governments and corporations have responded by proposing high-technology 'precision agriculture', with larger inputs of agrotoxics, genetically modified seeds and animals, 'climate-smart agriculture', suicide seeds with Terminator technology, and 'gene drives' aimed to eradicate entire species regarded as 'pests'. Faced with the energy and climate crises, the unsustainable systems of production and consumption based on fossil fuels are not questioned; instead, new technologies are put forward that would allow more intensive use of biomass, through synthetic biology and nanotechnology, thus promoting the expansion of giant monocultures of trees and genetically modified crops.

Industry invariably extols the potential benefits of these technologies while downplaying the risks or presenting them as uncertain or debatable. Networks of organizations, social movements and critical scientists are emerging as a response, with the goal of understanding and monitoring the complex technological horizon created by industry, while demanding application of the precautionary principle.[2]

One of the clearest and most extreme technological false solutions is geo-engineering, also called climate manipulation. Geo-engineering refers to a series of proposals for large-scale intervention in, and alteration of, ecosystems as a 'technofix' to climate change. It involves two main concepts, each with their own types of intervention: Solar Radiation Management (SRM), and Carbon Dioxide Removal (CDR), also known as Greenhouse Gas Removal (GGR). These proposals may involve interventions on land, sea or the atmosphere. None of them attempts to address the causes of climate change; instead, they just focus on managing some of its symptoms.

There is a variety of geo-engineering proposals, including injecting sulfates, or other chemicals into the stratosphere in order to block sunlight, aimed at producing a dimming effect; facilities to absorb carbon dioxide from the atmosphere and then burying it in marine or geological reservoirs; fertilizing the ocean with iron or urea to stimulate plankton blooms, hoping that they absorb greater volumes of carbon dioxide, altering the chemistry of the sea; and mega-plantations of transgenic crops that would reflect sunlight. All of the proposals carry huge risks, may have unpredictable synergistic negative effects and involve transboundary impacts.[3]

While each proposed geo-engineering scheme has specific risks and potential effects, they all share a number of negative impacts:

1. They propose to manipulate the climate – a global dynamic ecosystem essential for life on the planet – with the risk of creating larger imbalances than climate change itself.
2. To have an impact on the global climate, they must necessarily be mega-scale and so could also magnify the impacts.
3. Geo-engineering originated in military attempts to alter the climate as a weapon of warfare, the risk of weaponization is always present.
4. Schemes can be deployed unilaterally: a group of countries or economic actors could implement and deploy them for hostile or commercial interests.
5. Impacts will be felt unevenly among regions, severely affecting many countries in the global South that have contributed the least to climate change.
6. Experimental phases are not possible. Given the scale and time span necessary to differentiate effects from ongoing climate phenomena, experimentation would equal deployment.
7. Many schemes have been designed for profit and commercial use, particularly to earn carbon credits, which would increase the commercialization of climate crises.
8. Last but not least, technological solutions provide an excuse to continue the emission of greenhouse gases.

Along with governments of the global North, the main actors interested in geo-engineering are energy companies and other industries that have been among the major agents causing climate change. For them, geo-engineering is a good option because it enables them to continue emitting greenhouse gases while being paid for allegedly 'cooling the planet'.

The most active geo-engineering advocates include a small number of scientists, mostly in the United States and the United Kingdom, who have managed to convince their respective science academies to issue reports on geo-engineering. They also exert some influence on the Intergovernmental Panel on Climate Change (IPCC), which in its Fifth Assessment Report included a geo-engineering proposal – Bioenergy with Carbon Capture and Storage (BECCS) – as a component of most scenarios for the stabilization of the Earth's temperature. It is hoped that BECCS would provide the basis for 'net zero carbon emissions' or 'negative emissions'. These concepts, nevertheless, are highly speculative, and they create the illusion that greenhouse gas emissions can be increased since they would be compensated by BECCS or other technofixes. There is no independent proof or scientific study demonstrating their energetic, economic, or technological viability. Moreover, BECCS' impact on biodiversity and land and water usage could be enormous. It would also compete with agricultural land and pose a threat to indigenous and peasant territories.

Given the absence of a transparent and effective global scientific mechanism to deal with these technologies, in 2010 the Convention on Biological Diversity established a de facto moratorium on geo-engineering, upholding the precautionary principle due to the potential impacts of geo-engineering on biodiversity and the cultures that nurture it. Considering the seriousness of the impacts and their intrinsically unjust character, more than a hundred organizations and social movements across the world have demanded a ban on geo-engineering technologies since 2010.

Notes

[1] Since 2000, the ETC group referred to this convergence as BANG (bits, atoms, neurons, genes). In 2016, the World Economic Forum started to call the convergence the 'fourth industrial revolution'.

[2] The Red de Evaluación Social de Tecnologías en América Latina, RED TECLA, for instance, see http://www.redtecla.org/

[3] See http://www.geoengineeringmonitor.org/

Further Resources

Anderson, Kevin and Glen Peters (2016), 'The Trouble with Negative Emissions', *Science*. 354 (6309): 182–3, http://science.sciencemag.org/content/354/6309/182.

Biofuelwatch (2016), 'Last Ditch Climate Option or Wishful Thinking? Bioenergy
    with Carbon Capture and Storage',
    http://www.biofuelwatch.org.uk/wp-content/uploads/BECCS-report-web.pdf.
ETC Group (2010), 'Geopiracy: The Case against Geoengineering',
    http://www.etcgroup.org/content/geopiracy-case-against-geoengineering.
ETC Group and Heinrich Böll Foundation (2017), 'Climate Change, Smoke and
    Mirrors. A Civil Society Briefing on Geoengineering',
    http://www.etcgroup.org/content/civil-society-briefing-geoengineering.
Geoengineering Monitor, http://www.geoengineeringmonitor.org/.
SynbioWatch, http://www.synbiowatch.org/.

Silvia Ribeiro, originally from Uruguay, works in Mexico as the Latin America Director
for the international civil society organization Action Group on Erosion, Technology,
and Concentration (ETC Group), which is based in both Canada and the Philippines.

———◦·❖·◦———

GREEN ECONOMY

Ulrich Brand and Miriam Lang

Keywords: sustainability, growth, resource efficiency, valorization of nature,
externalization of costs

The Green Economy idea contains a threefold promise: to overcome the
economic crisis, the ecological crisis, and to alleviate poverty (UNEP
2011). Within the debate about a Green Economy, crucial commonalities
can be identified: its goals are a low-carbon production process, resource
efficiency, green investments, technological innovation and more recycling,
green jobs, poverty eradication, and social inclusion. The means to achieve
these goals are described as an 'adequate' political framework, one that is
able to internalize external costs, encourage sustainable consumption, green
businesses, and tax reforms. In 2011, the Organisation for Economic Cooperation and Development (OECD) developed a 'Green Growth Strategy'
that stresses innovation as a means to decouple growth from natural
capital depletion. The European Commission attempted to develop a plan
for sustainable growth promoting an ecological, yet competitive, market
economy by reducing resource use and increasing resource efficiency.

Proponents also argue that a wave of fascinating technological innovations promises a new economic growth period. One strategy against increasing environmental destruction consists in recognizing the economic value of nature by giving it a price. The assumption is that nature will be protected if it is included in the business calculation as 'natural capital' (Salleh 2012, Brand/Lang 2015, Fatheuer *et al.* 2016).

Promoted as a new global paradigm by the United Nations Conference on Sustainable Development (Rio+20) in 2012, the concept of a Green Economy is also strongly contested for its blind spots. Like sustainable development, Green Economy is an oxymoron used to legitimize international policy by bundling together quite different, even contradictory, interests and strategies for economic growth and the preservation of Nature.

A central critique of the Green Economy is its focus on growth, which actually means a material increase in resource extraction for the production of goods and services. This is measured in money and serves the logic of profits and capital accumulation. Who produces the products and under what conditions are secondary concerns. Moreover, governments of economically powerful countries do not question the western, imperial mode of production and living (Brand and Wissen 2012), holding to a form of capitalist globalization based on liberalization and deregulation. Competition for world-market share and the aim of economic growth prevail. For the business sector, a short lifespan for raw material–intensive products is often more profitable than the environment-friendly production of top-quality goods.

Strategies towards a Green Economy remain within capitalist rationality. The logic of being constantly oriented towards new investments, profit, and the dynamics of competition is not questioned. Green Economy does not counter capital-intensive mining, large-scale infrastructure projects, expensive offshore wind farms, or emissions trading. Very often, environmental problems are not solved but only displaced. An example would be when cars in Europe are run on 'renewable' agrofuels, while small farmers in Indonesia are expropriated of their livelihood as rainforests are cleared in order to establish palm oil plantations.

Gender perspectives and their focus on social reproduction and reproductive work are largely absent in the debate about a Green Economy, whose proponents usually mean the capitalist market economy, the goods and services produced as commodities to be sold. In contrast, feminist economists focus on everyday non-market activities which generate overall social well-being; and qualitative conditions such as the scope and capacity for self-determined action, or having more agency over one's time (Biesecker and Hofmeister 2010). In addition, ecofeminists emphasize the protection of global resources by choosing simple eco-sufficient lifestyles.

In sum, proposals for a Green Economy are at risk of intensifying the misleading capitalist valorization of nature. As a response to environmental destruction in some parts of the world, they remain aligned to the needs of corporations and the wealthy, and enable the stabilization of the capitalist, patriarchal, and imperial mode of production and living.

To understand the mentioned dominant dynamics in a critical manner, we suggest speaking of Green Capitalism instead of the Green Economy. This alternative concept points to the historical emergence of a new capitalist formation, replacing the old, crisis-ridden 'post-Fordist-neoliberal' mode of development, and its finance-dominated regime of accumulation. In countries such as Germany or Austria, 'green capital' might get stronger and 'traditional' fractions such as the automobile industry might become greener. A 'green' power bloc, a 'green' state and a 'green' corporatism – integrating workers and trade unions would be part of such a stabilization compatible with capitalist imperatives such as economic growth and competitiveness, and with distribution margins for enterprises and state institutions. For sure, a Green Capitalist social formation will evolve only in some countries and regions; it will be highly exclusive and – considering the externalization of costs into other world regions – not at all environment-friendly. In fact, Green Capitalism in some parts of the world means continued oligarchization of the imperial mode of living.

Further Resources

Biesecker, Adelheid and Sabine Hofmeister (2010), '(Re)productivity: Sustainable Relations Both between Society and Nature and between the Genders', *Ecological Economics*. 69 (8): 1703–11.

Brand, Ulrich and Markus Wissen (2012), 'Global Environmental Politics and the Imperial Mode of Living: Articulations of State–Capital Relations in the Multiple Crisis', *Globalizations*. 9 (4): 547–60.

———— (2015), 'Strategies of a Green Economy, Contours of a Green Capitalism', in Kees van der Pijl (ed.), *The International Political Economy of Production*. Cheltenham: Edward Elgar.

Fatheuer, Thomas, Lili Fuhr and Barbara Unmüßig (2016), *Inside the Green Economy: Promises and Pitfalls*. München: Oekom.

Salleh, Ariel (2012), 'Rio+20 and the Extractivist Green Economy', *Arena*. 119: 28–30.

United Nations Environment Programme (UNEP) (2011), *Towards a Green Economy: Pathways to Sustainable Development and Poverty Eradication*. Paris and Nairobi: UNEP.

Ulrich Brand is Professor of International Politics at the University of Vienna. His research and teaching focuses on global environmental and resource politics, social-ecological transformation, Latin America, and the 'imperial mode of living'. He was member of the German Bundestag's Enquete Commission 'Growth, Well-Being and Quality of Life' (2011–13) and is a current Member of the Permanent Group on Alternatives to Development established by the Rosa Luxemburg Foundation.

Miriam Lang teaches social and global studies at the Universidad Andina Simón Bolívar in Quito, Ecuador. She researches systemic alternatives, development critique, and the intersection of interculturality, gender, and societal relations to nature. She was coordinator of the Permanent Working Group on Alternatives to Development between 2011 and 2015 on behalf of the Rosa Luxemburg Foundation.

LIFEBOAT ETHICS

John P. Clark

Keywords: population control, neo-Malthusianism, tragedy of the commons

Lifeboat Ethics is a highly influential theory in contemporary applied ethics. It was developed by biologist Garrett Hardin and applied to issues such as world hunger, food aid, immigration policy and global population growth. In a 1968 article in *Science*, Hardin outlined his famous 'Tragedy of the Commons'; a situation in which individuals exploit a common resource for their exclusive personal benefit, resulting in degradation of the resource and serious harm for society in general. In a 1974 article in *Psychology Today*, he argued that such a tragedy occurs globally as an unintended functional outcome of food aid to those suffering from hunger and malnutrition.

Lifeboat Ethics asserts that the world is headed towards a catastrophic crisis in which global population will reach an unsustainable level, and that many countries have already reached such a level within their own borders. It claims that the primary cause of this crisis is the rapid rate of population growth, usually in countries of the global South. It holds that food aid from rich to poor countries is a major factor in producing unsustainably high fertility rates. It contends that food aid causes a 'ratchet effect' that prevents the population of a poor country from falling to a 'carrying capacity' that is posited as its 'normal' limit, and instead allows it to increase unsustainably. It predicts that the result of continued aid will be global economic collapse and population crash.

This standpoint carries on a long tradition of neo-Malthusian and Social Darwinist thought that has often been used to rationalize social inequality, economic exploitation and global imperialism as ways of maximizing the general good. As is typical of such ideologies, it is rife with theoretical inconsistencies and conflicts with empirical evidence.

To begin with, the core concept of carrying capacity is circular. No empirical evidence is offered that any specific level of population exhausts any actual capacity of a given geographical area to support human population; and no analysis is presented to demonstrate that any real-world case of decreasing population has been the result of exceeding such a capacity. Any concept of carrying capacity that is empirically based, as are some ecological footprint analyses, shows that affluent, industrialized societies consuming enormous quantities of fossil fuels and other resources will exceed capacity much more than poorer ones that consume relatively few resources per capita.

Furthermore, Lifeboat Ethics systematically ignores the fact that many poor and malnourished countries produce large quantities of goods, including agricultural products, that are exported to wealthy consumer societies, and that their domestic food scarcity is the result of global power relations and economic exploitation and trade, rather than high fertility rates exceeding carrying capacity.

Lifeboat Ethics rejects the possibility of a non-coercive 'benign demographic transition', but historical reality refutes this. Without the draconian population control measures that Hardin advocates, most countries in the world as of 2016, have fertility rates below replacement rate and three-fourths have rates below a modest 3.0. India's fertility rate at 2.45 is now far lower than the US's rate in the 1945–64 period, shortly before Hardin's Lifeboat Ethics manifesto was published.

Historical evidence also shows that contrary to the claims of Lifeboat Ethics the major causes of famine have been political and economic, not demographic. In such cases as Ukraine, Biafra, Bangladesh, Timor-Leste, and many others, famine was the result of deliberate state-policy goals, enforcing the authority of the ruling regime, protecting economic interests and most commonly, crushing dissident citizens and separatist movements.

In reality, the relationship between food security and fertility rates is precisely the opposite of that posited by Lifeboat Ethics. For example, much of sub-Saharan Africa has extremely low levels of social welfare in all areas, including food security. This should, according to Lifeboat Ethics, produce declines in fertility rates. Yet, the region also has the world's highest rates of population growth. Conversely, areas of the global South in which fertility is declining are those in which food production and other social welfare indicators have been improving.

The agenda of neo-Malthusianism is betrayed by inconsistent application of its own flawed ideological principles. Refusal to save human lives is not a path to the greater good of society. However, if saving lives in poor countries did in fact harm posterity, then saving lives in rich countries, in which each person consumes many times as much as a those in poor countries, would be much more harmful to future generations. Nevertheless, advocates of the Lifeboat Ethics argument never recommend sacrifice of the lives of affluent consumers to promote the general good.

Finally, Lifeboat Ethics is invalidated by the fact that it entirely ignores the intimate relationship between world hunger and colonial and neo-colonial development policies that have treated the global South as a source of cheap labour, raw materials, and agricultural products. Historically, colonial policies have gone through three stages that have generated widespread malnutrition and severe famine. These are:

1. forcible destruction of traditional subsistence economies based on the commons;
2. use of law, public policy, and coercive force to make indigenous labour subservient to the demands of imperial economic interests; and
3. refusal by authorities to allocate readily available food surpluses for famine relief.

Increasingly, today food scarcity is caused by economic and political factors combined with climatic conditions.

In view of its capacity to erase this history, to distort empirical realities and to disguise global exploitation as the normal course of nature, Lifeboat Ethics has functioned as a powerful tool of neo-colonial economics, and even genocidal models of development.

Further Resources

Clark, John (2016), *The Tragedy of Common Sense*. Regina, SK: Changing Suns Press.

Davis, Mike (2017), *Late Victorian Holocausts*. London: Verso.

Hardin, Garrett (1974), 'Living on a Lifeboat', *BioScience*. 24 (10): 561–68.

——— (1968). 'The Tragedy of the Commons', *Science*. 162 (3859): 1243–48.

Moore Lappé, Frances and Joseph Collins (2015), *World Hunger: Ten Myths*. New York: Grove Press.

Ostrom, Elinor (2015), Governing *the Commons*. Cambridge, UK: Cambridge University Press.

John P. Clark is a social ecologist; Director of the La Terre Institute for Community and Ecology; and Professor Emeritus of Philosophy at Loyola University, New Orleans. Widely published, his most recent book is *The Tragedy of Common Sense* (2016).

NEO-EXTRACTIVISM

Samantha Hargreaves

Keywords: accumulation, extractivism, resource nationalism, *progresismo*

A political discourse emerges in the late 2000s describing a new wave of extractivism in Latin American countries. It accompanies the global commodity price surge and coincides with the rise to power of some left-leaning Latin American governments. The related term, neo-extractivism, describes a variant of extractivism employed by these states to finance social reforms. Resource nationalism is a relative of neo-extractivism in Africa, understood as an assertion of a government's control over, and benefit from, natural resources in its territory.

Typically, neo-extractivist policy includes outright nationalization of some or all extractive industries, growth of public shareholding, a re-negotiation of contracts, efforts to grow resource rent through innovative taxation mechanisms, and value-adding beneficiation activities. Increasingly, neo-extractivist orientations have been promoted through global institutions and initiatives such as the United Nations, the Organisation for Economic Cooperation and Development (OECD), and the Extractive Industries Transparency Initiative (EITI). In the African context, the African Union, the African Development Bank (AfDB), and the Africa Progress Panel have played an active role.

Neo-extractivism and its weaker African counterpart, resource nationalism, have been held up by proponents as alternatives that support national development, safeguard the environment, and benefit local communities. Beyond this veneer of *progresismo*, however, the capitalist model of accumulation remains unchanged. 'Latin American neo-extractivism has demonstrated the limitations of this model of expecting exports and foreign investment to solve historical and structural problems of inequality, inequity, and above all, the destruction of the environment' (Aguilar 2012: 7).

Even though ownership patterns may shift in whole or part from the private sector to the state, productive processes continue to follow the standard rules of profit maximization, competitiveness, efficiency, and externalization of impacts. While the language of national self-determination accompanies neo-extractivism, rich countries remain in control, determining as they do what resources they import and where these are sourced from. The ideology of *progresismo* defends extractivism with a growth logic, arguing that the 'pie must grow' to fight poverty. Hence, foreign investment and productivism get to be promoted over protection of natural resources and the livelihood rights of indigenous and other communities.

Neo-extractivism induces natural resource conflicts, fails to create jobs, and externalizes social and environmental costs. Where nationalization has occurred, state-owned mining enterprises often operate no differently from private companies, as they continue to destroy the environment and disrespect social relations. Civic responsibilities of the state are compromised by the need to safeguard conditions for accumulation. The state defends neo-extractivism as being in the national interest, and movements and communities challenging its impacts are labeled 'anti-developmental'. Gudynas (2010) has noted that when surplus benefits are used by the state to finance social and public welfare programmes, the state gains a new source of 'social legitimacy'. The idea that extractivism is indispensable to development is heavily promoted in public discourse giving it hegemonic status.

The experience of Latin America is instructive for other regions seeking to progress via neo-extractivism and resource nationalism. The African Mining Vision (AMV) and its accompanying policy framework, Mining and Africa's Development (2011), is the most concrete expression of this tendency in Africa. The starting point of this framework is a claim that Africa is rich in untapped minerals that were misused historically and should now be exploited transparently, equitably, and optimally within the continent to achieve socio-economic development. Key to this modernist development strategy is minerals-based industrialization, expected to eradicate poverty and achieve sustainable growth as defined by the Millennium Development Goals. However, the AMV, widely supported by African civil society organizations through the African Initiative on *Mining* Environment and Society (AIMES), has failed in the face of a recent commodity price crisis, inter-state competition, and corporate interference in the national policy arena.

Returning to the question of externalization, the impacts of extractivism-based development are deeply sex-gendered ones. WoMin, a women-led women's rights alliance, fighting destructive natural resource extraction in

Africa, works on this alongside other feminist movements in Latin America, Asia, and the Pacific. These organizations make visible how cheap or unpaid labour of working class and peasant women in the global South contributes to capital accumulation. These often invisible relations of exploitation remain unchanged under state-led neo-extractivism. Women's reproductive labour roles mean that working class and peasant women work harder and longer hours than working men, to access safe drinking water and energy, and nurse children, workers, and other family members exposed to pollutants. Women's traditional responsibility for social reproduction is only partially relieved through state investment in social services. In addition to immediate social and environmental costs to communities, the long-term climate change impacts of extractivism will be carried mostly by poor women who labour to restore damaged eco-systems.

Neo-extractivism or resource nationalism is neither transformative nor emancipatory. It is, at best, a reformist trajectory, cloaked in the mantle of neoliberal development – *progresismo*. It may support some social reforms in the short to medium term, but will fail to resolve the deep contradiction between 'capital versus life', which is destroying humanity and the planet itself.

Further Resources

Aguilar, Carlos (2012), *Transitions towards Post-extractive Societies in Latin America*, https://womin.org.za/images/the-alternatives/fighting-destructive-extractivism/ C%20Aguilar%20-%20Post%20Extractive%20Societies%20in%20Latin%20 America.pdf.

Alternautas (Re)Searching Development, http://www.alternautas.net/about-us/.

Climate and Capitalism (2014), 'Progressive Extractivism: Hope or Dystopia?', http://climateandcapitalism.com/2014/06/24/progressive-extractivism-hope-dystopia/.

Gudynas, Eduardo (2010), 'The New Extractivism of the 21st Century: Ten Urgent Theses about Extractivism in Relation to Current South American Progressivism', *America's Program Report*. Washington, DC: Center for International Policy.

Petras, James and Henry Veltmeyer (2014), *The New Extractivism: A Post-Neoliberal Development Model or Imperialism of the Twenty-First Century?* London: Zed Books.

WoMin: African Gender and Extractives Alliance, http://womin.org.za/.

Samantha Hargreaves is the Founder and Director of the African Gender and Extractives Alliance (WoMin), a continent-wide African feminist NGO challenging destructive resource extraction. Her history of activism lies in land and agrarian reform and women's movement building.

REPRODUCTIVE ENGINEERING

Renate Klein

Keywords: misogynist science, reproductive engineering, genetic manipulation, surrogacy, eugenics

From test-tube babies to the erasure of women, Reprogen industries have gained increasing ground over the last forty years. Since the birth of Louise Brown by in vitro fertilization (IVF) in 1978, the twin industries of reproductive and genetic engineering, in tandem with population control advocates, have continued their global crusade to define:

- what parts of the world should be allowed to birth children
- what class, race, or age of women are acceptable as mothers
- what genetic qualities their children should have
- what sex these children should be.

Those scholars and activists among us who criticized these emerging technologies and policies in the 1980s were correct in combining the two industries, calling our network the Feminist International Network of Resistance to Reproductive and Genetic Engineering (FINRRAGE).

Since the 1980s, we have been witnessing industrial baby-making and genetic industries developing on parallel trajectories, but decades later these technologies are converging. The stated 'official aim' is to eliminate the pain and suffering of infertile people, as well as alleviating the pain and suffering caused by genetic diseases. The 'unofficial' aim is to make as much money for shareholders as possible. And indeed, they have: the capitalist genetic engineering and Reprotech industries are worth billions worldwide.

Pro-natalist and anti-natalist technologies are two sides of the same misogynist patriarchal coin. The strategy of Reprogen industries is to exploit the desire for a biological child and, more recently, to exploit women's fear of being judged 'unworthy' of reproduction. For those deemed 'inferior', such as disabled women or those belonging to a poor and marginalized ethnic group, the usual instruction is: 'contracept/abort or else be sterilized'.

In-vitro fertilization is a brutal, expensive and largely unsuccessful industry but IVF clinics proliferate globally. Their most recent offerings include the dangerous practice of egg 'donation' services for older women and 'surrogacy' services for gay men (Klein 2017). In addition, the population of sub-fertile men has increased dramatically over the last thirty years due to toxic pesticide use (Schafer 2014) and other environmental hazards. Half of all IVF 'treatments' use Intracytoplasmic Sperm Injection (ICSI); through this a single sperm is inserted into the egg cell of a fertile woman. The FINNRAGE data base indicates that taking into account every painful IVF cycle a woman undergoes until a live baby is born, real success rates are still only 20–30 per cent, although clinics around the world routinely claim a 70 per cent success rate.

While the Repro industry was dismembering women into wombs and egg cells and recombining them at will, the genetic engineering industry announced its Gene Revolution in the 1980s. Recombinant DNA technology can produce a plethora of genetically modified bacteria and viruses. Among these are hybrid seeds for a 'Gene Revolution' to right the wrongs of the failed 'Green Revolution' in the so-called Third World. By the year 2000, the Human Genome was mapped and since then 'personalized medicine' offered tests at $US 1,500–2,000 for any 'nasties' waiting to impair health in later life or impair the lives of our not-yet-conceived children. Despite the indeterminate nature of these future risks, this medicalization of people creates anxieties around their genetic-make up and a sense of responsibility to identify the carriers of 'bad' genes early.

Especially problematic for pregnant women is the recent introduction of Non-Invasive Pre-natal Tests (NIPTs) based on screening for cell-free placental DNA in maternal blood. A single blood test done as early as the tenth week of pregnancy can reveal up to 100 monogenetic (single-gene-determined) diseases a fetus might suffer from. The professional message is: abort and start again but next time use IVF and Pre-implantation Genetic Diagnosis (PGD). In PGD, a single cell of an early embryo is removed and quality-checked, including its sex, and only the 'best' embryo is implanted. This is medical eugenics in action, playing into people's fear of a disabled child. Iceland reports that 99 per cent of suspected Down syndrome pregnancies are aborted (Cook 2017). Medical exploitation of such anxieties is seriously interfering with people's enjoyment of life and pregnancy.

The advent of Reproductive Technology demands attention from the international community; a critical mass, especially among young people, must reconnect with the radical feminist slogan, 'Our Bodies–Ourselves'. We 'are' our bodies and must resist the regimental intrusion of techno-science into everyday life. Far from offering 'choice' or 'self-determination', the Reprogen industries lead to 'other-determination' through soulless and cut-and-paste ideologies.

Once the artificial womb 'ectogenesis' is perfected (Bulletti *et al.* 2011), and once 'enhancement' of the human race is made possible by germline embryo editing using clustered regularly interspaced short palindromic repeats (CRISPR-Cas9), the patriarchal erasure of women will be complete. The CRISPR is a guide molecule made of ribonucleic acid (RNA) and Cas9 is a bacterial enzyme. The CRISPR RNA is attached to Cas9 so as to work as molecular scissors. This fast new gene-editing technology can make changes to early embryos that will irrevocably be passed on to the next generation (Klein 2017).

Meanwhile, commercial surrogacy using poor women as breeders for rich people profoundly violates the human rights of the birth mothers, the egg 'donors' and the manufactured children (www.stopsurrogacynow. com).

It is imperative that left-wing progressive thinkers join radical feminists to stop the technological destruction of what might be the very definition of human life itself.

Further Resources

Bulletti, Carlo, Antonio Palagiano, Caterina Pace, Angelica Cerni, Andrea Borini and Dominique de Ziegler (2011), 'The Artificial Womb', *Annals of the New York Academy of Science*. 1221 (1): 124–28.

Cook, Michael (2017), 'Iceland: Nearly 100% of Down Syndrome Babies Terminated', https://www.bioedge.org/bioethics/iceland-nearly-100-of-down-syndrome-babies-terminated/12391.

Feminist International Network of Resistance to Genetic and Reproductive Engineering (FINRRAGE), http://www.finrrage.org.

Klein, Renate (2017), *Surrogacy: A Human Rights Violation*. Mission Beach: Spinifex Press [German edition: (2018), *Mietmutterschaft: EineMenschenrechtsverletzung*. Hamburg: Marta Verlag].

Schafer, Kristin (2014), 'Pesticides and Male Infertility: Harm from the Womb through Adulthood – and into the Next Generation', http://www.psr.org/environment-and-health/environmental-health-policy-institute/responses/pesticides-and-male-infertility.html.

Stop Surrogacy Now, http://www.stopsurrogacynow.com.

Renate Klein is a biologist and social scientist, co-ordinator of FINRRAGE (Australia) and publisher of Spinifex Press. A long time international feminist health researcher, she was an Associate Professor of Women's Studies at Deakin University, Melbourne. Widely published, her most recent book is *Surrogacy: A Human Rights Violation* (2017).

SMART CITIES

Hug March

Keywords: ICT, technological sovereignty, urban transformation

The Smart City is an ambiguous concept that is deeply shaping urban sustainability debates as well as urban competitiveness strategies, both in the global North and in the gobal South. Its cornerstone is the intensive and pervasive use of information and communication technology (ICT) to enhance urban management and improve its sustainability.

While it is impossible to present a comprehensive list of cities, regions or countries rolling out Smart City plans, it is worthwhile to mention some of the most paradigmatic cases. Europe has been a frontrunner in the Smart City, with cities such as Amsterdam[1] or Barcelona[2] topping the Smart City rankings in the past years. While most Smart City strategies aim to retrofit the existing built environment by adding a 'digital skin' to the city, one may mention the Smart Cities built from scratch, such as Masdar in the United Arab Emirates or Songdo in South Korea. The concept has not only impacted global North urbanism but is also deeply shaping urban debates in the global South. Remarkable is the magnitude and ambition of the Smart Cities mission in India, which includes over 100 projects developing across the country. Last but not least, it is also important to mention that the concept is beginning to influence African urbanism.

With the continuous capturing of fine-grained data on urban metabolism; through the pervasive use of mobile applications (apps), sensors, smart meters, smart grids, integrated management platforms and the like, the Smart City promises more efficient and optimal use of resources, decrease in urban pollution and a better quality of life. Large ICT conglomerates, utilities, and international consultancies emerge as key actors in the implementation at the urban level of the Smart City.

This corporate-led refurbished digital version of Ecological Modernization through ICT presents many perils for a post-development transition. First, mainstream understandings of the Smart City denote a high degree of technological determinism. The intensive use of ICT is acritically assumed as an obligatory passage-point that would automatically ensure a better quality of life for all. Therefore, under the Smart City imaginary, technological change spearheads social change. An ontological perspective that frames socio-environmental urban processes as engineering and technical challenges solvable through technological fixes usually characterizes this narrative. Fuelled by a depoliticized grandiloquence, it overestimates the transformative capacity of technology while overshadowing the structural political–economical dimensions of socio-environmental urban problems such as poverty, discrimination or inequality. By doing so, hegemonic Smart City deployments replace the pursuit of socio-environmental justice and the 'right to the city' with that of the democratization of technology. Second, Smart City technologies can deepen urban splintering, reinforce existing unequal power relationships and enhance social disparities and exclusion of some stakeholders. Third, the Smart City can be understood as an engine to accelerate capital circulation and rent extraction by and for private companies in times of post-crisis urban restructuring. Monopolistic private control over smart technologies may result in a socio-technical lock-in precluding the materialization of alternative, more egalitarian socio-technical transitions. Fourth, it could also be viewed as a step closer to an urban dystopia of complete surveillance and a shift towards authoritarian urban governance.

Beyond the political–economic implications of the Smart City, the environmental benefits of urban ICT should be subjected to critical scrutiny. Smart City solutions are aimed at reducing water and energy use, curbing emissions in an efficient and cost-effective way. On the one hand, efficiency improvements may lead to an unexpected increased use of resources, following Jevons paradox. On the other hand, the production of Smart City technologies may entail socio-environmental impacts derived from the manufacturing, operation and disposal of ICT (e.g. extraction of conflict-ridden scarce elements such as critical metals and rare earths).

In a nutshell, from a critical perspective the Smart City could be characterized as an empty, hollow, and depoliticized signifier built in the image of capital to extract urban rents and promote economic growth. In other words, the Smart City may be understood as a refurbished digital version of ecological modernization applied at the urban scale at odds with a post-development alternative. However, what is truly problematic about the Smart City are not ICT and smart technologies *per se* but the political economy underpinning technocratic and corporate, technological determinist, a-spatial and pro-growth Smart City imaginaries. Indeed, a progressive, bottom-up and emancipatory subversion Smart City technologies and ICT may be viable. If developed under an open-source logic by cooperatives, small and medium-sized enterprises (SMEs) or non-for-profit organizations, and held under democratic public control, many Smart City technologies, such as smart meters, sensors, smart grids or open platforms might be interesting for a post-growth transition. Indeed grassroots activists have shown through bottom-up experimentation with ICT, ranging from mapping apps to DIY sensors, that they have the ability to appropriate, enact and adapt those technologies and the capacity to produce new data to articulate a politics of urban socio-environmental contestation. Elsewhere, local administrations concerned with issues on technological sovereignty are beginning to articulate alternatives to hegemonic and corporate-led Smart Cities. These alternative discourses and practices revolve around the collaborative redistribution of 'intelligence' and open up the possibility of a progressive civic, democratic, cooperative, citizen-based and community-led urban transformation, neither controlled by the technocratic elites and capital nor subsumed to the fetishism of perpetual economic growth.

Notes
[1] Documentary 'Smart City – In search of the Smart Citizen' (2015, directors. Dorien Zandbergen and Sara Blom) documents and discusses the conversion of Amsterdam into a Smart City world benchmark. It is published in creative commons and it can be watched at https://gr1p.org/en/documentary-smart-city-in-search-of-the-smart-citizen/.
[2] Barcelona Digital City (2017–20) is a roadmap drawn by the Barcelona town council to use technology and data to provide affordable and better services to citizens while making the government more participative, transparent, and effective. It also has a focus on digital empowerment to fight inequality and digital innovation to address social challenges. More about this can be read at https://ajuntament.barcelona.cat/digital/en.

Further Resources
Glasmeier, Amy and Susan Christopherson (2015), 'Thinking about Smart Cities', *Cambridge Journal of Regions, Economy and Society* (Special Issue on Smart Cities). 8: 3–12.
March, Hug (2018), 'The Smart City and Other ICT-Led Techno-Imaginaries: Any Room for Dialogue with Degrowth?' *Journal of Cleaner Production.* 197 (2): 1694–1703.
March, Hug and Ramon Ribera-Fumaz (2016), 'Smart Contradictions: The Politics of Making Barcelona a Self-sufficient City', *European Urban and Regional Studies.* 23 (4): 816–30.

Hug March teaches at the Faculty of Economics and Business, Universitat Oberta de Catalunya, Spain and is researcher at its Urban Transformation and Global Change Laboratory, Internet Interdisciplinary Institute (IN3). He is an urban political ecologist interested in critically understanding the role of technology and finance in socio-environmental transformation. He has extensively researched the political ecology of the water cycle.

SUSTAINABLE DEVELOPMENT

Erik Gómez-Baggethun

Keywords: eco-development, Brundtland Report, free trade, green growth

Sustainable development was defined in the report 'Our common future' – widely known as the Bruntland report – as 'development that meets the needs of the present without compromising the ability of future generations to meet their own needs' (WCDE 1987). Ever since, the term is the guiding principle to harmonize environment and development policies worldwide and it has recently gained renewed momentum with the launch of Sustainable Development Goals.[1]

Despite its popularity in policy circles, environmentalists have criticized sustainable development as green washing of conventional growth and development policies. It has also been blamed for restoring international consensus on growth in the 1980s, after the Club of Rome report *Limits to growth* (1972) had convinced many political leaders around the world (including the fourth president of the European Commission, Sicco Mansholt) that perpetual growth in a finite planet was not feasible.

In those early days of international sustainability policy, consumerist societies of rich industrialized countries were pointed to as the main threat to the global environment. Given limits to growth, redistribution of wealth was the favoured option to harmonize environmental protection and social justice. This spirit was captured in the term 'eco-development' and in the 1972 Stockholm Conference of the Human Environment. The influence of eco-development peaked at the 1974 Cocoyoc Symposium, which final declaration stated that 'the hope that rapid economic growth benefiting the few will "trickle down" to the mass of the people has proved to be illusory' and rejected the idea of 'growth first, justice in the distribution of benefits later' (UNEP/UNCTAD 1974: Article 1). However, eco-development soon met fierce opposition from powerful actors like Henry Kissinger who, as chief of US diplomacy, entirely rejected the Cocoyoc declaration in a cable sent to UNEP and UNCTAD directors (Galtung 2010). Sustainable development followed in the 1980s as the new guiding principle in sustainability policy, turning upside down previous framings of environmental problems and solutions. Growth was no longer presented as the cause of environmental problems but as the remedy. The Bruntland report states that 'the international economy must speed up world growth'

(WCED 1987: paragraph 74) and advocates 'more rapid economic growth in both industrial and developing countries' (ibid: paragraph 72). Anticipating the idea of 'green growth', the reports claimed that more rapid growth could be sustainable if nations shift the content of their growth towards less material and energy-intensive activities and develop more resource-efficient technologies (ibid: paragraph 32). This expectation of a 'dematerialized' economy where growth is decoupled from pollution and resource use was formalized some years later in the so-called Environmental Kuznets Curve (EKC) hypothesis, used by economists and bureaucrats since the 1990s to claim that growth and free trade are good for the environment.

Sustainable development effectively reshaped sustainability principles to fit economic imperatives of growth and shift in emphasis from social justice to 'poverty alleviation' also fitted dominant economic ideas favouring 'trickle down' over redistribution of wealth. In addition, the Bruntland report shifted the responsibility of environmental decline from the rich to the poor, referring to a 'downward spiral of poverty and environmental degradation' and claiming that 'poverty place unprecedented pressures on the planet's lands, waters, forests, and other natural resources' (p. 7).

Furthermore, the report's support of 'expansionary policies of growth, trade, and investment' (article 24) paved the way for a harmonious relation between sustainable development and the neoliberal globalization agenda of economic deregulation and free trade. Since the Bruntland report was launched, all Earth summit declarations have endorsed growth and trade liberalization in the name of sustainable development (Gómez-Baggethun and Naredo 2015). Sustainability policy, earlier watchdog and counterbalance of dominant economic ideas, had been turned by sustainable development into a docile servant.

Surprisingly (or may be tellingly), international sustainability policy keeps promoting growth despite empirical data that prove green growth and dematerialization to remain a myth. Although some environmental indicators have improved at local and urban levels, on a planetary scale, GDP remains highly coupled to resource use and carbon emissions. Some countries have dematerialized in relative terms (per unit of GDP) but there are no symptoms of absolute dematerialization (Wiedmann *et al.* 2015). The ECK hypothesis of dematerialization with GDP growth has come true only in developed countries that have outsourced their industry to developing countries with cheaper labour force and softer environmental regulations (Jackson 2017). Empirical data have proven the claim that economic growth is a precondition for environmental sustainability equally problematic. Evidence shows that per-capita carbon and material footprint of rich nations remains, in average, far larger than per-capita footprints of poor nations (Martínez-Alier 2005).

Three decades since the launch of the global sustainable development agenda, scientists argue that humanity has never been moving faster nor further from environmental sustainability than now. It is time for global sustainability policy to leave behind its subordination to the precepts of dominant economic ideology, including the technological dream of dematerialization and the case for an expansionary economy premised on the axiomatic necessity of growth. Meeting basic needs for all within planetary boundaries is humanity's biggest challenge for the twenty-first century and achieving this goal requires a radical change in our economic mindset (Raworth 2017). Whether we call it sustainable development or something else, the organizing principle to navigate this challenge needs to recognize the importance of redistributing our existing wealth and abandon *growth ideology*, that obsolete idea that growth must be placed at the core of economic and sustainability policy.

Note
[1] Sustainable Development Goals (SDGs) are a set of 17 global Goals and 169 targets created by the United Nations that define the guidelines for policies on sustainable development from 2015 to 2030.

Further Resources
Galtung, Johan (2010), 'The Cocoyoc Declaration', TRANSCEND Media Service, https://www.transcend.org/tms/2010/03/the-cocoyoc-declaration/.

Gómez-Baggethun, Erik and José Manual Naredo (2015), 'In Search of Lost Time: The Rise and Fall of Limits to Growth in International Sustainability Policy', *Sustainability Science*. 10: 385–95.

Jackson, Tim (2017), *Prosperity without Growth*. London: Earthscan.

Martínez-Alier Joan (2002), *The Environmentalism of the Poor*. Cheltenham: Edward Elgar.

Raworth, Kate (2017), *Doughnut economics: seven ways to think like a 21st-century economist*. Vermont, US: Chelsea Green Publishing.

WCED (World Commission on Environment and Development) (1987), *Our Common Future*. Oxford: Oxford University Press.

Wiedmann, Thomas O., Heinz Schandl, Manfred Lenzen, Daniel Moran, Sangwon Suh, James West and Keiichiro Kanemoto (2015), 'The Material Footprint of Nations', *Proceedings of the National Academy of Sciences*. 112 (20): 6271–76.

Erik Gómez-Baggethun is a professor of environmental governance at the Norwegian University of Life Sciences (NMBU), a senior scientific advisor at the Norwegian

Institute for Nature Research (NINA) and a senior visiting researcher at the University of Oxford. His research covers topics in environmental policy, ecological economics, and sustainability science.

———◆◆◆◆◆◆———

TRANSHUMANISM

Luke Novak

Keywords: transhumanism, artificial intelligence (AI), progress, singularity, existential risk

Transhumanists believe that human nature can evolve through the application of science to increase life span, intellectual and physical abilities, even emotional control (Bostrom 2007). By replacing cells and organs with genetically enhanced or machine-operated equivalents, people would be able to move more quickly and process information faster. The tools of transhumanism include technologies such as genetic engineering, IVF, cloning, germline therapy, artificial intelligence (AI), as well as the ultimately complete merging of machines and humans, known as singularity. Transhumanists aim to eliminate suffering and achieve a divine wisdom that far surpasses the capabilities of even the most intelligent human beings today.

Transhumanism is a small yet powerful clandestine movement. Its leading proponent, Ray Kurzweil, is employed by Google in machine learning and language processing and is a close friend of Larry Page, the CEO of Alphabet Inc, parent company of Google. He is the most radical of transhumanists, steering its agenda and inviting others to follow his project of human enhancement. Kurzweil (2005: 9) states:

> Singularity will represent the culmination of the merger of our biological thinking and existence with our technology, resulting in a world that is still human but that transcends our biological roots. There will be no distinction, post Singularity, between human and machine or between physical and virtual reality. If you wonder what will remain unequivocally human in such a world, it's simply this quality: ours is the species that inherently seeks to extend its physical and mental reach beyond its current limitations.

This merging of human and machine into a Singularity will foster what is called the 'post-human' age, anticipating a transcendence towards the Omega Point envisaged back in 1965 by philosopher Teilhard de Chardin, as a collective consciousness resembling God. At the Omega Point, humans will able to live indefinitely, conquer the cosmos and form consciousness with the universe. This transformation is believed to commence by uploading the human mind on to a computer. Further, transhumanists believe this world has meaning because it is mystical and magical and no one can calculate what the universe will look like after Singularity is achieved.

Transhumanism begins with an implicit belief in rational scientific progress. That is, progress for progress's sake. One scholar has suggested that 'transhumanism is "a secularist faith" that secularises traditional religious motifs, on the one hand, and endows technology with salvific meaning, on the other hand' (Wolyniak 2015: 63). This stems from the aspiration to transcend the human condition as the current biology of human beings is seen as weak and failing to meet the needs of the future. The highly political question of who is entitled to define future needs is not asked here.

Why are some scholars and scientists advocating transhumanism and why have so many people in late capitalist societies adopted this narrative? Humanists traditionally separate the categories of human and animal. In addition, they highlight 'reason and individual autonomy' in decision making. This idea of progress can be viewed from two lenses. German sociologist, Max Weber, points to its cultural affinity with scientific knowledge and the effect this has on the development of society and individuals. Conversely, the other lens is a historical one and based on a conviction that humans have always been on a project of technological advancement; that it is in the nature of being human to 'improve' one's life, akin to picking up the first tool (Toffoletti 2007).

Transhumanists hope that by transcending biology and taking control of natural processes of evolution, they will realize the end goal of becoming post-human. The World Transhumanist Association (2006) defines the post-human as a hypothetical future being 'whose basic capacities so radically exceed those of present humans as to be no longer unambiguously human by our current standards'. This specific usage of the noun 'post-human' should not be confused with the more cultural–philosophic usage of the term. The latter post-humanists claim that positioning oneself as a de-centred 'post-human subject' allows one to challenge the limitations of humanist anthropocentrism, masculinist nomenclature and binary relationships such as nature/culture, machine/human.

Transhumanists believe that by the year 2045, the combined human brainpower of all humans will be surpassed by computers (Kurzweil 2005: 70). Yet they fear a world dominated by super-intelligent AI inventions. Some even argue that the only way to alleviate such risk is for humans themselves to become transhuman. For example, Kurzweil feels we must prevent the rise of super-intelligent AI by encouraging humans to merge with machines. However, the risks transhumanists want to protect the world from are the same risks as their solutions would encourage.

Humanist and Christian scholars alike, and even Fukuyama, believe that enhancement will pose moral questions if it provides one individual with an unfair advantage over another. Theologians reject transhumanism on the basis of natural law; any attempt to alter the human condition is viewed as a sinful affront to God. Sociologist, Nick Bostrom, has introduced the term 'existential risk' in the context of dangerous technologies such as AI. He thinks that super-intelligence is one of several existential risks 'where an adverse outcome would either annihilate Earth-originating intelligent life or permanently and drastically curtail its potential' (Frankish and Ramsey 2014: 329). Either way, this world of the post-Singularity is so remote that it is impossible to calculate the risks.

Further Resources
Bostrom, Nick (2007), *In Defense of Posthuman Dignity*.
    https://nickbostrom.com/ethics/dignity.html.
Frankish, Keith and William Ramsey (2014), *The Cambridge Handbook of Artificial
    Intelligence*. Cambridge: Cambridge University Press.
Kurzweil, Ray (2005), *The Singularity Is Near: When Humans Transcend Biology*. New
    York: Viking.
Toffoletti, Kim (2007), *Cyborgs and Barbie Dolls: Feminism, Popular Culture and the
    Posthuman Body*. London: I. B. Tauris.
Wolyniak, Joseph (2015), '"The Relief of Man's Estate": Transhumanism, the
    Baconian Project, and the Theological Impetus for Material Salvation', in Calvin
    Mercer and Tracy J. Trothen (eds), *Religion and Transhumanism: The Unknown
    Future of Human Enhancement*. Santa Barbara: Praeger.
World Transhumanists Association, http://transhumanism.org/index.php/wta/hvcs/.

Luke Novak recently completed a BA Honours in sociology and anthropology. His thesis looked at existential risks posed by transhumanism, Singularity, and Artificial Intelligence (AI). He is currently studying for his Juris Doctor, at the University of New South Wales in Sydney, hoping to research the intersection of law with advanced technologies, and the risks they pose.

A PEOPLE'S PLURIVERSE: TRANSFORMATIVE INITIATIVES

AGACIRO

Eric Ns. Ndushabandi and Olivia U. Rutazibwa

Keywords: dignity, development, Rwanda, decoloniality

Being deprived of self-worth for so long taught us the full meaning of *Agaciro*. *Agaciro* is about creating self-worth. It is only achieved if we all value one another. Unity, standing together and holding one another accountable means we are fulfilling the responsibility of *Agaciro*. . . .

– H.E. Paul Kagame, President of Rwanda

'*Agaciro*' is a concept with multiple meanings, depending on the historical and geographical context and whether it refers to things, people or their interaction. Mostly translated as 'worth', 'dignity', and 'self-respect' (Behuria 2015), *Agaciro* speaks to the concrete lived experiences of Rwandans.

This essay explores the creative ways in which the concept of dignity is re-centered in development thinking and social practices through *Agaciro*. It looks at the extent to which it constitutes a post-development or decolonial alternative to hegemonic international development. Our preliminary conclusion is that in Rwanda, the symbolic power associated with *Agaciro* enables Rwandans to project themselves as principal agents of development, rather than recipients or beneficiaries. Considering the different policies for which *Agaciro* is invoked, we conclude that *Agaciro*'s potential for radical alternative thinking is real but not automatic.

Most consider *Agaciro* as a cultural attitude dating from precolonial times. *Agaciro* is seen as having been temporarily 'lost' during colonization – first by Germany, then Belgium – and even after independence (1962) that culminated in its complete suppression during the genocide against the Tutsi (1994). Ever since, the concept has not only been revisited and revalorized in popular, urban and elite culture, but has also been consciously translated into political ideology and policies, like the Agaciro Development Fund, after the genocide by the ruling party, the Rwandan Patriotic Front (RPF).

Agaciro reveals itself through deliberate public policy making aimed at both Rwandans in the country and the diaspora, as well as an indirect instrument of communication with the rest of the world. The Agaciro Development Fund was proposed by Rwandans at the ninth *Umushyikirano* or National Dialogue Council in 2011. Rwandans in the country and the diaspora were invited to make voluntary contributions to the Fund. The

Fund was launched officially by President Kagame on 23 August 2012 as a means to build up public savings to achieve self-reliance, maintain stability in times of turbulence in the national economy and accelerate Rwanda's socio-economic development goals, including the financing of key national projects under the country's Vision 2020. President Kagame said that the establishment of the Fund should be perceived as a sign of dignity – something earned and not given by others. What is important is not the amount of money, but the 'idea' itself.

As both a philosophy under construction and a public policy pushing for home-grown solutions, *Agaciro* does not automatically embody a radical alternative to international development thinking and practice. But it can.

When asked how the idea of *Agaciro* influences her thinking and behaviour, former Minister of East African Community (EAC) Affairs, Valentine Rugwabiza, gave the example of Rwanda's desire to do away with the second-hand clothing market in the region which, for her, runs counter to the idea of dignity in *Agaciro*. This example showcases the challenges that lie in the path of *Agaciro* automatically constituting a post-development alternative. In such a context it might constitute a simple transposition of liberal market preferences pushed by the West to one pushed and carried out locally.

On the other hand, the mere contestation of the leftover mentality, does, even if at this stage merely at a discursive or symbolical level, fundamentally challenge the idea of 'catching-up-with-the-West' through the deployment of the idea of dignity.

What, then, needs to be foregrounded for *Agaciro* to be deployed as a decolonial, or otherwise radical, alternative philosophy? Here we suggest three avenues related to epistemology, ontology and normativity as invitations for further research and engagement.

The first concerns the need to 'de-silence' knowledge on and from the global South. A systematic mapping, building a living archive of *Agaciro*, would greatly contribute to such an imperative. The aim should be to document its different interpretations and enactments related to the intrinsic value of dignity and self-reliance.

Secondly, *Agaciro* can and should be deployed to 'de-mythologize' some of our ontological understandings of both development and international relations or solidarity. *Agaciro* allows us to challenge our deep-grained suspicion with which home-grown solutions tend to be met and invariably held to an idealized Western yardstick; our near inability to conceive of the good life for the 'Rest' in the absence of the 'West'. *Agaciro*, as a renewed focus on dignity and ensuing self-determination, invites a radical rethinking of what international solidarity might look like.

Finally, there is an imperative to normatively connect *Agaciro* to an (im)material anti-colonial project of 'decoloniality' and unabatedly pitch its deployments against such a standard, avoiding its glorification simply because of its 'home-growness'. It requires a constant investigation of the extent to which the invocation of *Agaciro* constitutes a breach or reproduction of inequality, oppression, exclusion or violence. In the spirit of *Agaciro*, this exercise is first and foremost to be led by the people concerned. This includes vigilance against *Agaciro* being reduced to a neo-liberal strategy of individual responsibility for entrepreneurial success, when the concept is also – perhaps first and foremost – a radical re-affirmation of the dignity and undeniable value of every individual, embedded in history, community and environment. For external actors, deploying *Agaciro* decolonially might mean that their primary task is to apply the anti-colonial test to their (im) material engagement with the country.

Since the end of the genocide, Rwanda's reconstruction process has proceeded by denouncing the role of colonial governance and the two last regimes, which destroyed the relationship of Rwandan citizens with each other.

Agaciro stands for value, self-worth and dignity, and in different ways it is deliberately being enacted through public policy in contemporary Rwanda. At the heart of this idea lies an invitation for constant co-creation, appropriation, subversion, reinterpretation, and negotiation of what the re-centering of dignity in domestic and international, communal and inter-communal relations might and should entail. This is a call that transcends the Rwandan context, but a lot can be learnt from the Rwandan experience in this regard.

Further Resources

Agaciro Development Fund, http://www.agaciro.rw/index.php?id=2.

Behuria, Pritish (2016), 'Countering Threats, Stabilising Politics and Selling Hope: Examining the "Agaciro" Concept as a Response to a Critical Juncture in Rwanda', *Journal of Eastern African Studies*. 10 (3): 434–51.

Hasselskog, Malin, Peter J. Mugume, Eric Ndushabandi and Isabell Schierenbeck (2016), 'National Ownership and Donor Involvement: An Aid Paradox Illustrated by the Case of Rwanda', *Third World Quarterly*. 38 (8): 1816–30.

Rutazibwa, Olivia Umurerwa (2014), 'Studying Agaciro: Moving beyond Wilsonian Interventionist Knowledge Production on Rwanda', *Journal of Intervention and Statebuilding*. 8 (4): 291–302.

Rwanda Days, http://rwandaday.org/.

Uwizeye, Annette (2011), 'Agaciro Documentary Film', https://www.youtube.com/watch?v=1oBN_qmlQMU.

Eric Ns. Ndushabandi is the Director of the Institute of Research and Dialogue for Peace, (IRDP), in Kigali, Rwanda, an independent think tank, which includes a Youth Debate School. He teaches political science at University of Rwanda; his research focuses on homegrown initiatives in rebuilding the nation-state in a post-genocide context, and political and civic education through *ingando* (solidarity camp) and *Itorero* (traditional leadership).

Olivia U. Rutazibwa is a senior lecturer at the University of Portsmouth, UK. Her research centres around vernacular international relations, decolonial understandings of global solidarity, building on philosophies and practices of self-determination through Agaciro and Black Power, and Autonomous Recovery in Somaliland. She is the former Africa Desk editor and journalist at the Brussels-based magazine, *Mondiaal Nieuws*.

AGDALS

Pablo Dominguez and Gary J. Martin

Keywords: communal resource management, heritage, resilience, Maghreb, Mediterranean

'*Agdal*' – plural *Igudlan/Igdalen* or more commonly '*agdals*' in English – is a term of Tamazight origin, that is, of the Amazigh/Berber language family of North Africa. It primarily refers to a type of communal resource management in which there is a temporary restriction on the use of specific natural resources within a defined territory, with the intention of maximizing their availability in critical periods of need (Auclair *et al.* 2011).

Pastoral *agdals* in the High Atlas mountains of Morocco, where pastoralists may bring their herds only during a communally agreed period of time, are a salient example. *Agdals* may also be arboreal areas, such as the argan – *Argania spinosa* – woodlands in western Morocco, where access is prohibited as fruits ripen before harvest. *Agdal* restrictions are more broadly applied to agricultural fields, algal beds, areas of forage and medicinal plants, orchards and sacred sites, among other places in coastal lowlands, mountains, oases, and even urban areas.

Analysis of *agdals* reveals both biological and social advantages over other approaches to landscape management. They favour relatively equitable distribution of benefits from natural resources through collective decision-making and recognition of common access rights, and are governed by local institutions that establish use rules. *Agdals* maximize the annual production of culturally important biodiversity and maintain the viability of resource extraction over time, resulting in ecosystems that are more biodiverse and resilient than others in similar geographical contexts that are not communally managed. For example, pastoral *agdals* are typically opened to herds of domesticated animals in late spring to early summer when most grazed plants have completed their reproductive cycle, thus avoiding the degradation witnessed in the Maghrebian open access pastures.

At the same time, focusing solely only on the productive side of *agdals* would not present a complete picture. Beyond an agro-economic tool, the *agdal* is a cultural element around which turns a whole system of sacramental, ethical, aesthetic and other symbolic references that transforms the *agdal* in a faithful reflection of a mountain Amazigh culture and raises it to a total social fact. Indeed, ritual regulation used to comprise a strong share of the *agdal* system. Such rituals would include having the annual *agdal* prohibitions often announced and legitimized by saint descendants or even offerings such as bags of grain, butter, couscous, and cattle sacrifices given to them, who, in turn, would share the products with all the visitors and distribute *baraka* among the ritual holders.

The existence of *agdals* is putatively ancient, with its origins being traced back to the Amazigh societies that have been present in the region for many thousands of years (Auclair *et al.* 2011; Navarro Palazón *et al.* 2017). They have presumably persisted and recreated themselves continuously over time in North Africa and the Sahara, having adapted to diverse economic, environmental and political situations. The term '*agdal*' has also been used in new contexts, even in urban centres of rapid demographic expansion such as Rabat or Marrakech. In the latter, the term could have started to be applied to the real estate located south of the medina after the Alawite restoration (seventeenth century), which in turn was founded by the Almohade dynasty in the twelfth century (Navarro *et al.* 2017).

Although the *agdal* garden of Marrakech has attained an enhanced public profile – it was listed as a World Heritage Site in 1985 – rural *agdals* have greater contemporary relevance despite a general lack of awareness about their existence. They are the prototypical Indigenous Peoples' and Community Conserved Territories and Areas (ICCAs) of the Maghreb. They play a fundamental role in maintaining both intangible and tangible cultural heritage, ranging from ritual ceremonies to traditional agro-ecosystems, while enhancing the resilience of social-ecological systems in diverse environments. Following intensive academic study in the early twenty-first century, they are regularly evoked as a model of conservation and sustainable use based on knowledge, innovations, and practices of indigenous and local communities embodying traditional lifestyles, as defined in the Convention on Biological Diversity and other international environmental agreements.

Adgals are significant biodiversity hotspots, not only as rich reservoirs of fauna and flora, but also as areas highly threatened by expansion of privatized agricultural frontiers, grazing intensification, rural exodus, transition in cultural beliefs, and climate change. Pastoral and forest *agdals* have gained particular prominence as Important Plant Areas in the Moroccan High Atlas as they harbour a large number of endemic, endangered and useful plant species.

Increased awareness of such importance is leading to their inclusion in community-based projects of cultural and ecological restoration. This is piloted, albeit with funding and support from external institutions, by community associations that seek to either defend existing *agdals* or apply the principle of communal management to degraded areas, based on temporal prohibition of resource use. The motivation for this incipient renaissance is drawn from practical concerns about erosion and other environmental challenges, and broader interest in recovering cultural identity and landscapes.

Although *agdals* typically lack explicit governmental recognition – and are scarcely considered in national agricultural and conservation plans – awareness of the biocultural benefits of these communally managed areas has the potential of influencing public policy on sustainable land use.

In their multiple manifestations as cultural fact, communally managed territory and even urban green space, *agdals* remain anchored in a specific human and geographical context. Initiatives to recreate or impose them outside this realm would be inappropriate, but they do remain a viable alternative in specific areas across the Maghreb, and are an inspiration to advocates of community management of natural resources in the Mediterranean Basin and worldwide.

Further Resources

Agdal, Voices of the Atlas, https://www.youtube.com/watch?v=PMOfQXzlmDI&feature=youtu.be, a documentary, through a prior request for authorization to Pablo Domínguez.

Auclair, Laurent and Mohamed Alifriqui (eds) (2012), *Les Agdals du haut Atlas marocain: savoirs locaux, droits d'accès et gestion de la biodiversité*. Rabat: IRCAM/IRD, http://horizon.documentation.ird.fr/exl-doc/pleins_textes/divers13-07/010059469.pdf.

Auclair, Laurent, Patrick Baudot, Didier Genin, Bruno Romagnyand and Romain
    Simenel (2011), 'Patrimony for Resilience: Evidence from the Forest Agdal in the
    Moroccan High Atlas Mountains', *Ecology and Society*. 16 (4): 24,
    http://dx.doi.org/10.5751/ES-04429-160424.
Domínguez Pablo, Francisco Zorondo-Rodríguez and Victoria Reyes-García (2010),
    'Relationships between Saints' Beliefs and Mountain Pasture Uses', *Human
    Ecology*. 38 (3): 351–62, http://dx.doi.org/10.1007/s10745-010-9321-7.
Navarro, Julio, Fidel Garrido and Íñigo Almela (2017), 'The Agdal of Marrakesh
    (Twelfth to Twentieth Centuries): An Agricultural Space for Caliphs and Sultans',
    *Final Report of the Research Project Almunias of Western Islam: Architecture,
    Archeology and Historical Sources*. 106.
The Global Diversity Foundation, www.global-diversity.org.

Pablo Dominguez, an environmental anthropologist, has studied High Atlas Amazigh
populations and their systems of communal natural resource management. He now
focuses on how the concept of heritage can promote or devalue *agdals* and other
commons in Spain, France, Italy, East Africa, and Latin America. He teaches in the
Geography of the Environment Laboratory, Université Toulouse, France, and at
the Institute of Environmental Science and Technology, Universitat Autònoma de
Barcelona, Spain.

Gary Martin has been involved in conservation and ethno-botanical practice for over
thirty-five years in more than fifty countries. He is the Director of the Global Diversity
Foundation, University of Oxford, and was a lecturer in the School of Anthropology
and Conservation at the University of Kent, UK.

AGROECOLOGY

Victor M. Toledo

Keywords: agroecology, sustainable agriculture, food systems,
socio-ecological crisis

Agroecology is an emerging field of knowledge that offers solutions to the
serious environmental and food production problems caused by modern or
industrialized agriculture and agribusiness in the entire world. Agroecology
is a 'hybrid discipline', because it combines knowledge coming from the
natural and the social sciences. It adopts a multi-disciplinary scope along the

lines of 'post-normal science'. It seeks not only to be an applied knowledge but also an example of participatory research. As a form of critical thinking, agroecology engages in contesting not only social inequality, but also environmental disturbances. Agroecology scholars identify three spheres or dimensions as a sort of 'holy trinity': ecological and agricultural scientific research; empirical agricultural practices; and the need to elaborate an approach with and for rural social movements. In the last two decades, the number of publications and initiatives that people describe as agroecological has increased exponentially. Similarly, the number of social and political movements embracing agroecology as their main goal has been growing at a high rate.

While the scientific and practical dimensions of agroecology refer to cognitive and technical fields respectively, the third dimension is linked to social movements and the political actions of peasant communities. Many actors – including peasants, rural households, indigenous people, rural landless workers – men and women – are using agroecology as a tool for the contestation and defence of their territories and natural resources, their lifestyles and their biocultural heritage. Examples are the numberless national-level peasant unions principally in Latin America, India and Europe. The most well-known is La Via Campesina[1], a global alliance of 200 million farmers, comprising about 182 local and national organizations in 73 countries from Africa, Asia, Europe, and the Americas. It defends small-scale sustainable agriculture as a way to promote social justice and dignity. It strongly opposes corporate-driven agriculture that destroys people and nature.

In Latin America, the practice of agroecology involves scientific and technological research carried out in close association with rural social and political movements, a trend that has experienced an unprecedented expansion in many countries of the region. Agroecology is practised by tens of thousands of peasant households as a result of either social movements or the implementation of public policies, with extraordinary advances in Brazil, Cuba, Nicaragua, El Salvador, Honduras, México, and Bolivia, and with moderate achievements in Argentina, Venezuela, Colombia, Peru, and Ecuador.

In its early stages, agroecology was conceived as a mere technical field focused on applying ecological concepts and principles to the design of sustainable agricultural systems. However, as it grew, this approach was followed by a more explicit integration of concepts and methods from the social sciences addressing cultural, economic, demographic, institutional, and political issues. The dominant trend of agroecology in Latin America is 'political agroecology', understood as a practice of agroecology that acknowledges that agricultural sustainability cannot be achieved simply by technological innovations of environmental or agronomical nature, but by much-needed institutional change in power relations, that is, by taking into account social, cultural, agricultural, and political factors.

The evolution of agroecological thinking constitutes a very interesting process from an epistemological perspective. Its main epistemological innovation has been the 'intercultural dialogue', through which scientific researchers recognize the forms of knowledge embedded in the minds of traditional farmers. It is considered that this local, traditional, or indigenous knowledge constitutes a 'biocultural memory or wisdom', orally transmitted through hundreds of generations. This non-scientific knowledge has been utilized by indigenous peoples for thousands of years to produce foods and other raw materials. Because of this, some authors define agroecology as a trans-cultural, participatory, and action-oriented approach; some also consider agroecology as a new expression of Participatory Action Research (PAR), a movement that emerged from critical social scientists throughout the so-called Third World in the 1970s as an innovative approach promoting emancipatory change. Because agroecologists recognize traditional *cosmovisiones* or worldviews, knowledge and practices as a basis for scientific and technological innovation, it puts into practice the concepts of intercultural dialogue and the co-production of knowledge.

Most of the world's agricultural production continues to be generated by peasants, or small-scale traditional farmers with an estimated population between 1,300 and 1,600 millions. Their farming knowledge and practices are the product of over 10,000 years of tradition and experimentation. This fact has been recently corroborated by the Food and Agriculture Organization (FAO), showing that the bulk of the foodstuffs destined to feed the nearly seven billion humans are produced by small household producers, a recognition leading FAO to declare 2014 as International Year of Family Farming. A study made by the international non-governmental organization GRAIN in 2009 confirmed that peasants or small-scale farmers indeed produce most of the food globally consumed by humans, but adding that they accomplish this feat in only 25 per cent of the total agricultural land surface in plots averaging 2.2 hectares. The remaining three quarters of the total agricultural land is owned by 8 per cent of agricultural producers including medium-, large-, and very large-scale landowners such as owners of *haciendas* or *latifundia*, companies and corporations, which usually adopt the agro-industrial production model.

Agroecologists work predominantly, but not exclusively, with small-scale farmers, peasant communities and indigenous peoples in the amelioration of food systems, agrarian justice and the emancipation of rural peoples. In fact, to overcome the crisis of the modern industrial and technocratic contemporary world, we require food production systems that are compatible with the environment, rural cultures, and human health. Agroecology is, therefore, a crucial scientific, technological, intercultural, and socio-political instrument that confronts the ecological and social crises of the contemporary world, as one searches for a post-industrial, alternative modernity.

Note

[1] Via Campesina (from Spanish La Vía Campesina, the campesino way, or the peasants' way) was founded in 1993 by farmers organizations from Europe, Latin America, Asia, North America, and Africa. It describes itself as 'an international movement which coordinates peasant organizations of small- and middle-scale producers, agricultural workers, rural women, and indigenous communities'. It is a coalition of over 182 organizations, advocating family-farm-based sustainable agriculture and was the group that coined the term 'food sovereignty'. La Vía Campesina, https://viacampesina.org/en/index.php.

Further Resources

AgriCultures Network, https://magazines.agriculturesnetwork.org/.

Altieri, Miguel A. and Victor Manuel Toledo (2011), 'The Agroecological Revolution in Latin America: Rescuing Nature, Ensuring Food Sovereignty and Empowering Peasants', *Journal of Peasant Studies*. 38 (3): 587–612.

Méndez, V. Ernesto, Christopher M. Bacon, Roseann Cohen and Stephen R. Gliessman (eds) (2015), *Agroecology: A Transdisciplinary, Participatory and Action-oriented Approach*. Boca Raton: CRC Press.

The Latin American Scientific Society of Agroecology (SOCLA) promotes the development of the science of agroecology as the scientific basis of a sustainable rural development strategy in Latin America, https://www.socla.co/en/.

Toledo, Victor Manuel and Narciso Barrera-Bassols (2008), *La Memoria Biocultural*. Barcelona: Icaria Editorial. See also a Portuguese version in Editora Expressào Popular, 2015.

Victor M. Toledo works at the Research Institute on Ecosystems and Sustainability, National University of Mexico (UNAM) with a focus on the study of the relationships between indigenous cultures and their natural ambience (ethno-ecology), sustainable societies, and agroecology. He is author of over 200 scientific publications including twenty books.

ALTER-GLOBALIZATION MOVEMENT

Geoffrey Pleyers

Keywords: World Social Forum, prefigurative activism, counter-expertise, social movements

The uprising of the Zapatista indigenous movement against the Mexican government and the North American Free Trade Agreement on 1 January 1994 symbolizes the birth of the 'alter-globalization' or 'global justice movement'. La Via Campesina, which now has gathered as its followers over 200 million small farmers worldwide, had been founded three months earlier. Unexpectedly, indigenous people and small peasants became the frontrunner of the global movement that denounces the neoliberal order and explores or renews the paths to achieve emancipation.

While globalization was perceived as the end of the Second World sovietism and as the 'final victory' of the West's 'market democracy', the alter-globalization movement rather points to the end of the 'Third World', as it was understood in the twentieth century, and the rise of the global South among progressive movements. It is significant that the start of the movement, its main encounters, the World Social Forums, and the major steps of its 'ecological turn' mostly found place in the global South. The 'transformative encounter' between activists from the North and struggles, activists, epistemologies, and *cosmovisiones* from the South is the constitutive essence and the aim of the alter-globalization movement and the roots of emancipation struggles, practices, and epistemologies in the twenty-first century.

Local and national mobilizations dominated the first period of the alter-globalization movement. But its globality was more and more readily apparent, notably during mobilizations organized around global events such as the protest against the 'Millennium Summit' of the World Trade Organization (WTO) in 1999 in Seattle, USA. Committed intellectuals also played a major role in raising public awareness on the social consequences of free trade and challenging the hegemonic Washington Consensus. The movement gained impetus with protests against Free Trade Agreements and the World Social Forums held since 2001. It gathered up to 120,000 (Mumbai, 2004) and 170,000 (Porto Alegre, 2005) activists to share ideas and experience showing that 'another world is possible'. Since 2011, a new global wave of protests in all continents has prolonged alter-globalization movements, notably by denouncing austerity politics, rising inequalities, and the collusion between political, economic, and media elites.

Alter-globalization activists point to financial speculation, tax heavens and the concentration of resources in the hands of the super-rich as the main cause of social and ecological damages. Since its beginning, the alter-globalization movement has also mobilized against the extreme-right, nationalism, borders wars, and in support of migrants. Environmental issues gained increasing attention inside the movement. Global justice and green activists united in the 'Climate Justice Network', founded in Bali, Indonesia, in 2007. They stated that avoiding global warming requires structural changes in the current capitalist economy and political system.

Against the dominant idea that 'there is no alternative' to neoliberal policies, alter-globalization activists claim that ordinary citizens can have an impact on local, national, and global politics. They build on the following three cultures of activism.

The way of reason: Citizens' counter-expertise. In the 'way of reason' alter-globalization activists consider that a fairer world requires citizens to engage in the debates about global issues. They show that neoliberal policies are socially unfair, undemocratic, scientifically irrational, and economically inefficient. They develop alternative and 'more rational' policies oriented towards the common good.

These activists consider the major challenge to be the way that economy is bound to social, cultural, environmental, and political standards. Following a rather top-down concept of social change, these activists urge policy makers and international institutions to regulate the economy under the monitoring of committed experts and citizens. Based on clear scientific evidence, efficient campaigns, and citizen mobilization, they have had a major impact in raising public awareness and prompting policy actions on major global issues, such as global warming and tax evasion. However, the rise of greenhouse gas emissions in spite of scientific reports urging governments to tackle climate change, and the continuous strength of neoliberal policies that have lost all scientific legitimacy, also point to the limits of clear scientific arguments to foster global change.

The way of subjectivity: pre-figurative activism. Rather than expecting policy-makers to solve their problems, the second trend opts for a bottom-up concept of social change. The 'new world' starts with oneself and at a local scale. The guideline lies in the consistency between one's practices and values (democracy, participation, sustainability, gender equality, and so on). Indigenous and rural communities, autonomous movements, social centres, or Occupy movements, all seek to create 'spaces of experience', understood as 'autonomous spaces distanced from capitalist society which permit actors to live according to their own principles, to knit string and convivial social relations and to express their subjectivity and creativity'. Concrete practices in daily life are significant far beyond the local scale as they challenge the capitalist way of life, opposing the hold of a consumerist culture. Activism is thus 'prefigurative' – it prefigures in concrete actions the elements of a sustainable and more democratic world. Activism is also 'performative' – the 'other world' starts here and now, in concrete and local practices.

Alliances with progressive governments. A more classic component of the movement believes that social change occurs through the alliance between progressive national governments and popular movements. In 2005, progressive leaders and movements managed to stop the Free Trade Agreements of the Americas. Alliances between movements and progressive governments placed *buen vivir* in the Ecuadorian constitution, and Bolivian President Evo Morales even borrowed the Social Forum repertoire to organize the 2010 Peoples' World Conference on Climate Change in Cochabamba. A few years later, Latin American progressive governments however left alter-globalization activists disappointed on the socio-economic and environmental fronts.

On other continents, Occupy and the 'post-2011' movements took their distance from party politics and implemented participatory democracy on the squares, denouncing the collusion of political, media, and economic elites. After 2013, some activists opted for launching new parties or combining party politics with more horizontal activism.

Overall, however, the relation between movements and governments is increasingly dominated by repression, suggesting that the use of force is often the only way that neoliberal and extractivist development can be implemented.

Taken together, these three cultures of activism offer concrete guidelines for a global and multidimensional approach to social change and to more sustainable processes of meeting human needs and well-being which simultaneously acknowledge the key roles to be played by local communities and grassroots actors, citizens' activism, international institutions, and political leaders. Beyond the North–South divide, the alter-globalization movement challenges the centrality of economics, fosters international solidarity and provides concrete answers to tackle global challenges, starting with global warming and rising inequalities.

Further Resources
Holloway, John (2002), *Change the World Without Taking Power*. London: Pluto Press.

Juris, Jeffrey S. (2008), *Networking Futures*. Durham, NC: Duke University Press.
Pleyers, Geoffrey (2010), *Alter-Globalization: Becoming Actors in the Global Age*.
  Cambridge: Polity Press.
Sen, Jai, Anita Anand, Arturo Escobar and Peter Waterman (eds) (2004), *World Social
  Forum Challenging Empires*. New Delhi: Viveka Foundation.
Smith, Jackie (2008), *Social Movements for Global Democracy*. Baltimore: Johns
  Hopkins University Press.

Geoffrey Pleyers is a National Fund for Scientific Research (FNRS) Professor at the University of Louvain, Belgium. He is the president of Research Committee 47 (Social Movements) of the International Sociological Association. He is the author of *Alter-Globalization. Becoming Actors in the Global Age* (Polity Press, 2010). Other research focuses on environmental movements, critical consumption, and social movements in Europe and Latin America.

ALTERNATIVE CURRENCIES

Peter North

Keywords: alternative currencies, localization, transition initiatives

Community, alternative or complementary currencies (CCs)[1] are forms of money created by non-state actors as alternatives to and remedies for the perceived pathologies of state-created money and growth-focused development.

Four types of CCs can be identified. Local Exchange Trading Schemes or LETS are local currencies called either Green Dollars or a locally significant name such as 'Bobbins' in Manchester, UK. LETS credits may be valued in alignment with national money, at an hourly rate, or a mix of the two. They have no physical form: users meet each other through a directory and then pay each other by cheque denominated in LETS units, which they back with their personal commitment to earn enough credits repay this commitment in the future. Account balances are kept on a computer.

Many critique any alignment of CCs with state-created money on the basis that it reproduces the pathologies of capitalism. Instead, they advocate money denominated by time, so that bankers will be paid no more than cleaners. Time Banking users help each other out and keep score with an electronic time credit, with each credit valued as one hour irrespective of the work done. 'Hours' are local time-denominated paper currencies which can be used as easily as cash, with no need for any central records. After the financial crisis in Argentina in 2001 millions of people supported each other using community-created paper currencies denominated in 'credits' – unrelated to either state money or time.

Activists wanting to avoid dangerous climate change and resource depletion have developed 'transition currencies' – for example, Totnes, Lewes, or Stroud Pounds. These are locally circulating paper currencies denominated in units aligned with and backed by state money. Finally, the development of personal computers and smartphones has led to experiments with electronic currencies such as BitCoin and FairCoin. Worldwide, a bewildering variety of CCs can be found (North 2010).

There has been a long history of contestation over money (North 2007). At the dawn of capitalism in the UK, the utopian socialist, Robert Owen, advocated and experimented with labour notes, as did nineteenth century anarchist communities across the United States. In the 1880s and '90s, US Populists organized for more liberal issuance of silver money in conflict with banking interests who defended the gold standard to enforce labour discipline, often violently. In the 1930s local authorities in the United States, pre-Nazi Germany, Switzerland, Austria, and revolutionary Spain issued their own paper currencies. The Swiss Business Ring exists to this day, connecting small business owners. Widely-available information technology has catalysed creation of CCs by non-state groups to unprecedented levels, for example, Kenya's Bangla-Pesa.

The first wave of LETS and Time Banks was created by green activists in Anglo-Saxon countries – people who were deeply aware of the unsustainability of global capitalism. They argued that in order to survive the regular crises of capitalism a diversity of local currencies should be created, so that if one form of currency – usually state-created – was unavailable, alternatives would be. Being created by users themselves and backed by their commitment to repay this currency in the future, CCs would be more widely available than state-issued currencies, thereby generating local demand (North 2005). Second, CCs would be available in amounts necessary to meet the needs of those who created them, irrespective of the pre-existing availability of money. Finally, given that everyday services such as babysitting, gardening, and so on were widely available in CCs networks – as opposed to highly remunerated professional services – CCs valued work done by those the capitalist market did not value. As a result of making personal credit money widely available to everyone willing to agree to the networks' value of reciprocity, CCs advocates argued that they had created a social change mechanism that would bring into being a localized, convivial sustainable economy, prioritizing the needs of people above either buying 'stuff', maximizing GDP or accumulating money for its own sake.

These schemes worked well enough if what people wanted to buy from each other could be produced within the resources owned or controlled by their members. The initiatives were more than what Marx had described as the 'dwarfish cooperation' of the utopian socialists of the nineteenth century, but often did not meet basic needs – food, shelter, power – or more complex desires generated by the capitalist system. The organizers struggled to keep accounts and would be overwhelmed in an economic crisis if millions of people started using them. Therefore, CCs advocates started to issue local paper currencies called 'hours' in the US, *creditos* in Argentina, transition currencies in the UK, thereby abolishing any need to keep accounts. Paying attention to the design of the notes meant they would be more likely to be taken seriously by conventional market actors. Unlike LETS and Time Credits though, these were not personal credit currencies – 'new money'. The issue then was why would users change 'universal' state-backed money into a locally circulating currency that could not be used for everyday needs, unless they shared the values of the network? As a result, the networks often stayed small, restricted to enthusiasts.

Are Alternatives Radical or false solutions? It depends. On the one hand, the assumption is that CCs support locally owned businesses and these are of value in themselves. The existence of a CC is believed to generate more local production and more convivial ways of creating more localized economies. On the other hand, transition and electronic currencies have been promoted by local business interests who see them as a means of supporting locally owned, but perhaps otherwise unsustainable, forms of economic activity. For instance, a local grocery that does not sell anything produced locally. These projects may still exclude those who do not have much ordinary money. Thus the extent that CCs go beyond mere visions of post-development to actualizing it remains – 'work in progress'.

Note

[1] The diversity of existing alternative currencies across the world is logged by the Complementary Currency Resource Center, http://complementarycurrency.org. Other good sources for recent research on alternative currencies is the online and open access *International Journal of Complementary Currency Research*, the *Journal of the Research Association on Monetary Innovation and Complementary and Community Currencies* (RAMICS), https://ijCCsr.net/.

Further Resources

North, Peter (2005), 'Scaling Alternative Economic Practices?' *Transactions of the Institute of British Geographers.* 30 (2): 221–33.

—— (2007), *Money and Liberation: The Micropolitics of Alternative Currency Movements.* Minneapolis: University of Minnesota Press.

—— (2010), *Local Money: Making It Happen in Your Community.* Dartington: Green Books.

Peter North is Reader in Alternative Economies in the Department of Geography and Planning, University of Liverpool, UK. His research focuses on social and solidarity economies as tools for constructing and rethinking alternative geographies of money, entrepreneurship, and livelihoods. This forms part of a project constructing strategies for local economic development given resource constraint, dangerous climate change, and economic crisis.

●•••●

ARBITRATION FOR SOVEREIGN DEBT

Oscar Ugarteche Galarza

Keywords: international monetary agreements, international borrowing, debt, bankruptcy, settlements

Let us recognize that external debt is, at all times, the most visible expression of an evolution that goes well beyond mere financial and economic concerns. That is why it is not enough to simply state that external debt and its management have caused repeated economic crises in many countries of the world. Debt crises themselves have been, on numerous occasions, yet another manifestation of the crises of the capitalist system. These happen recurrently, with a series of new elements each time and a repetition of previous elements. So, time and again, debt has carried out and continues to carry out an important role as leverage by imposing the will of lending countries, almost always large imperial powers, upon the indebted. It is an imposition that covers diverse characteristics, including violence.

The necessity of a mechanism to respond to the problem of global debt has been contemplated based on a long, accumulated history, highlighting the fact that there is no system of law for an unbiased, transparent, and equitable treatment of the problems of external debt.

The idea of a mechanism of this type has a long history, since the grave debt crisis of the 1980s. Alberto Acosta and Oscar Ugarteche designed an International Board of Arbitration for Sovereign Debt (IBASD), whose fundamental elements were debated within the United Nations General Assembly, twice[1] and the process was blocked due to the opposition of six large economies of the planet.

The principle of the Board is that receivership is a regular part of long economic cycles. Therefore, a mechanism is needed to determine what the components that motivate receivership are and how to proceed against them. A recurrent concern in the documentation on this question is when and who declares that a debtor can no longer be charged, a standstill, that payment has been suspended and that a process of reordering the debts has been entered into. Ideally, all creditors would simultaneously suspend debt collection while the remaking of the calendar of payments is agreed upon. This implies that no creditor may leave this list, meaning there are no free riders.

Who decides that a country is unable to fulfil its obligations? This is a fundamentally ethical question. If a contract exists which commits the debtor to pay and if there is the principle of *pacta sum servanda*, then, faced with a stormy economic forecast, the lending state may ask for a reordering of the debt even before it actually faces a payment crisis. This option allows a worsening crisis to be foreseen, which seriously affects the economic, social, and even environmental conditions of the indebted country. Based upon this decision, the way to resolve this situation transparently, without bias and adhering to the minimal norms of the rule of law, is the International Board of Arbitration for Sovereign Debt.

The basic principles for the operation of the Board are the following:

1. The payment of external debt cannot be, at any moment, an impediment to human development or a threat to environmental balance. The objective is the creation of a more stable and equitable economic system that results in benefits for all of humanity, which should incorporate the Tobin tax to curb speculative fluxes, as well as the demise of tax havens, to briefly mention two indispensable actions for the construction of another economy for another civilization.

2. It is unacceptable, within international law, that external debt agreements be instruments of political pressure, enabling a lending state or a case controlled by lending states, the IMF and/or World Bank, to impose unsustainable conditions on an indebted state.

3. The conditions of any agreement, internationally negotiated, must be based on human rights and rights of nature.
4. Based on the stipulated principles, it is necessary to create an international financial code that embraces all countries without exception.
5. The starting point of any solution, including arbitration, abides in the identification of legally acquired debt that can be paid, distinguishing this from illegitimate indebtedness based on the doctrine of hateful and corrupt debt.
6. This requires the active participation of civil society, where neither governments nor lenders intervene, only auditing guilds, legal and accounting associations accompanied by other organizations who will deliver their findings directly to the Board.
7. Proportionality – *paripassu* – must be established amongst all the lenders. According to this, multilateral, bilateral, and private lenders must be subject to debt negotiation, not solely private lenders.
8. For the service of contracted and renegotiated debt with definitive agreements in conditions of legitimacy, clear parameters must be established in fiscal terms so that debt servicing that still remains to be paid affects neither social investments, nor the capacity of internal savings, so assuring the payment capability of the country.
9. The causes of payment suspension must be established so that cases of *force majeure* are treated differently than cases of poor administration.
10. The space for this board must be established in Geneva due to the proximity of the United Nations Conference on Trade and Development (UNCTAD) offices and United Nations units specialized in external debt, outside of the IMF.
11. The IMF must retake its original role. This controller must be supervised and it must be national and international civil society, in each case, who have the capacity to monitor it. The IMF must be accountable to the General Assembly of the United Nations, along with a system of sanctions.
12. The creation of an international network of civil society organizations that monitor, in each country, the running of this international organization. They will deliver reports to their governments and to a permanent supervisory of the United Nations. An evaluation of the work of the IMF will be carried out on a yearly basis.

The subject of sovereignty is complex. A country that accepts arbitration would be, to some extent, recognizing its inability to pay and would be subject to the conclusions of the Board. Nevertheless, by accepting the dogma of no bankruptcy, in order not to weaken their sovereignty, a country will passively assume the necessity of conditions to guarantee debt payment and lose *defacto* its sovereignty through the policies that stem from agreements imposed by their lenders.

In a truly globalized world, the Board would be made up of representatives from all countries.

Note

[1] UN General Assembly (2013) Draft resolution A/68/L.57/Rev.1, adopted by 124 votes to 11, with 41 abstentions (resolution 68/304). In 2015, the text was refined and the Assembly then adopted it by a recorded vote of 136 in favour to 6 against (Canada, Germany, Israel, Japan, United Kingdom, United States) with 41 abstentions. https://www.un.org/press/en/2015/ga11676.doc.htm.

Further Resources

Ugarteche, Oscar and Alberto Acosta (2006), 'Los problemas de la economía global y el tribunal internacional de arbitraje de deudasoberana', *Polis*.13: 30, http://polis.revues.org/5393.

United Nations (2015), 'State of Palestine Flag to Fly at United Nations Headquarters Offices, as General Assembly Adopts Resolution on Non-Member Observer States', https://www.un.org/press/en/2015/ga11676.doc.htm.

Oscar Ugarteche and Alberto Acosta (2007), 'Global Economy Issues and the International Board of Arbitration for Sovereign Debt (IBASD)', *El Norte: Finnish Journal of Latin American Studies*. 19 (2), www.elnorte.fi/archive/2007-2/2007_2_elnorte_ugarteche.pdf.

Dr. Oscar Ugarteche Galarza is a Peruvian economist, at the Institute of Economic Research of the National Autonomous University of Mexico. He is Member of the National System of Researchers/National Counsel of Science and Technology. He was earlier the systems consultant to the United Nations in matters of external debt. He is author of 27 books, 62 academic articles, and 56 chapters in edited volumes, and has been a guest researcher at half a dozen universities in Europe.

AUTONOMY

Gustavo Esteva

Keywords: autonomy, radical democracy, patriarchy, modernity

Autonomy today alludes to attitudes, practices, and positions across the entire ideological spectrum, from the self-rule of sovereign individuals to real movements that adopt radical democracy as an emancipatory horizon beyond capitalism, the industrial mode of production, western modernity and patriarchy. Rather than autonomy, there are autonomies, both in reality and as political projects, as mobilizing myths and as horizons – as what is not yet.

Consequently, I exclude from this essay two schools of thinking and action that in my view are not real alternatives to the dominant regime:

- The individualist school, sometimes called 'libertarian', and its voluntary unions of egoists (Stirner), which usually operate within capitalist pseudo-anarchism.
- The socialist school, Leninist and supposedly anti-capitalist, which reduces autonomy to a decentralized form of administering the vertical powers of the state within structures of domination justified as requirements for the transition to socialism. Autonomy as the self-activity of the multitude (Negri, Virno) belongs to this school, as all approaches dealing with masses, not people.

Let us turn to the heart of the matter and to alternatives that offer real possibilities.

The word 'autonomy' is very old. In the seventeenth century, in Europe, the Greek term could either be used to allude to the liberty granted to Jews living according to their own laws, or to discuss the Kantian autonomy of the individual will. Several European schools of thinking and action adopted the word in the twentieth century to characterize their positions and aspirations. In the rest of the world, other notions, attitudes, and practices that today would be called autonomic have existed since time immemorial.

To understand current debates, we can differentiate between *ontonomy*, the traditional, endogenous norms still in force everywhere; *autonomy*, referring to the processes by which a group or community adopts new norms; and *heteronomy*, when the rules are imposed by others. Autonomic movements attempt to widen as much as possible the spheres of ontonomy and autonomy.

A new semantic constellation arising from emancipatory social and political movements shares, at least in part, the following elements:

It goes beyond formal democracy. Both Greece, which coined the word 'democracy', and the United States, which gave to it its modern form, were societies with slaves. During the last 200 years, softened forms of slavery were fostered or hidden in regimes that the great black intellectual W.E.B. Dubois correctly characterized as democratic despotism. Participatory democracy fails to eliminate the verticality of democratic societies, ruled by professional dictatorships in which professionals assume legislative, executive, and judiciary powers in each field and prevent the participation of common people in the functions of government.

Disenchantment with democracy is today universal. The wake-up call of the Zapatistas, in 1994, put autonomy at the centre of the political debate. 'Enough! All of them should go!' said the Argentinians in 2001. 'My dreams don't fit into your ballot box', affirmed the *Indignados* in Spain. Occupy Wall Street, in the US, enabled millions of people to finally acknowledge that their system is at the service of the 1 per cent. There are still attempts to reform it, but many struggles try instead to widen, strengthen, and deepen the spaces in which the people can practise their own power. They are literally constructing democracy from the roots, in which common people can assume the power of the Leviathan, free to speak, to choose and to act (Lummis 1996). Attempts of this kind are innumerable and all over the world. On January 1, 2017, for instance, the National Indigenous Congress of Mexico, with the support of the Zapatistas, launched a proposal to create a Council of Government based on both Indigenous and Non-Indigenous autonomies. Instead of trying to seize the state apparatus, conceived and operating for control and domination, they are attempting to dismantle it and create institutions where the practice of commanding by obeying can thrive.

Beyond economic society. Autonomic movements, widely visible in Latin America, are not only challenging neoliberal globalization, but are also acting explicitly against capitalism without becoming socialist. Some are not only attempting to end their dependence on the market or the State, but are also breaking with the 'premise of scarcity' that defines economic society: the logical assumption that human wants are great, not to say infinite, while his means are limited. Such assumption creates an economic problem par excellence: resource allocation through the market or the plan. These movements, by contrast, adopt the 'principle of sufficiency', thus avoiding the separation of means from ends in both economic and political terms. Their struggles adopt the shape of the outcome they want to bring about.

Beyond western modernity. An increasing number of people painfully disassociate themselves from the truths and values that define western modernity and in which they came to believe. Most of these people cannot yet find a new system of reference. Confronted with such a loss of values and orientation some may become fundamentalists. Others, however, may acknowledge the relativity of their previous truths, immerse themselves in different forms of radical pluralism, and practise new forms of knowing and experiencing the world, participating in the insurrection of subjugated knowledge. Inspired by Raimon Pannikar, they substitute nouns creating dependence – education, health, food, home, and so on – for verbs that bring back their personal agency, their autonomy: learning, healing, eating, dwelling. They acknowledge the individual as a modern construction from which they disassociate themselves, in favour of a conception of persons as knots in nets of relationships, which constitute the many real we's defining a new society.

Beyond patriarchy. Several feminist schools participate in autonomic movements that gobeyond conventional visions of post-patriarchal societies. A clear example is the Zapatista society, where politics and ethics, and not the economy, are at the centre of social life, and caring for life, women and Mother Earth has the highest priority. In these societies, autonomous practices characterize all areas of daily life, ruled through democratic processes that organize communally the art of hope and dignity.

Further Resources

Albertani, Claudio, Guiomar Rovira and Massimo Modonesi (Coord.) (2009), *La autonomía posible: Reinvención de la política y emancipación*. México: Universidad Autónoma de la Ciudad de México.

Dinerstein, Ana Cecilia (2015), *The Politics of Autonomy in Latin America: The Art of Organising Hope*. Hampshire, England: Palgrave MacMillan.

Enlace Zapatista, http://enlacezapatista.ezln.org.mx.

Linebaugh, Peter (2006), *The Magna Carta Manifesto*. Berkeley: University of California Press.

Lummis, Douglas (1996), *Radical Democracy*. Ithaca: Cornell University Press.

Pannikar, Raimon (1999), *El espiritu de la politica*. Barcelona: Peninsula.

Gustavo Esteva is an activist and public intellectual. Columnist in *La Jornada* and occasionally in *The Guardian*, he participates in local, national, and international grassroots organizations and is an author of numerous books and essays.

BIOCIVILIZATION

Cândido Grzybowski

Keywords: civilizational crisis, care, commons, social and environmental justice, human rights and responsibilities

The notion of biocivilization, or life civilization, refers to the search for a new civilization paradigm, a concept that is still in an embryonic stage. Biocivilization indicates a direction to move towards. However, instead of presenting a uniform model for the world, it seeks to be a concept that is extremely diverse, like the planet and life itself. For biocivilization to occur, we must become reintegrated with life and the dynamics and rhythm of ecological systems, adjusting to them, enriching them and facilitating their renovation and regeneration. In lieu of the free market logic and the search for private interests, the core guiding principle should be caring for the ethics of both collective and individual responsibility, with regard to all relations and processes, in the economy and in terms of power, in science and in technology. The following are the pillars of biocivilization: doing the best that is possible locally, following the principle of subsidiarity in relation to other levels; keeping the commons at the centre; creating decent work shared among all men and women; ensuring human rights, equality, freedom, happiness, and the fulfilment of people's potentialities, in all their diversity and according to their will.

We are facing a very serious philosophical and political challenge. It is serious because the challenge is to dismantle those presumptions of thought and action that we have internalized and which, for this reason, often without our conscious knowledge shape our minds and, thereby, organize economy and power in society. We are led to believe that lack of development, non-development or underdevelopment are at the root of society's ailments. Development is the dream and the ideology that dominates planet Earth; it is understood as a rising gross domestic product (GDP), involving the ever-increasing possession and consumption of material goods, no matter what.

Faced with the crisis of the dominant capitalist civilization, an issue that emerges as a *sine qua non* condition is the need to restructure and rebuild our relationship with nature. After all, we are part of the biosphere. We are nature ourselves, living nature, gifted with a conscience. Future generations have the same right to healthy natural conditions that we have. Moreover, the planet's integrity is a value 'in itself', and it is our duty to preserve it. To interact with nature is, by definition, to be living. From a biocivilization perspective, it is in this relationship with nature that we define the sustainability of life and the planet.

Environmental destruction should be viewed as an aspect of growing social inequality. After all, environmental destruction is socially unequal, with some groups and societies being more responsible for it than others. Linking the fight for social justice to the fight against environmental destruction is vital, since one depends on the other. This radically redefines the social struggles of our time from the perspective of life's civilization.

To become sustainable again, modern human civilization must give up anthropocentrism and radically change its vision and relationship with nature (Calame 2009). All life forms, as well as the complex inter-related ecological systems that regulate planet Earth, have the fundamental right to exist. This must be the founding principle, condition, and limitation of human intervention in the relationship with nature and in building flourishing societies.

The values of caring, living together and sharing refer to the basics of an economy focused on life. They are actually the core economy, since human life is based on these values (Spratt *et al.* 2009). Care is the essential activity of daily life. This vital work is done mostly by women, who carry the burden of the double journey and suffer from male domination. We are, in fact, suffering an inversion, where what is essential – care – is considered private and worthless by the dominant economy. Care refuses to abide by the market capitalist principle of value, expressed by GDP. We must locate care at the centre of the economy, as a principle for the management of the symbiosis of human and natural life that makes up the planet, and as the basis of life of communities, where we ineluctably live together and share with others.

De-commoditization and de-commercialization of the commons are crucial conditions to overcome the civilization crisis and walk towards sustainability. Commons are one of the core principles of biocivilization. Reclaiming, expanding, and creating anew the commons are all tasks that comprise the creation of a new civilizational paradigm.

The principles of living together and sharing are the corollary of care. Care flourishes with community life and friendship. Cultural life, dreams, imagination, beliefs, knowledge, and cooperation all flourish with care.

Here is where a fundamental issue arises, one that is present in the current political culture of human rights, but not stressed enough. There are no human rights without human responsibilities. In order to be seen as holder of rights, the condition is to acknowledge the same entitlement for everyone. That is to say, in order to have rights we must, at the same time, be responsible for the rights of all others. This is a shared relation and, as such, a relation of co-responsibility. The growing awareness of human rights and responsibilities, both within and between societies, as well as in the relation with the biosphere, sheds light on the fundamental issue of interdependence from the local and territorial to the planetary levels.

The proposal for a biocivilization was the main issue discussed first at a workshop in 2011, organized by IBASE, World Social Forum Organizing Committee and the Forum for a New Global Governance, in Rio de Janeiro. About 60 social activists from Brazil, South America, South Africa, India, China, and Europe participated. This was an activity in preparation to the Thematic Social Forum (2012) and the Rio+20 Summit. Since then, the idea of biocivilization has been a main theme of the Europe-Latin America-China dialogues.

Further Resources

Bollier, David and Silke Heilfrich (eds) (2012), *The Wealth of the Commons: A World Beyond Market and State*. Armherst, MA: Levellers Press.

Calame, Pierre (2009), In Charles Léopold Mayer (ed.), *Essai sur l'oeconomie*. Paris: IRE.

Grzybowski, Cândido (2011), *Caminhos e descaminhos para a biocivilização*. Rio de Janeiro: IBASE, http://www.ibase.br/pt/wp-content/uploads/2011/08/Caminhos-descaminhos.pdf.

Grzybowski, Cândido (2015), 'Biocivilization for Socio-Envoironmental Sustainability: A Brazilian View on the Hard but Necessary Transition', in Michel Reder, Katharina Hirschbrunn and Georg Stoll (eds), *Global Common Good*. Frankfurt-on-Main: Campus Verlag GmbH.

Spratt, Stephen, Andrew Simms, Eva Neitzert and Josh Ryan-Collins (2009), *The Great Transition: A Tale of How It Turned Out Right*. London: New Economics Foundation.

Cândido Grzybowski is a philosopher, sociologist, and social activist who is strongly engaged with the World Social Forum. He is former director and currently the Management Advisor of the Brazilian Institute for Social and Economic Analysis (IBASE), Rio de Janeiro, Brazil.

———•••———

BODY POLITICS

Wendy Harcourt

Keywords: embodiment, feminism, racism, queer activism, heteronormativity

'Body politics' has been an important political project among feminist and queer activists transnationally since the 1980s. Within body politics, bodies are considered sites of cultural and political resistance to the dominant understanding of the 'normal' body as white, male, Western, and heterosexual from which all 'other' forms of bodies differ. Body politics, then, ranges from liberal economic justice demands for recognition of the integrity of sexual orientation rights for all peoples. For example, body politics was a disruptive and critical force in queer and feminist interventions in the 1990s United Nations global conferences on human rights (1993), population (1994), and women (1995). Activists working in these different UN conferences brought to international attention issues such as domestic violence, rape as a weapon of war, sexual and reproductive rights of women, and the rights of indigenous, homosexual, and transgender people. Their campaigns spoke out not only against gender inequalities but also against racism, ageism, and heterosexual norms. In this way, body politics has linked different forms of bodily oppression with radical forms of democracy (Harcourt 2009).

Examples of body political actions and campaigns include performative protests such as those in India against Miss World pageants to large direct actions and long-term campaigns, online and offline; the 1980s marches against the criminalization of abortion in the US, Australia, and Europe; and the multiple global campaigns to end violence against women including rape, forced sterilization, femicide, and paedophile sex trafficking. Examples range from feminist campaigns to recognize domestic violence in marriage to plays such as *The Vagina Monologues* that began in New York and is now held all over the world. Other examples would be the demands in the 2000s for the right of same-sex marriage and for transgenders to be recognized as a third gender alongside male/female in official documentation in public bathrooms and in schools. Sexuality Policy Watch (SPW),[1] an online institution, based in Brazil, produces data that documents many of these actions and campaigns illustrating how important body politics is in the global political landscape. A recent outstanding example of global body politics would be the hundreds of women's marches held around the world on 21 January 2017, the day after the 45th Presidential inauguration in the

US. Millions marched in protest against Donald Trump's misogynist, racist, and homophobic statements and his behaviour towards women.

Body politics is undoubtedly controversial because it makes visible intimate and often taboo issues. Body politics in this sense 'speaks' the unspoken in political and economic spaces as it challenges norms that condone and institutionalize gender and other inequalities. For instance, campaign for sex workers' rights to a fair wage demand that sex workers be seen as workers like all others. Body politics challenges homophobia even in places such as Africa where homosexuality is penalized and criminalized. It speaks out about racialized discrimination also within feminist movements, where the markings of historical erasure associated with colonialism can be seen in white privilege (Harcourt, Icaza and Vargas 2016). Decoloniality in the last few years has become an important part of global body politics. For example, the campaigns to end the violent discrimination, including sterilization, of indigenous women in Central America challenge assumptions about development and progress. Another example is the Latin American and Caribbean Feminist Encounters in the early 1990s which opposed countered liberal gender and development ideas of success through intercultural dialogues that included non-white, queer, and indigenous peoples (Harcourt, Icaza and Vargas 2016). As these last examples suggest, body politics is critical of forms of Western power expressed in the sexism, racism, misogyny, and heterosexism that accompany imperial and colonial knowledge systems informing development practices (Mohanty 2003).

Body politics is not only about struggles to end oppression but also about ways to re-imagine and remake the world. This includes understandings of sexuality, diversity and well-being from the perspective of the marginalized 'other.' One example is the European Feminist Forum (EFF), held from 2004–08. The EFF engaged feminist and queer activists from twenty European countries. It created a digital space where Roma women, young queer feminists from Central and Eastern Europe, domestic migrant and migrant sex workers could meet and debate the future they envisaged, one that was not informed by the givens of dominant EU politics based on individualistic ideas of success and progress. Linked to the World Social Forum, this experience contributed alternative ways of organizing based on a pluriversal understanding of care in non-exploitative, non-heteronormative familial groups.

Body politics activism, then, challenges the narrative of modernity that defines gender, bodies, and sexuality through the lens of progress. In its activism and vision for alternatives, body politics opens up other ways of seeing politics, beyond social and economic development based on individual human rights and alleged economic equality, to be delivered by the State according to the rule of law. Instead, body politics unpacks the racial/sexual/gendered system that modernity has imposed on society. It challenges the modern classification of the norm of privileged white male heterosexual bodies. It brings to post-development theory and practice the challenge to question and unmake the ways bodies are shaped by social relations embedded in neoliberal capitalism. It invites post-development to build on the multiple resistances and rebellions expressed in feminist and queer struggles for the bodily integrity of the many 'others' to white male privilege.

In its claim to otherness, body politics is an essential entry point for post development rethinking. It is also a space for transformative collective action which connects the body to radical alternatives in social movements, thus fashioning strategies for diversely embodied transformation. The challenge for post-development is to take seriously the notion that there are a multiplicity of bodies and forms of embodiment in order to go beyond the normalizing histories and practices of modern development. And for feminists and queers there is the challenge to understand the multiple and diverse ways of connecting with the spiritual dimension of life in order to relate to non-human bodies or 'earth others'.

Note
[1] Sexuality Policy Watch (SPW) is a global forum comprising researchers and activists from a wide range of countries and regions of the world, see http://sxpolitics.org.

Further Resources
CREA, http://www.creaworld.org.

Harcourt, Wendy (2009), *Body Politics in Development: Critical Debates in Gender and Development*. London: Zed Books.

Harcourt, Wendy, Rozalba Icaza and Virginia Vargas (2016), 'Exploring Embodiment and Intersectionality in Transnational Feminist Activist Research', in Kees Biekart, Wendy Harcourt and Peter Knorringa (eds), *Exploring Civic Innovation for Social and Economic Transformation*. London: Routledge.

Mohanty, Chandra Talpade (2003), *Feminism without Borders: Decolonizing Theory, Practicing Solidarity*. Durham: Duke University Press.

Wendy Harcourt is Professor in Critical Development and Feminist Studies at the International Institute of Social Studies of the Erasmus University, The Hague, Netherlands. From 1988 to 2011 she was Editor of the journal *Development* and Director of Programmes at the Society for International Development in Rome, Italy. She has published widely on post-development, political ecology, and feminism.

BUDDHISM AND WISDOM-BASED COMPASSION

Geshe Dorji Damdul

Keywords: compassion, happiness, education, social engagement

Since the day of our birth, we have engaged in actions such as suckling milk from the mother's breast and crying, all of which are driven with the desire to shun miseries and acquire happiness. This fact pervades all beings, with no exception to anyone, whether believer or non-believer, rich or poor, educated, or uneducated. Directly or indirectly, we all seek the greatest happiness.

During one of the public lectures by His Holiness the Dalai Lama, at Delhi University in 2008, a girl asked him, 'What is the meaning of life?' His Holiness, without a second thought, said, 'We live on hope, and we hope for genuine happiness. Therefore, genuine happiness is the meaning of life.'

The ultimate source of happiness is within us. Like the sound of a clap which invariably requires two hands coming together, all of the miseries we go through arise from the coming together of two conditions – the external and the internal. The number of external factors is often so high that we are hardly able to remove any part of it. On the contrary, like removing one hand, if the internal factor is removed, no matter how strongly the other hand moves, there is no clap or sound. The sound of the miseries stops. The Buddha indicated that the self-referential ego, which is triggered by ignorance, is the worst of the internal factors, the eradication of which stops all miseries.

Means for happiness through eradicating the egoistic ignorance. Just as one requires light to eliminate the darkness, it is through introducing the light of wisdom that the darkness of ignorance can be eliminated, said the Buddha or Dhammapada.

The wisdom of interdependence to engender loving kindness. Division and hatred arise out of the ignorance which sees the family of humanity as not one, but divided into different ethnicities, religions, and countries. Worldwide phenomena such as global warming do not know national borders. The birds in the air and the fish in the ocean do not know these virtual lines either. Bird flu, mercury poisoning from fish, and radioactivity spread across the globe without border checkpoints. In actuality, we are all interdependent, and the recognition of this fact should give rise to love and affection, similar to how a young child feels incredible love towards its mother when it realizes his or her dependence upon the mother's love and kindness. This philosophy is very much in line with quantum physics' presentation of how the observed exists only in dependence of the observer.

Ego versus self-confidence. Oftentimes people confuse the terms 'ego' and 'self-confidence' as interchangeable. This is the gravest mistake. The 'ego' is the one that gives birth to all destructive emotions and, thereby, not only does it give rise to all forms of unrealistic and harmful actions, but, in turn, also brings miserable experiences upon oneself. In contrast, 'self-confidence' keeps one calm. The person with 'self-confidence' acts realistically and thus accomplishes all the desired results.

It is very saddening to see that some have strong self-hate. It is very unhealthy and harmful to oneself. By making the distinction between 'self' and 'ego', one should learn to be kind towards the self and hate only the self-referential ego.

Buddhism applied in society. It is on the basis of what has just been stated that the Buddha's message has contributed and can further contribute immensely to ecological preservation and social equality. The Buddha prohibited his followers including monastics from polluting the rivers and harming vegetation (Max Muller 1974).

In line with this, His Holiness the Dalai Lama has for many years advocated strongly against harming animals and using animal skin and fur. Following him, the Tibetans in Tibet instantly boycotted the use of skin, fur, and elephant tusk on a massive scale by burning them, including those inherited many generations ago. This act was a courageous one, especially because of the existing ban on the burning imposed by the Communist Chinese, who feared to see the enormous influence of His Holiness, the Dalai Lama on Tibetans in Tibet.[1]

Also, a massive conversion to vegetarianism happened in the Tibetan communities in India some twenty years ago on the advice of His Holiness the Dalai Lama. Around 1997, a poultry with 87,000 chickens in a Tibetan settlement in South India was closed down when he said that he would surely live for a minimum of 87 years if the settlement released all the chickens.[2]

The world, despite being at the height of modern education, is torn apart by deep crises such as corruption, terrorism, the gap between the rich and the poor, gender discrimination. His Holiness the Dalai Lama recognizes the fact that the modern education system, which favours the development of the brain and not the heart, is responsible for these crises. He has been working hard to bring 'universal ethics' as a subject into the modern education system. He is confident that such ethics are the most important of all, not only for one ethnicity or religion or country or region, but for the entirety of humanity. At the heart of universal ethics lies the value of warm-heartedness based on the appreciation of the concept of interdependent nature, among human beings, among nations, between humans and nature.[3] This subject was introduced by Delhi University in 2012 and is now being introduced in the Tata Institute of Social Sciences, Mumbai, and other institutes. For the Dalai Lama, education is the final answer to the current global crises.

Notes

[1] See, for instance, Tibetans in Yunnan Give Up Wearing Animal Skins, Burn Valuable Furs, https://www.rfa.org/english/news/tibet/animal-03042015170302.html.
[2] Public teaching of HH the Dalai Lama, 1990, recorded by author.
[3] Public talk on 'Universal Ethics', by HH the Dalai Lama in Thayagraj Stadium, 2016, as recorded by author.

Further Resources

Dalai Lama (2001), *Ethics for the New Millennium*. New York: Riverhead Books.
——— (2011), *Towards a True Kinship of Faiths*. London: Penguin Random House.
——— (2012), *Beyond Religion*. New Delhi: Harper Collins Publishers.
Engaged Buddhism, https://en.wikipedia.org/wiki/Engaged_Buddhism.
Max Muller, Friedrich (ed.) (1974), *Vinaya, Mahavagga (Vinaya Texts, Sacred Books of the East)*. Delhi: Motilal Banarsidas.
Ricard, Matthieu (2007), *Happiness: A Guide to Developing Life's Most Important Skill*. New York: Little, Brown and Company.
Thanissara (2015), *Time to Stand Up: An Engaged Buddhist Manifesto for Our Earth: The Buddha's Life and Message through Feminine Eye*. Berkeley, CA: North Atlantic Books.

Geshe Dorji Damdul is Director at Tibet House, New Delhi, the Cultural Centre of His Holiness the Dalai Lama. He finished his Geshe Lharampa Degree (PhD) in 2002 from Drepung Loseling Monastic University in Karnataka. He has been the official translator to His Holiness the Dalai Lama since 2005. He travels widely, for instance, Mumbai, USA, UK, and Singapore, teaching Buddhist philosophy, psychology, logic, and practice.

———•••———

BUEN VIVIR

Mónica Chuji, Grimaldo Rengifo, Eduardo Gudynas

Keywords: alternatives to development, South America, modernity, indigenous movements

The category of *buen vivir* or living well expresses an ensemble of South American perspectives which share a radical questioning of development and other core components of modernity, offering at the same time alternatives beyond it. It is not akin to the Western understanding of well-being or the good life, nor can it be described as an ideology or culture. It expresses a deeper change in knowledge, affectivity, and spirituality, an ontological opening to other forms of understanding the relation between humans and non-humans which do not imply the modern separation between society and nature. It is a plural category under construction that takes specific forms in different places and regions. It is heterodox in that it hybridizes indigenous elements with internal critiques of modernity.

References to the ideas of *buen vivir* have been registered since the mid-twentieth century, but its current meanings were enunciated in the 1990s. Important in this regard were, the contributions of the Andean Project of Peasant Technologies (PRATEC in Peru; Apffel-Marglin, 1998); of the Andean Centre for Agriculture and Livestock Development (CADA; Medina 2001) in Bolivia; and of diverse intellectuals, as well as social and indigenous leaders, among which Acosta (2012) stands out, in Ecuador. A wide range of social movements have supported these ideas, which affected political changes in Bolivia and Ecuador, attaining constitutional recognition.

Buen vivir includes different versions specific to each social, historical and ecological context. These come about through the innovation, linking, and hybridization of concepts stemming from indigenous traditions with critical postures within modernity itself. Among these are the Aymara's *suma qamaña*, the Bolivian Guarani's *ñande reko*, *sumak kawsay*, the Ecuadorian Kichwa's *allin kawsay*, and the Peruvian Quechua's *allin kawsay*. The Ecuadorian/Peruvian *shür waras* and the Chilean Mapuche's *küme morgen* (Rengifo 2001 and Yampara 2011) are analogous concepts.

Among the Western contributions are the radical critiques of development, including post-development; the recognition of the coloniality of power and knowledge; feminist critiques of patriarchy; alternative ethics that recognize the intrinsic value of the non-human; and environmental visions such as deep ecology.

There is no single *buen vivir*; for example, Ecuador's *sumak kawsay* is different from Bolivia's *suma qamaña*; approximate translations to Western categories in the first case refer to the art of good and harmonious living in a community, although defined in social and ecological dimensions at the same time, while the latter also address conviviality in mixed communities but in specific territories. At the same time, it is as incorrect to say that *buen vivir* is exclusively an indigenous proposition, as to say that it implies returning to a pre-colonial condition, although those contributions are essential to its construction.

There do exist shared components beyond this diversity (Gudynas 2011). All of them question the concept of progress and the notion of a single universal history. They are open to multiple, parallel, non-linear, and even circular, historical processes. They question development because of its obsession with economic growth, consumerism, commodification of nature, and so on. This critique spans capitalist as well as socialist varieties of development. From this perspective, a socialist *buen vivir* makes no sense. The alternatives are both post-capitalist and post-socialist, disengaging from growth, and focusing on the complete satisfaction of human needs from the standpoint of austerity.

Buen vivir displaces the centrality of humans as the sole subject endowed with political representation and as the source of all valuation. This implies an ethical opening (by recognizing the intrinsic value of non-humans, and the rights of nature), as well as a political opening (the acceptance of non-human subjects). It confronts patriarchy, even at the heart of rural and indigenous domains, postulating feminist alternatives that revive the key role of women in the defence of communities and nature.

The modern separation betweeen humanity and nature is also challenged. *buen vivir* acknowledges extended communities made up of humans and non-humans, animals, plants, mountains, spirits, and so on, in specific territories – as with the Andean concept of *ayllu*, mixed socio-ecological communities rooted in a specific territory.

Buen vivir rejects all forms of colonialism and keeps a distance from multiculturalism. It upholds, instead, a type of interculturality that values each tradition of knowledge, thus postulating the need to refound politics on the basis of plurinationality.

Buen vivir bestows substantial importance upon affectivity and spirituality. Relationships in extended communities are not restricted to market exchanges or utilitarian links; instead, they incorporate reciprocity, complementarity, communalism, redistribution, and so on.

The ideas behind *buen vivir* have been the subject of a harsh critique. Some consider that they reflect an indigenous reductionism, while others affirm that, in actuality, they are a New Age invention. Intellectuals from the conventional left have maintained that they are a distraction from the true objective, which is not alternatives to development, but alternatives to capitalism; they also reject the intrinsic value of non-humans. Despite these arguments, *buen vivir* ideas have achieved strong and widespread support within Andean countries; from there, they have spread rapidly throughout Latin America and the global scene, providing the basis for concrete alternatives to development, as in the constitutional recognition of the rights of Nature and of the Pacha Mama; moratoria on Amazon drilling; models for transitions to post-extractivism; or the cosmopolitics based on the participation of non-human actors.

The sharp contradiction between these original ideas of *buen vivir* and the development strategies of the Bolivian and Ecuadorian governments, who have promoted extractivism like mega-mining or Amazonian oil extraction, has become evident. The progressive regimes have attempted to surmount these contradictions via new definitions of *buen vivir*, whether as a type of socialism in Ecuador, or as integral development in Bolivia, thus placing it again within modernity. These positions have been supported by some state agencies, by intellectuals, and by non-South American intellectuals who despite their intentions, only rehearse the coloniality of ideas.

Despite everything, the original ideas of *buen vivir* have been maintained. These continue to nourish social resistance to conventional development, for instance, in the case of the indigenous and citizen demonstrations in Bolivia, Ecuador, and Peru in defence of territory, water, and Mother Earth. This demonstrates that *buen vivir* in not limited to a few intellectuals and NGOs, but that it has garnered a high level of popular support.

To sum up, *buen vivir* is an on-going proposal, nourished by different movements and activists, with its advances and setbacks, innovations and contradictions. It is inevitably under construction as it is not easy to go beyond modernity. It must necessarily be plural as it encompasses positions that question modernity while opening up other ways of thinking, feeling, and being – other ontologies – rooted in specific histories, territories, cultures, and ecologies. However, there are clear convergences within this diversity that distinguish it from modernity, such as the delinking from modernity's belief in progress, the acknowledgement of extended communities stemming from relational worldviews, and an ethics that accepts the intrinsic value in non-humans.

Further Resources

Acosta, Alberto (2012), *Buen vivir. Sumak kawsay: Una oportunidad para imaginar otros mundos*. Quito: AbyaYala.

Apffel-Marglin, Frédérique (ed.) (1998), *The Spirit of Regeneration: Andean Culture Confronting Western Notions of Development*. London: Zed Books.

Gudynas Eduardo, (2011), 'Buen Vivir: Today's Tomorrow', *Development*. 54 (4): 441–47.

Medina, Javier (ed.) (2001), Suma Qamaña: *La comprensión indígena de la Buena Vida*. La Paz: Gesellschaft für Technische Zusammenarbeit (GTZ) and Federación de Asociaciones Municipales de Bolivia (FAM).

Rengifo, Grimaldo (2002), *Allin Kawsay: El bienestar en la concepción andino amazónica*. Lima: Andean Project of Peasant Technologies (PRATEC).

Yampara H., Simón (2011), 'Cosmovivencia Andina: Vivir y convivir en armonía integral – Suma Qamaña', *Bolivian Studies Journal /Revista de Estudios Bolivianos*. 18: 1–22.

Mónica Chuji is an Amazonian Kichwa intellectual; she was a part of the Constitutional Assembly of Ecuador; a former Minister of Communication; and former government spokesperson for the first administration of Ecuador under President Correa.

Grimaldo Rengifo is a Peruvian educator and promoter of Andean culture, especially that which is linked to the Earth. He is a team member of the Andean Project for Peasant Technologies (PRATEC).

Eduardo Gudynas from Uruguay, is a researcher at the Latin American Centre for Social Ecology (CLAES) and an associate researcher in the Department of Anthropology; University of California, Davis.

CHINESE RELIGIONS

Liang Yongjia

Keywords: Confucianism, Taoism, ecology

The term 'Chinese religions' refers here to the religious beliefs and practices originating in and transmitted from China over centuries. The most influential ones that have survived are Confucianism and Taoism.

Confucianism was founded by Confucius (551–479 BCE), who reinvented the royal rituals into a system of thought and practices around the idea of filial piety and loyalty. It became the imperial ideology in the second century BCE and was followed in China until the early twentieth century. Taoism was institutionalized around the first century CE by syncretizing different philosophies about the universe and salvation; it continued to be one of the most influential religions in China over two millennia, largely due to its diffuse and immensely popular practices.

The expression 'Chinese religions' appeared first in the West from the pens of the seventeenth-century Jesuits, and was used by Max Weber in *The Religion of China: Confucianism and Taoism* (1915). Weber's elitist construction leaves the impression that Confucianism and Taoism are the authentically operative religions in China. But actually one cannot understand them without taking Buddhism and other religions into consideration. Moreover, both Confucianism and Taoism significantly declined in the early twentieth century, undergoing radical changes so as to better address contemporary issues. This lead to a reinterpretation of concepts and traditions relating to harmony, nature, justice, and ecology.

By the early twentieth century, Confucianism had lost its orthodox authority and institutions because of the demise of the Chinese empire. In the first three decades of the People's Republic of China (PRC), it was further denounced as a feudal remnant. Since the turn of the twenty-first century, Confucianism has experienced a strong revival with state support. The major players are the 'mainland neo-Confucianists' who are keen to build a state-recognized religion and direct the country towards Confucian constitutionalism.

The revivalists generally believe that Confucianism can overcome the crisis of Western modernity, with alternative social and bodily techniques relating to inner peace, communal solidarity, civility, and self-constraint (Tu 2010). It offers remedies to political corruption, economic disparity, social instability, and ecological disaster. They believe that modernity, characterized by liberal democracy and global capitalism, is no longer desirable for China. Instead, Confucianism will teach China and the world about the spiritual, moral and ritual life, a sort of 'civil religion' (Jensen 1997:4). Many promote reading Confucian classics in educational institutions and offer trainings to entrepreneurs, politicians, professionals, and spiritual seekers. There are also environmentalists who re-orient the Confucian anthropocentric outlook in line with ideas of harmony with nature, yearning for longevity, sympathy to hardship and desire for sustainable justice. The International

Confucian Ecological Alliance (ICEA) tries to combine Confucian wisdom and ecological science to foster a global network to raise awareness of the severe ecological crisis of the world.

The contemporary revival of Confucianism remains largely rhetorical. Lacking institutional power, it is mostly active through philosophical writings, commercial endeavours, grassroots activism, and political lobbying. The movement resonates with various syncretic, redemptive and salvationist movements that flourished a century ago, such as the Consistent Way (Yiguandao) and the Confucian Church (Kongjiaohui) (Goossaert and Palmer 2011: 91–122). Most of the ideas are speculative invocations of a better world, but the movement is also institutionalizing. Its potential to offer alternative ways of development lies in its ethical outlook on the harmony of human society and the wider world (Fei 1992[1948]).

In the early twentieth century, Taoism was severely attacked by Chinese Enlightenment thinkers who dismissed it as superstitious, pseudo-scientific, and selfish. Taoism was weakly institutionalized in 1912 (Goossaert and Palmer 2011: 43–66), although its practices continue to influence Chinese social life.

At the turn of the twenty-first century, the Taoist elites began to promote the value of human–nature harmony. Taoism's founding canon, *Tao Te Jing*, is celebrated as one of the earliest teachings of ecological ethics in human history. Many scholars argue that Taoism had discovered that humans are part of the universe and they should 'return to the innocence' (*fanpuguizhen*) by keeping harmonious relations with nature rather than overpowering it. Other ideas along this vein include 'no-action' (*wuwei*), 'life-nourishment' (*yangsheng*), 'think less, desire less' (*shaosiguayu*), all promoting limited exploitation of natural resources and constrained use of pleasure (Girardot *et al.* 2001).

Taoist scholars and practitioners write about the virtues of altruism, simplicity, and compassion contributing to the health of bodies, spirits, societies, and nations. They celebrate how the old teachings can cure modernity's ills: excessive consumerism, energy crisis, pollution, food insecurity, income disparity, and social injustice. The Chinese Taoist Association promotes environmentalism in Taoist architecture, body techniques, rituals, greenery, water-drainage, and incense burning. It declares that the founder of Taoism, Lao Zi, is 'the God of Ecological Protection' (Duara 2015: 43–44).

Taoist practices in everyday life are full of practical knowledge about food, body techniques, geomancy, and communal rituals. Ordinary Chinese have no difficulty appreciating the ideas of constrained usage of resources, balanced flow of cosmic energy and the art of being non-aggressive. Village temples are filled with activities of proper reciprocity between humans and the cosmic force. Urban parks are full of elders who practise the proper flow of life energy. Food nourishment, breath-exercises, and spiritual cultivation reflect the Taoist ideas of balance and constraint. Independent of any institutional promotion, Taoism in China embodies the true locus of alternative ways to human well-being.

Further Resources

Duara, Prasenjit (2015), *The Crisis of Global Modernity*. Cambridge: Cambridge University Press.

Fei, Hsiao-t'ung (1992 [1948]), *From the Soil: The Foundations of Chinese Society*. Translated by Gary G. Hamilton, and Zheng Wang, Berkeley: University of California Press.

Girardot, Norman J., James Miller and Xiaogan Liu (eds) (2001), *Daoism and Ecology: Ways within a Cosmic landscape*. Cambridge, MA: Harvard Divinity School.

Goossaert, Vincent and David Palmer (2011), *The Religious Question in Modern China*. Chicago: The University of Chicago Press.

Jensen, Lionel M. (1997), *Manufacturing Confucianism: Chinese Traditions and Universal Civilization*. Durham: Duke University Press.

Tu, Weiming (2010), *The Global Significance of Concrete Humanity: Essays on the Confucian Discourse in Cultural China*. New Delhi: Centre for Studies in Civilizations and Munshiram Manoharlal Publishers.

Weber, Max (1951 [1915]), *The Religion of China: Confucianism and Taoism*. Glencoe, Illinois: The Free Press.

Liang Yongjia is Professor of Anthropology in the Department of Sociology, China Agricultural University. He specializes in religion and ethnicity in China. His recent publications include *Reconnect to Alterity: Religious and Ethnic Revival in Southwest China* (Routledge, 2013) and journal articles on cultural heritage, kingship, and the gift.

CHRISTIAN ECO-THEOLOGY

Fr. Seán McDonagh

Keywords: industrialization, neocolonial development, earthcare, creation theology, technological costs

Modern Catholic Social teaching has its roots in Europe's industrial revolution. Priests such as Adolph Koloping in Cologne observed that industrialization was adding to the misery of the poor. Bishop Kettler of Mainz was also aware of the negative social impacts of liberal capitalism. This was captured by Pope Leo XIII in *Rerum Novarum* (1891), the encyclical which criticized the exploitation of factory workers, called for a living wage, and the right to form unions. In 1967, Pope Paul VI moved beyond economic development in his ground-breaking *Populorum Progressio*. This encyclical, 'On the Development of Peoples', created a framework for evaluating authentic human development. Many countries, especially Africa, had won political freedom from their colonial masters, but at the same time were experiencing neocolonialism though dependent economic ties to Europe or North America.

Almost all the social encyclicals written in the twentieth and early twenty-first centuries are weak on ecological concerns. The Second Vatican Council coincided with the publication of Rachel Carson's *Silent Spring* (1962), and although it transformed Catholicism in many ways, there was little discussion of ecology. However, the Council's final discussion piece, *The Pastoral Constitution of the Church in the Modern World*, referred to the vision of the great Jesuit mystic and scientist, Pierre Teilhard de Chardin. Condemned by the Vatican in 1957, he had trained as a paleontologist and geologist and took part in the discovery of Peking Man. These experiences culminated in influential books such as the *Phenomenon of Man* and the *Divine Milieu*.

Fr Thomas Berry developed Teilhard de Chardin's work into viable theology of creation especially in two books *The Dream of the Earth* and *The Universe Story.* He sets out to capture the proper place for humankind in the evolutionary story of the emerging universe, of the planet Earth and of all life on the planet. Berry argues that this story, drawing on a variety of sciences and religious reflection, must now become a touchstone for the dawning ecological age. For Berry, the task of humankind is to design a new way of living with the rest of the natural world.

I became involved in ecological issues in 1978 while living among the T'boli people in the highlands of South Cotabato on the island of Mindanao, Philippines. Here I witnessed the dire impacts of deforestation. In 1980, I studied with Thomas Berry at the Riverdale Center for Religious Research in New York. The clarity of his insights into the 'technological trance' that grips global culture made sense of the disasters I had seen in the Philippines. My own first book, *To Care for the Earth: A Call to a New Theology*, was published in 1986. Following this, I helped draft the first pastoral letter by Filipino Bishops: *What Is Happening to Our Beautiful Land?*

In the 1970s, Leonardo Boff in Brazil and Peruvian theologian Gustavo Gutierrez laid the foundations for liberation theology. During the early years, liberation theologians often dismissed 'creation theology' as middle class, concerned more about the planet than the poor. But Boff's book *Cry of the Earth, Cry of the Poor* (1997) links liberation theology with ecological issues such as climate change, extinction of species and poisoning of oceans and fresh water. Focusing on the Amazon region, Boff traces connections between the fate of the rainforest, indigenous peoples and other poor people, and this work had an extraordinary impact. The Australian priest and systematic theologian Denis Edwards is another figure who has helped ground ecology in the Catholic faith. In *Ecology at the Heart of Faith* and other books, Edwards helps the general reader, the preacher, the spiritual director, the student and the theologian tear down the walls that too often separate mysticism, theology, prophecy, poetry, and science.

On June 18, 2015, Pope Francis published his encyclical *Laudato Si': On Care for Our Common Home.* In line with his mentor St Francis, he appeals: 'The urgent challenge to protect our common home includes a concern to bring the whole human family together to seek a sustainable and integral development, for we know that things can change' (No. 13). In No. 57, the Pope challenges denialists with the following words: 'Such evasiveness serves as a license to carry on with our present lifestyles and models of production and consumption' (No. 59). In fact, the 'gravity of the ecological crisis demands . . . patience, self-discipline and generosity, always keeping in mind that "realities are greater than ideas"' (No. 201). Pope Francis writes, 'We are not faced with two separate crises . . . but rather with one complex crisis which is both social and environmental' (No. 139). Further, in No. 48: 'The deterioration of the environment and of society affects the most vulnerable people on the planet'.

Human technologies cause major damage to the fabric of life on Earth. In *Laudato Si'*, No. 128, Pope Francis maintains that work is a necessary part of the meaning of life on earth, a path to growth, human development, and personal fulfilment. Unfortunately, recent history has made it clear that technological change, automation, robotics, and machine learning mean the end of self-fulfilling work as we have known it. The development artificial intelligence and big data will lead to serious unemployment in the service sector of the economy. The Catholic Church and other religions need to start campaigning now as a matter of urgency for a basic income scheme to protect the well-being of citizens under this regime of technological 'development'.

Further Resources

Berry, Thomas (1988), *The Dream of the Earth*. San Francisco: Sierra Club Books.
Boff, Leonardo (1997), *Cry of the Earth, Cry of the Poor*. New York: Orbis Books.
Carson, Rachel (1962 [2002]), *Silent Spring*. Boston and New York: Mariner Books.
de Chardin, Pierre Teilhard (1957), *Divine Milieu*. New York: Harper Torchbooks.
——— (1965), *The Phenomenon of Man*. London: Fontana Books.
Edwards, Denis (2006), *Ecology at the Heart of Faith*. New York: Orbis Books.
His Holiness, Pope Francis, '*Laudato Si': On Care for Our Common Home*',
    http://w2.vatican.va/content/francesco/en/encyclicals/documents/papa-
    francesco_20150524_enciclica-laudato-si.html.
McDonagh, Sean (1986), *To Care for the Earth: A Call to a New Theology*. London:
    Chapman.
Populorum Progressio, http://w2.vatican.va/content/paul-vi/en/encyclicals/documents/
    hf_p-vi_enc_26031967_populorum.html.
Rerum Novarum, http://w2.vatican.va/content/leo-xiii/en/encyclicals/documents/
    hf_l-xiii_enc_15051891_rerum-novarum.html.
Swimme, Brian and Thomas Berry (1992), *The Universe Story: From the Primordial
    Flaring Forth to the Ecozoic Era – A Celebration of the Unfolding of the Cosmos*. San
    Francisco: Harper Collins.
The Pastoral Constitution of the Church in the Modern World, http://www.vatican.
    va/archive/hist_councils/ii_vatican_council/documents/vat-ii_cons_19651207_
    gaudium-et-spes_en.html.
What Is Happening to Our Beautiful Land? http://www.catholicsocialteaching.org.uk/
    wp-content/uploads/2010/11/What-is-Happening-to-our-Beautiful-Land.pdf.

Fr Sean McDonagh is a Columban missionary priest, writer, and lecturer on ecology and theology. He now lives in Ireland, but worked for over twenty years with the T'boli people on the island of Mindanao, Philippines. He has written *To Care for the Earth: A Call to a New Theology* (1986), one of the first books in English on creation theology, followed by *On Care for Our Common Home: Laudato Si* (2016).

CIVILIZATIONAL TRANSITIONS

Arturo Escobar

Keywords: Western civilization, capitalist patriarchal modernity, ontologies, pluriverse

The notion of civilizational transition(s) designates the complex movement from dominance of a single, allegedly globalized, model of life – often designated as 'capitalist hetero patriarchal modernity' – to the peaceful, though tense, co-existence of a multiplicity of models, 'a world where many worlds fit', a pluriverse. It originates in claims that the current multi-headed crisis of climate, energy, food, poverty, and meaning is the result of a particular *modelo civilizatorio* or civilizational model, namely that of 'Western civilization'.

This thought finds echo in a variety of social spaces, from indigenous, Afro descendant, and peasant struggles in Latin America to alternative science and futures research, Buddhism, spiritual ecology and anti-capitalist, ecological, and feminist writing and activism in both the global North and the global South. Anticipated by anti-colonial thinkers such as Aimé Césaire: 'A civilization that proves incapable of solving the problems it creates is a decadent civilization. . . . A civilization that uses its principles for trickery and deceit is a dying civilization' (Césaire 1972[1955]: 9). It is echoed today in many quarters. In the words of the revered Buddhist teacher, Thich NhatHanh, we need to contemplate actively the end of the civilization that is causing global warming and pervasive consumerism: 'Breathing in, I know this civilization is going to die. Breathing out, this civilization cannot escape dying' (NhatHanh 2008: 55).

The origins of the Western civilizational model – as a project of economic, military, sex-gender, racial, and cultural dominance – are variously placed in the Conquest of America; the Peace of Westphalia (1648) that ended intra-European religious wars by providing the foundation for the modern nation-state; the Enlightenment; or the French Revolution that inaugurated the Rights of Man. However, its ultimate roots lie in the historical soil of Judeo-Christian patriarchal monotheism. From a critical perspective, it is characterized by the following:

- the hierarchical classification of differences in terms of racial, gender-based, and civilizational ladders (coloniality)

- economic, political, and military dominance over most world regions
- capitalism and so-called free markets as its mode of economy
- the secularization of social life
- hegemonic liberalism based on individual, private property, and representative democracy
- knowledge systems based on instrumental rationality, with its sharp separation between humans and nature (anthropocentrism)

Every civilization is based on a particular system of beliefs and ideas (epistemic and ontological premises), often deeply embedded in foundational myths. Civilizations are not static, and inter-civilizational relations are always changing and are subject to power. All major historians and theorists of civilization agree that they are plural – in other words, there cannot be a single civilization.[1] Nevertheless, the West has acquired a high degree of civilizational dominance, based on a measure of economic and political unification. The same cannot be said of the cultural domain, despite the inroads made by modernization into non-modern societies and, in more recent decades, globalization as the universalization of a 'higher civilization'.

However, the project of a global civilization has not come to pass. Nations and civilizations refuse to assemble neatly into a single order, despite the fact that the global experience is deeply shaped by a Eurocentric, trans-Atlantic model. In México, for instance, after more than five centuries of imposition of the Western colonial project, the Mesoamerican indigenous civilization continues to be alive and culturally vibrant. Perhaps the same could be said of other countries and world regions. It is increasingly evident that democracy cannot be exported by force; this is truer in the case of civilizations. The irrationality and violence of the dominant model are everywhere in sight. Some critics underline the spiritual and existential poverty of modern life, given the spread of the patriarchal and capitalist ontology of hierarchy, domination, appropriation, control, and war that has come to characterize it.

A diverse and pluralistic movement calling for the end of Eurocentric and anthropocentric dominance is arising as a result of its drawbacks, failures and even horrors, despite its huge technological achievements (increasingly questionable on ecological and cultural grounds). This movement involves a range of creative transition visions as well as concrete actions. In the global North, the call for civilizational change can be gleaned in ecofeminist subsistence economies, proposals for degrowth, the defence of the commons, inter-religious dialogue, and strategies for the localization of food, energy, and transport, among other areas. In the global South, visions of transition are grounded in ontologies that emphasize the radical interdependence of all that exists. This biocentric view finds clear expression in notions of *buen vivir* collective well-being according to one's *cosmovisions*, the Rights of Nature, and transitions to post-extractivism, all instances of post-development.

It is too early to say whether these loosely assembled heterogeneous visions and movements will achieve a degree of self-organization capable of ushering in significant transformations and perhaps large-scale transitions. For most transition theorists, while the outcome is by no means guaranteed, the move to a different civilizational model – or set of models – is not foreclosed. For many it is already happening, in the multiplicity of practices that embody, despite limitations and contradictions, the values of deeply ecological, non-capitalist, non-patriarchal, non-racist, and pluriversal societies.

The notion of civilizational transitions establishes a horizon for the creation of broad political visions beyond the imaginaries of development and progress and the universals of western modernity such as capitalism, science, and the individual. It does not call for a return to 'authentic traditions' nor for forms of hybridity to be arrived at through the rational synthesis of the best traits of each civilization, as if the seductive but harmless liberal language of 'best practices' could be applied to civilizations. Far from it, this call adumbrates a pluralistic co-existence of 'civilizational projects' through inter-civilizational dialogues that encourage contributions from beyond the current Eurocentric world order. It envisions the reconstitution of global governance along plural civilizational foundations, not only to avoid their clash, but to constructively foster the flourishing of the pluriverse.

Note
[1] This is the case of Arnold Toynbee, Fernand Braudel, and even Samuel Huntington who coined the celebrated concept of 'the clash of civilizations', based on an objectified notion of multiple, yet separate civilizations.

Further Resources

Bonfil Batalla, Guillermo (1987), *México Profundo. Una civilización negada*. México, DF: Grijalbo.

Césaire, Aimé (1972[1955]), *Discourse on Colonialism*. New York: Monthly Review.

EboussiBoulaga, Fabien (2014), *Muntu in Crisis: African Authenticity and Philosophy*. Trenton, NJ: Africa World Press.

Great Transition Initiative (GTI), http://www.greattransition.org/.

Nandy, Ashis (1987), *Traditions, Tyranny, and Utopias*. Delhi: Oxford University Press.

Nhat Hanh, Thich (2008), *The World We Have*. Berkeley: Parallax Press.

Arturo Escobar is Professor Emeritus of Anthropology at the University of North Carolina, Chapel Hill, and is associated with several Colombian universities. His most well-known book is *Encountering Development: The Making and Unmaking of the Third World* (1995). His recent books include *Otro possible es possible: Caminando hacia las transiciones desde AbyaYala/Latino-America* (2018); and *Designs for the Pluriverse: Radical Interdependence, Autonomy, and the Making of Worlds* (2017). He has worked with Afro-Colombian social movements for more than two decades.

COMMONS

Massimo De Angelis

Keywords: commons, commoning, tragedy of the commons, capital, social movements

By commons, I mean social systems formed by three basic interconnected elements: 1) a commonwealth, that is, a set of resources held in common and governed by 2) a community of commoners who also 3) engage in the praxis of commoning, or doing in common, which reproduces their lives in common and that of their commonwealth. In this sense, all forms of non-hierarchical human cooperation are different forms of commons. This definition is more general and comprehensive than the conventional wisdom – which sees the commons simply as resources shared by a set of individuals.

The notion of the commons has its own history and its own diverse interpretations. Historian, Peter Linebaugh (2008), sees the origin of this word in Medieval peasant practices to collectivize the King's land, a practice that was called commoning. With the advance of capitalism and its successive waves of enclosures of the commons, the term became less and less used, while political/theoretical language started to focus on other terminology to refer to alter-capitalist practice. This state of affairs lasted until the movements of the 1960s and 1970s forced communitarianism and 'sharing' back on the agenda. It is precisely in the midst of this era that the modernist critique of commons emerged: In 1968 ecologist Garrett Hardin published a paper in *Science* on 'The Tragedy of the Commons'. This was a seminal article claiming that sharing land or any other resource among a group of farmers always results in resource depletion. To demonstrate this,

Hardin deployed methodological individualism to assume that different farmers only aim at maximizing their individual utility. They seek to achieve this by allowing their cows to graze longer in the meadow or by bringing in more cows. Obviously, within this increasing competition among farmers in the context of a fixed resource pool, the latter will deplete, hence the tragedy of the commons. The solution proposed by Hardin was twofold: either going ahead with the privatization of the commons into different property rights or by monitoring and enforcing rules over the commons by the state.

It was Elinor Ostrom (1990) who produced an effective and simple critique of Hardin's thesis. After having studied thousands of different cases of existing commons around the world, some of which lasted hundreds of years, Ostrom argued that Hardin did not talk about the tragedy of commons, but the tragedy of free access. What the Hardin parable does not consider is that commons are governed by the commoners who decide collectively the rules of access, and continuously monitor them. Thus, the commoners are not only looking after their own interests, but also taking care that their own collective interaction with shared resources is sustainable, otherwise they would all lose the resources they depend on. Ostrom thus made a clear connection between commonwealth or wealth held in common, a community of commoners, and their governance system.

Ostrom's original definition of the commons, however, suffers from an important limitation. In her approach, commons appear as goods that are rival, or subtractable, and with a low degree of excludability. This implies that only the resource systems are commons – say a fishing ground, an irrigation channel, a ground water basin or a grazing area – while not the resource units derived from those systems such as 'the tons of fish harvested from a fishing ground, the acre-feet or cubic metres of water withdrawn from a groundwater basin or an irrigation canal, the tons of fodder consumed by animals from grazing area' (Ostrom 1990: 31).

This distinction, however, runs counter to both historical experience and the complexity of contemporary forms of commons, as they are reclaimed by plural social movements around the world. In the first place, both historically and in current experience, there are indeed myriad examples in which communities pool excludable resource units into a 'common pot' and then establish rules or customs for their individual appropriation: from toy libraries to communal kitchens. Second, in recent decades there has been growing interest in non-rival common goods such as knowledge, music or software codes commons. Similar to Ostrom's common resources, it is difficult to prohibit people from enjoying these latter resources. Also, the 'stock' of the resource is not reduced when one uses it. Rather, the real issue posed by these resources is the fact that capital imposes their privatization, making them artificially scarce. The open access movement is one that, with different nuances, is founded on the refusal of privatization of non-rivalrous goods such as information and knowledge. In academia and in cyberspace, it is a social movement dedicated to the principles of information sharing, open source, copyleft, creative commons and anti-privatized knowledge commons (Benkler 2003).

All these cases point at the definition of commons not just as a type of resource shared – which could be anything – but a social system comprising three elements (De Angelis 2017): commonwealth, a community of commoners and a praxis of commoning, of doing in common, including the act of governing the relation with the commonwealth and nature and with one another among the commoners themselves. In recent decades, we have witnessed indigenous communities and new commons systems emerging and becoming more visible and innovative pretty much everywhere: from the streets of Cochabamba to those of New York, Johannesburg, Athens, and Mumbai. They subtract resources from capital systems and insert them into processes of collective production and cultures based on value practices, which are participatory and democratic, and whose horizons are the welfare of the commoners and of environmental sustainability. This emergence of collective action occurs first as a survival motive to the many enclosures and crises of neoliberalism and as a refusal to submit to its exploitative technologies on social reproduction issues such as food, houses, energy, health care, education, arts and culture, or even the 'global commons' of the biosphere. Second, it acts as an innovative participatory exploration of new technologies and forms of cyber cooperation open source software production, peer-to-peer cooperations such as Wikipedia, open-source machines. These multisided spaces of cooperation are injecting hope for a post-capitalist transformation, in that they represent the emergence of socio-ecologies for an alternative production model to capitalism and authoritarian state systems. There is, however, the danger that these latter systems are able to co-opt the commons in such a way so as to shift further the cost of social reproduction onto them. This requires the weaving of emergent and traditional commons and social movements into commons movements that rebuild the social fabric of social reproduction and set an expanding limit to capital's drive for endless growth.

Further Resources

Benkler, Yochai (2003), 'The Political Economy of Commons', *The European Journal for the Informatics Professional*, 4 (3): 6–9, http://www.boell.org/downloads/ Benkler_The_Political_Economy_of_the_Commons.pdf.

Creative Commons, https://creativecommons.org/.

De Angelis, Massimo (2017), *Omnia Sunt Communia: On the Commons and the Transformation to Postcapitalism.* London: Zed.

Hardin, Garrett (1968), 'The Tragedy of the Commons', *Science.* 162 (3859): 1243–48.

Linebaugh, Peter (2008), *The Magna Carta Manifesto: Liberties and Commons for All.* Berkeley: University of California Press.

Ostrom, Elinor (1990), *Governing the Commons: The Evolution of Institutions for Collective Action.* Cambridge: Cambridge University Press.

The Commoner, http://www.thecommoner.org.

The Digital Library of the Commons, http://dlc.dlib.indiana.edu/dlc/.

Massimo De Angelis is Professor of Political Economy at the University of East London. In 2000, he founded *The Commoner,* a web-based journal. He is the author of many publications on critical political economy, neoliberal globalization, social movements, and the commons, among which are *The Beginning of History: Global Capital and Value Struggles* (Pluto, 2007) and *Omnia Sunt Communia: Commons and Post-Capitalist Transformation* (Zed Books, 2017).

COMMUNITY ECONOMIES

J.K. Gibson-Graham

Keywords: diversity, commoning, ethical negotiation, habitat

The term 'community economy' denotes a space of reflection and action. Community economies are made up of diverse, ethically negotiated practices that support the livelihoods of humans and non-humans to build flourishing habitats. Through their enactment of other worlds, here and now, they challenge and sidestep the dominance of capitalism.

In 1996, in a world in which the capitalist economy had come to stand for the only kind of economy possible, feminist economic geographer J. K. Gibson-Graham proposed that how 'the economy' is represented, also constrains our actions to change it. With the demise of state socialism as a counter pathway, it seemed that the only political response was to critique capitalism and pursue strategies of resistance. But what of the positive project of building more equitable and sustainable economies?

Gibson-Graham pointed to the incredible diversity of economic practices such as working, doing business and transacting goods and services that were excluded from mainstream economic theories, or included only as subordinate practices unable to 'drive' economic dynamism. As theorists of the global South, feminists and anthropologists know all too well that the unpaid work of women and family members, subsistence farmers, small traders, indigenous land-carers, and worker and producer cooperatives 'hold up half the sky'. Gibson-Graham argued that a whole range of economic models was being ignored by those interested in changing the world. What if these diverse economic activities were to become a new basis for collective understanding and action?

The language of a 'community economy' was proposed to describe the diversity of efforts aimed at building more ethically responsible ways of negotiating survival – that is, basic needs provision – and of generating and distributing the surplus that enables life to flourish. By enrolling the freighted concept of 'community' to qualify the master term 'economy' – and removing the term 'capitalist' as primary – Gibson-Graham aimed to highlight the fact that, despite social and cultural differences, cohabitation is the starting point from which people, 'we', must begin to negotiate and manage our earthly home, our *oikos*.

In *Take Back the Economy*, Gibson-Graham and colleagues encapsulate the concerns of a community economy in the following sequence of questions:

- What do we really need to live healthy lives both materially and psychically? How do we survive well?
- What do we do with what is leftover after we've met our survival needs? How do we distribute surplus?
- What type of relationship do we have with the people and environments that enable us to survive well? How do we encounter others as we seek to survive well?
- What materials and energy do we use up in the process of surviving well? What do we consume?
- How do we maintain, restore and replenish the gifts of nature and intellect that all humans rely on? How do we care for our commons?
- How do we store and use our surplus and savings so that people and the planet are supported and sustained? How do we make futures?

Such questions guide many existing innovations, practical experiments and social movements that focus on making 'other worlds' possible. Think, for example, of movements around basic income grants, transition towns, the solidarity economy, *buen vivir*, fair trade, sustainable consumption, community land trusts, ethical banking and community finance, worker-owned cooperatives, and land care to name just some.

Everywhere there are instances of people forging more considered ways of meeting their needs and the needs of others without savaging the environment or disregarding future generations. These movements are usually grounded in particular place-based concerns of people, species, and landscapes. But they are also connected into assemblages that have political force at national and global levels. Consider the multi-scalar impact of the fair trade movement. Here, new standards for ensuring people and environments are not degraded by commodity transactions continually renegotiated under international regulations. One may also consider the common will of international alliances that led to agreements on the regulation of chloro-fluoro carbons, thus healing our atmospheric commons of its ozone hole. The language of a 'community' – as distinct from 'capitalist' economy – offers a transformative reframing of what lies at the core of economic reason. It pushes to the foreground the profound interdependence humans have with each other and non-human others, be they plant and animal species, fungal and bacterial communities, or geological and climatological earth systems.

Can the logic of community economies displace the hegemony of capitalist economic thinking and behaviour? Certainly, a different way of knowing, representing and talking about economic 'realities' is a prerequisite to this displacement. Certainly, on-the-ground community economic experimentation provides inspiration for what comes next. A growing network of thinkers and activists is now theorizing community economy interactions. The aim is to model complex adaptive systems, driven not by capital, but by ethical negotiation attuned to the needs of the 'more than human' world. By sharing our capacities, acting and thinking together, we are setting ourselves on a path 'beyond development' as it has been known.

Further Resources

Community Economies Research Network, www.communityeconomies.org.

Gibson-Graham J. K. (1996), *The End of Capitalism (As We Knew It): A Feminist Critique of Political Economy*. Oxford: Blackwell.

Gibson-Graham, J. K., Jenny Cameron and Stephen Healy (2013), *Take Back the Economy: An Ethical Guide For Transforming Our Communities*. Minneapolis: University of Minnesota Press.

——— (2016), 'Commoning as a Post-Capitalist Politics', in Ash Amin and Philip Howell (eds), *Releasing the Commons*. London and New York: Routledge.

Take Back the Economy: An Ethical Guide for Transforming Our Communities Website, http://takebackeconomy.net/.

J.K. Gibson-Graham is a pen name shared by the late Julie Graham, who passed away in 2010, and Katherine Gibson, Research Professor at the Institute for Culture and Society, Western Sydney University. J.K. Gibson-Graham co-founded the Community Economies Collective, which now hosts an internationally growing Community Economies Research Network.

——•••••——

COMUNALIDAD

Arturo Guerrero Osorio

Keywords: Communality, Oaxaca, post-development, The We, Originary peoples

Comunalidad, or Communality, is a neologism that names a mode of being and living among the peoples of the Sierra Norte of Oaxaca, and in other regions in this state located in southeastern Mexico. The term was coined at the end of the 1970s by two Oaxacan thinkers: Floriberto Díaz Gómez and Jaime Martínez Luna. It expresses a stubborn resistance to all forms of development that have arrived in the area, which has had to accept diverse accommodations as well as a contemporary type of life that incorporates what arrives from afar without allowing it to destroy or dissolve what is its own – *lo propio*. Communality appeals to the best of our peoples' traditions that has persisted, that of changing traditions traditionally so as to continue being who they are despite the pressures to dissolve, marginalize, or convert them into something else, that is, develop them.[1]

Communality is the verbal predicate of the We. It names its action and not its ontology. Incarnated verbs such as eat, speak, learn are collectively created in a specific place. It only exists in its execution. The We is realized in the 'spiral of experience'. We can distinguish three moments within it.

Recognition/Exchange/Evaluation. The exercise and understanding of the We are not epistemological activities but lived ones. They entail 'the recognition of the ground' on which one walks. 'You recognize yourself with the people on that ground. We recognize what we do and what we achieve.' That is to say, we recognize our potential and limit.

We recognize that our existence is only possible with the others by constructing a We, thereby distinguishing ourselves from the Others. We open ourselves to all beings and forces, because even if the We manifests itself in the actions of concrete women, men, and children, yet in that same movement, all that is visible and invisible below and in the Land also participates, following the principle of 'complementarity' among all that is different. The communal is not a set of things, but an 'integral' fluidity.

After recognition comes an exchange of experiences, tools, and knowledge within the We and/or with Others. It is a 'mutual hospitality'. We harbour the Truth of the Other while the Other hosts ours. We encounter one another in 'sharing',[2] that is, *guelaguetza* in Zapotec, a communal aesthetic principle: to be with the other in key moments of life, sharing the experience. Homeomorphic equivalents of communality could be the Quechuan *sumak kawsay* and the Tzeltal's *lekil kuxlejal*. All are tilled by an ethic of 'reciprocity'. The exchange entails both rational critique as well as trust and faith. This learning culminates in an evaluation of the recognition and exchange that have taken place. It creates within us a new recognition towards another exchange and new evaluation.

We/Orality/Sediment. The 'We' is recreated first of all in the mental space of 'orality', the 'image', even if today these are mixed up with textual and cybernetic mentalities. Within orality, the We is produced on a concrete ground – *un suelo* – and under a concrete sky, which is a place where the bodies of all those present and disappeared are, each with the unique appearance they have precisely at the moment of recognition and exchange. *Guelaguetza* occurs on a 'sediment' of life and death. All that has occurred since Mother Earth was born is deposited there: it is on that bundle of traces where one speaks and listens.

Everyday/Remembrance/Hope. Experience lives in its duration; it is not measured by linear time. For the We, it is an extended present. In the everyday we remember, having the sediment as our grip and trigger. It is there that we shelter our hopes for the future.

The experience of the We takes place on the horizon of an 'inside spiral'. In it we distinguish two dimensions: 'Agreement' and the 'Root'. Agreement is the rationalization and verbalization of the root. It establishes the ordering of the We in its internal relations and with the exterior. Experience sediments itself in the agreement, and the agreement determines the experience. The norms establish the We's forms of sharing and sets limits to individualism and envy. From the agreement emerge the communal institutions of the 'assembly', 'cargos', posts or communal obligations, and *tequio* or collective labour for the common good, without remuneration.

The assembly is the form brought forth by the We to come to a consensus and make agreements. It is 'communalocracy' that operates there, rather than democracy, between the different people that share in the We, instead of equal and free individuals competing with each other. In the assembly, the authorities are 'named' – not elected – 'grievances' are resolved, and it is decided collectively which common path will be followed. The authorities do not govern: they provide a 'service' as ordered by the assembly: it is the 'leading by obeying' of the Ejército Zapatista de Liberación Nacional (EZLN). The duties associated with positions of authority, like services, are carried out in an obligatory manner, without payment, and willingly – although normally people avoid them; they are onerous. One activity that the authority of each We organizes is the *tequio*.

By definition, the Root is invisible, unknowable. It is origin and sustenance. Jaguar and serpent. It is the communal myth, its horizon of intelligibility.[3] We sense the form of the Root and not its contents, since each community has its own which is different from that of others – with four directions or pillars. These are the mentioned recognitions: the ground, the people, their endeavours, and their achievements. In other words: land, authority, labour, and communal celebration.

Meanwhile, communality can only be understood in its relation with the non-communal exterior, which is to say, with the economic society. This is the 'outside spiral': it begins with an external 'imposition', which unleashes, or not, an internal 'resistance' and develops into an 'adaptation'. This result is *lo propio* – what is one's own – and the We.

Notes

[1] I express my gratitude to Gustavo Esteva for revising this text and his support in situating it within the framework of post-development.

[2] Translation note: The author uses another neologism, *compartencia*, impossible to translate fully. Here it has been translated as 'sharing'.

[3] Panikkar 1999: p. 45. Homeomorphic equivalencies are deep correspondences between words and concepts belonging to distinct religions or cultures; see http://www.raimon-panikkar.org/english/gloss-homeomorphic.html.

Further Resources

Guerrero Osorio, Arturo (2013), 'La comunalidad como herramienta: una metáfora espiral', *Cuadernos del Sur*. 34: 39–55, http://pacificosur.ciesas.edu.mx/Images/cds/cds34.pdf.

———— (2016), 'La comunalidad como herramienta: una metáfora espiral II', *Bajo el Volcán*. 23: 113–29, http://www.redalyc.org/articulo.oa?id=28643473007.

Martínez Luna, Jaime (2013), *Textos sobre el camino andado*. México: CMPIO/ CAMPO/ CEEESCI/ CSEIIO.

Panikkar, Raimon (1999), *El espíritu de la política*. Barcelona: Península.

Robles Hernández, Sofia and R. Cardoso Jiménez (2007), *Floriberto Díaz. Escrito: comunalidad, energía viva del pensamiento mixe*. México: UNAM.

Arturo Guerrero Osorio was born in Mexico City in 1971. For two decades he has worked with intellectuals and activists of Oaxaca on the idea of communality; and been involved with community radio in southeast Mexico and Colombia. He is a Collaborator of the Universidad de la Tierra in Oaxaca and the Fundación Comunalidad and is currently a doctoral candidate in rural development at the Universidad Autónoma Metropolitana-Xochimilco.

———•✦•———

CONVIVIALISM

Alain Caillé

Keywords: conviviality, political philosophy, convivialist manifesto, cosmopolitanism

'Convivialism', the term, emerged as the only natural choice during a conference held in 2011 in Tokyo under a heading referencing Ivan Illich: 'Is a convivial society possible?' Among those taking part were Serge Latouche, degrowth economist, Patrick Viveret, a leading theorist of alternative wealth indicators and Alain Caillé, editor of the monthly journal of the Anti-Utilitarian Movement in Social Science, *Revue de MAUSS*. One of the conclusions drawn at the event was the need to focus more on the points of convergence than divergence. A term was needed to refer to these commonalities. It was 'convivialism' – in other words, to be brief, the philosophy of living together, of conviviality. It was a way of paying homage to Ivan Illich.

Two years later, the name and the idea was gaining ground. To such an extent that Edgar Morin was soon able to write: 'Convivialism is a key concept without which there can be no civilisation policy.' Following a good year of debate, a booklet, the *Convivialist Manifesto*, was published in 2013, signed by 64 well-known French-speaking alternative intellectuals. The authors were located on a spectrum ranging from the left of the left to centre-left. Support from the right was also possible, as it was inspired by the view that, given the range of threats we were faced with on a global scale, only a massive shift in international public opinion towards the torrent of hubris could save us. But to make that possible, we first have to agree, from a 'pluriversal' standpoint, on a number of core values that a large portion of humanity is likely to adhere to, regardless of religious or political tradition. It may seem like an audacious gamble, but it is questionable whether there are others worth the risk.

What made this convergence between such diverse authors possible? Their agreement, be it explicit or implicit, is based on at least six points:

1. First and undoubtedly foremost, a very strong sense of urgency. There is very little time left to avoid disaster.

2. The belief that a share of the current threats stems from the global hegemony of rentier and speculative capitalism, which has become the primary enemy of humanity and the planet, through its generation and embodiment of a paroxysmal crystallization of excess hubris and corruption.

3. The fact that capitalism's omnipotence rests on the powerlessness of those who are suffering its effects, and who aspire to another way of life, to realize what they have in common and to name it.

4. The certainty that adherence to democratic values – and not least their universalization – can no longer be based on indefinite and significant GDP growth. Developed nations will not see a return to such growth, for structural reasons, and it is therefore futile to see it as the solution to all our ills. The planet, moreover, would not be able to survive a proliferation of the Western or American way of life.

5. The certainty that, rather than technical, economic, and environmental solutions, what is most crucial to the process of embarking in earnest on the task of building a post-growth world is a political philosophy that is wide enough to grasp today's world in its entirety. The ideologies that form our legacy – liberalism, socialism, communism, and anarchism – no longer enable us to consider a possible present or future, for two key reasons: because they remain too nationally centred, and because all four of them share the belief that humanity's fundamental problem is that of material scarcity, and that human beings are beings of need. As such, they fail to address the issue of desire and thus make the very notion of a post-growth world inconceivable.

6. Finally, the certainty that the only hope of thwarting, in a civilized manner, all the threats we are faced with, is to develop and radicalize the democratic ideal. Our hopes, therefore, need to be turned not only towards the Market and/or the State, but also towards Society itself, that

is, self-managed, associationist, civil society, or what convivialists refer to as 'civic society'.

The consensus among the 64 authors of the Convivialist Manifesto, soon joined by around fifty world-renowned intellectuals was built on the identification of four principles:

A principle of common humanity: respect for differences. This precludes all forms of exclusion and stigmatization.

A principle of common sociality: the wealth of social ties. This lays down the absolute need to ensure the quality of social relations.

A principle of legitimate individuation: self-actualization. This specifies that social relations must be organized in a way that allows everyone to be recognized in their uniqueness.

A principle of peaceable and constructive opposition. This establishes the primary political goal of allowing humans to cooperate by 'opposing each other without slaughtering each other (and to be able to fully dedicate themselves to a cause without sacrificing their lives for it)'.

It should be noted that these are the principles that have been and continue to be opposed by all totalitarian powers and dictatorships, including financial dictatorships. The first principle expresses the central aspiration of communism, the second, that of socialism, the third, that of anarchism, and the fourth, that of republican liberalism. It is probably not too audacious to assume that these four key values are shared by all religions. A State or a government, or a new political institution cannot be deemed legitimate unless it adheres to these four principles. Convivialism could be considered the art of combining them.

The *Convivialist Manifesto* (2013) was initially written or endorsed by around 100 well-known French-speaking intellectuals and academics, and then by almost 4,000 people. It has been translated into ten languages.

Further Resources
Caillé, Alain/The Convivialists (2016), *Éléments d'une politique convivialiste*.
    Lormont: Le Bord de l'eau.
Convivialism Transnational, http://convivialism.org/.
Convivialist Manifesto (2013), *A Declaration of Interdependence*, Le Bord de l'eau.
Journal du MAUSS, www.journaldumauss.net.
La Revue du M.A.U.S.S., http://www.revuedumauss.com.fr/
Les Convivialistes, http://www.lesconvivialistes.org/.

Alain Caillé is an Emeritus Professor of Sociology at Paris-Ouest-Nanterre University. He also directs an interdisciplinary social science and political philosophy review, *La*

Revue du MAUSS (Anti-Utilitarian Movement in Social Science), www.revuedumauss.com.

⟡

CONVIVIALITY

David Barkin

Keywords: conviviality, tools, convivial reconstruction

In its modern form, 'conviviality' became a word of common usage with the publication of Ivan Illich's *Tools for Conviviality* in 1973.[1] Although not a new concept, he chose to endow it with a different connotation, 'a technical term to designate a modern society of responsibly limited tools' (p. xxiv); he is explicit in differentiating his meaning from its common use as 'joyfulness', by applying the term to 'tools' rather than to relations among people. In this context, he also introduced another fundamental characteristic of this society: austerity, understood as 'a virtue which does not exclude all enjoyments, but only those which are distracting from or destructive of personal relatedness' (p. xxv). What Illich proposed was a steady process of convivial reconstruction, offering concrete guides for it that are even more relevant today than they were then.

Conviviality is a platform for the forging of a new society, one that transcends the profound limitations of our present world, to move towards a socialism that would require 'an inversion of our present institutions and the substitution of convivial for industrial tools' (p. 12). This new framework 'will remain a pious dream unless the ideals of socialist justice prevail' (p. 12). He goes on to highlight a position common to many of today's social movements:

> The present crisis of our major institutions . . . abridge basic human freedom for the sake of providing people with more institutional outputs . . .. A convivial society would be the result of social arrangements that guarantee for each member the most ample and free access to the tools of the community and limit this freedom only in favor of another member's equal freedom. (p. 12)

This is crucial: although a convivial world does not lead to an equal society, traditional as well as new economic devices would be needed to 'keep the net transfer of power within bounds' (p. 17).

In this world, a balance must be sought between people, their tools and the collectivity. Key to this discussion is a different understanding of tools, instruments, and institutions. In his presentation, Illich carefully traces a process for redesign so that it works for people and society, satisfying needs rather than the opposite, as occurs at present. Advancing this agenda, he insists on the ethical centrality of freedom rooted in interdependence, rather than the atomistic value of today's society, so closely related to the competitive dynamic imposed by today's class society.

Conviviality, however, has to take into account our current social structures and planetary limits. The emerging communities where it is taking root are not groups that simply decide to separate from the nation-state. The process involves much more, an effort to forge a measure of independence, of autonomy to govern themselves, to create new institutions that enable genuine democratic participation and a sharing of the tasks of governance. This is facilitated by assuming control over a territory, an area they identify with and, ideally, to which they can lay an historical claim. The consolidation of the collective administration of the territory involves the claim and/or recognition of the commons as an institution, a tool, if you wish, that transforms the individuals into a deliberative collectivity, groups taking control of their lives and the sources of their livelihoods.

The commons are an important source of support for this transformation. A physical area where the community can support itself, a historical space with which it can identify, and an institutional space that allows it to formulate the new relations, facilitating people's ability to support each other, extending the possibilities of conviviality. In their defense of the commons, communities are facing the challenges of reversing historical trends of social disintegration and environmental devastation as well as the advance of the new models of expropriation or appropriation. These function through the market or outright theft such as land and water grabbing by the perverted use of dominant institutions. These conflicts are a growing part of the anti-capitalist dynamic that 'commoning' is defining and advancing. The same processes of defence are strengthening their commitment to a common purpose, creating a renewed ability to forge alliances with other communities in expanding areas of influence with a greater capacity to propose new forms of governance. Of course, these communities and their evolving institutions are also implementing new forms of living, more compatible with the exigencies of the planetary boundaries and the possibilities of the bioregional ecosystems.

Today, conviviality is far more likely to emerge than when Illich first formulated his vision. Across the world, a growing number of groups are taking the reins of their future into their own hands. No longer fooled by the promises of a prosperous future of perpetual growth, myriad initiatives are sprouting in the search for alternatives. While some are the illusory product of idealist abstractions, many are firmly grounded in more realistic attempts to learn from their heritage, adjusting these lessons to the concrete circumstances they face today. The communities are actively engaged in productive dialogues with others, locally and globally, strengthening alliances and networks to create the possibilities of overcoming the limitations of small size and individual ecosystems. In this way, they are 'reappropriating' community, integrating a history of collective work and knowledge, assuring the well-being of their societies, replacing the concern for profit with programmes to strengthen their institutions and conserve their ecosystems.

The numerous initiatives to build and rebuild community, moving beyond resistance against the forces of the global marketplace, are examples of people searching for and forging new alternatives. In place of scarcity, they define new convivial objectives: placing production and institutions at the service of the collectivity while ensuring the health of their environment.

Note
[1] In this text, I am citing the internet version of the 1973 book, published by Calders and Bacon in London, freely available at http://www.preservenet.com/theory/Illich/IllichTools.html. The page references are to this version.

Further Resources
Esteva, Gustavo (2014), 'Commoning in the New Society', *Commons Sense: New thinking about an Old Idea*, special issue of the *Community Development Journal*. 49: 144–159.
Illich, Ivan (1973), *Tools for Conviviality*. London: Calder and Bacon, http://www.preservenet.com/theory/Illich/IllichTools.html.
McDermott, Mary (2014), 'Introduction', *Commons Sense: New thinking about an Old Idea*, special issue of the *Community Development Journal*. 49: 1–11.
Shaw, Mae (2014), 'Learning from the Wealth of the Commons: A Review Essay', *Commons Sense: New thinking about an Old Idea*, special issue of the *Community Development Journal*. 49: 12–20.

David Barkin is Distinguished Professor at the Universidad Autónoma Metropolitana, Xochimilco Campus, Mexico City. A founding member of the Ecodevelopment Center, he received the National Prize in Political Economy and is an emeritus member of the National Research Council. His latest book is *De la Protesta a la Propuesta* (México and Buenos Aires: Universidad Autónoma Metropolitana y CLACSO). He is presently working on *Food Sovereignty* to be published by Zed Books, London.

COOPERATIVE ECOSYSTEMS

Enric Duran Giralt

Keywords: grassroots movements, commons, alternative economy

We understand a cooperative ecosystem as a process of building a post-capitalist society grounded in the creation of cooperative relationships in all aspects of life – economy, politics, ecology, culture, and human needs. Cooperative ecosystems are antagonistic to the capitalist system and to nation states; they aim to spread solidarity relationships amongst participants, and autonomy in as many domains of social life as possible.

Cooperative ecosystems are based on principles, codes, connections, and actions that allow each project to support as many needs as possible. Each part has a role in that the whole is only possible with the participation of all parts. We say they are cooperative because all initiatives and individuals involved are based on mutual support, solidarity and equity, generating cooperative practices in opposition to the competitive experiences that predominate in capitalist systems.

Actually, the phrase is adopted primarily from FairCoop, which introduced the idea of a 'cooperative ecosystem' in its 2016 slogan: 'Earth cooperative ecosystem for a fair economy.'

A cooperative ecosystem can become a synergistic meeting point for different alternative economy frameworks; joining practices that any of the networks agree on. These ecosystems can work locally or globally in an interrelated way, while consistent with their values, since local action and decision processes have autonomy.

The following are some *key elements* of a cooperative ecosystem:

Openness: A fundamental aspect is that individuals, collectives or even networks and other ecosystem processes enjoy cooperative ecosystem tools. That inclusiveness allows initiatives to be fluid and versatile, without authoritarian decisions that close spaces or generate division amongst participants.

Assemblies and direct democracy: Decision-making by consensus is key to maintaining process freshness during the early stages of development. In the long run, a successful process may need to introduce a voting process, due to the overcrowding of participative spaces, but it is important that in the early stages, consensus should be the way to decide on common issues. Consensus avoids the temptation to privilege decision-making only by those who are registered. It avoids conflict between diverse territories or sectors.

Dynamic construction: It is important for the interaction between means and ends to be fluid and flexible. Transition change needs a shared idea about where that change will go; even though this plan may change at each assembly. A future plan cannot be an untouchable document, because that will take freedom and constituent power from those who are building it and learning in the process. In this way present practices, and the theory that is bringing us towards them, interact and transform each other through time.

Sustainable self-management of the transition process: Building a series of institutions and productive capacities requires investment in materials, time, learning, experimentation. In the case of economic cooperatives, costs coming from errors can mean a significant economic expense before enough income is generated to recover such costs. To cover this difference between incomes and expenses generated by the productive process – and without being dependent on external factors such as the state or capitalist business that could deactivate the project, we need to be clear about how to access resources from the start. Examples to learn from might be: economic disobedience towards political institutions; reinvestment of tax money into communal autonomy such as the 'Catalan Integral Cooperative'; or recovery of the value of a digital cryptocurrency like Faircoin,[1] supported by a global cooperative grassroots movement.[2]

On the other hand, for these transition processes to work, we need strategic tools, such as the following *autonomous tools*:

Social market: The confusion between the market and capitalism has historically made it difficult, given antagonism towards the neoliberal system, to debate the role of markets in the development of alternatives. A market with values related to the process being built has a very important role in generating cooperative relationships between diverse communities, both at a local level and with the commerce between different regions, taking into account the priority for local circuits to be aligned with nature.

Generating resource distribution processes outside of markets demands consensus within the political community. A market allows that wherever such consensus does not exist, participants still have access to diverse day-to-day material and service needs. A fair or social market can have many shapes, from a street one to an online platform.

Currency: Currency has been one of the key inventions in history. It has enabled complex societies to exchange products and services in a quick and efficient manner without the need for participants to have products of the same value and utility. Having one or many currencies is a key element for everyday practices in a cooperative ecosystem. In addition to exchange, these days decentralized systems such as the blockchain technology facilitate secure online transactions. Currency is important for the storage of value allowing us to incorporate savings and investments as strategic elements in building ecosystems.

Commons projects and autonomous projects: By analogy with the public and private spaces present in modern states, cooperative ecosystems have basically two types of relationships. The majority of projects are autonomous, meaning one person or a group of people take their own decisions to offer their product or services. From that, they generate their own income and with it they provide a progressive contribution to the commons budget. The Catalan Integral Cooperative has successfully applied this methodology since 2010.

A commons project dependant on participant decision-making may serve to meet participant needs or have a longer-term strategic function.

Notes
[1] See https://www.fair-coin.org.
[2] See https://www.fair.coop.

Enric Duran Giralt comes from Vilanova i la Geltrú in Catalonia. He is also known as 'Robin Banks' or Robin Hood of the Banks, being a Catalan anti-capitalist activist and founding member of the Catalan Integral Cooperative (CIC: Cooperativa Integral Catalana) of FairCoop and Bank of the Commons.

COUNTRY

Anne Poelina

Keywords: Indigenous First Water Law, colonization, rights and responsibilities in country

My Indigenous heritage is Nyikina; in my language *ngajanoo Yimardoowarramarnil* means 'a woman who belongs to the river'. This centres me as property belonging to the Mardoowarra, Fitzroy River country. We are the traditional custodians of this sacred river in the Kimberley region of Western Australia. We were given the rules of Warloongarriy Law from our ancestor Woonyoomboo. He created the Mardoowarra by holding his spears firmly planted in Yoongoorroonkoo, the Rainbow Serpent's skin. As they twisted and turned up in the sky and down in the ground, together they carved the river valley track as sung in the Warloongarriy river creation song. This is the First Law from Bookarrarra, the beginning of time. This is inherent to what we call 'country'.

On 17 November 2015, the Western Australia *Constitution Act of 1889* was amended to recognize the state's Aboriginal inhabitants for the first time as the First People of Western Australia and traditional owners and custodians of the land – what we ourselves call 'country'. The amendment promotes the view that the state parliament should seek reconciliation with Western Australia's Aboriginal people. Although the amendment was a gesture of support, neither state nor federal governments have yet recognized the full extent of Indigenous rights.

On 2 and 3 November 2016, Aboriginal leaders met in Fitzroy Crossing to showcase to the world the recognition that the National Heritage Fitzroy River is our living ancestor from source to sea. The Fitzroy Declaration claims:

> Traditional Owners of the Kimberley region of Western Australia are concerned by extensive development proposals facing the Fitzroy River and its catchment and the potential for cumulative impacts on its unique cultural and environmental values. The Fitzroy River is a living ancestral being and has a right to life. It must be protected for current and future generations, and managed jointly.

Building on the *United Nations Permanent Forum on Indigenous Issues Background Guide (for) 2017* launched in 2016, we recognize this as an important model for cultural governance of our natural and cultural resources. The UN framework grounds the Fitzroy River Declaration and the final resolution of Kimberley traditional owners and custodians, allowing us to 'investigate legal options . . . strengthening protection under the Commonwealth Environmental Protection and Biodiversity Act (1999)', along with protection under the Western Australian Aboriginal Heritage Act (1972), whilst exploring legislation in all its forms to protect the Fitzroy River Catchment.

The British colonial invasion and occupation of our country and peoples was violent and brutal. It resulted in subjugation and slavery of the people, but invasion was defined as 'development'. The colonial states were established to create wealth for private and foreign interests at the expense of Indigenous people, our lands and waters. Since the historical discourse regarding development from the Anglo-Australian perspective has been in terms of the process and impacts of invasion, this begs the questionas to how it benefits First Nation peoples, Aboriginal or Torres Strait Islanders.

As a traditional owner of 27,000 square kilometres of Nyikina country, I am a witness to, and share in, the struggle to reconcile conditions imposed on Aboriginal people across the continent with the fulfilment of traditional law. The focus of federal and state government policy and private investment is on the development of Northern Australia within a Western economic framework. So, the Anglo-Australian settler society disregards the value of our human 'capital' grounded in traditional knowledge systems and the rights of nature. Foreign interests view our country as a resource for investment: from the pastoral industry and intensive agriculture to mining for diamonds and gold, and from pearls to fracking for gas and oil. None of these industries is sustainable: each has an adverse effect on air, land, water, and biodiversity; each brings poverty to local people.

The experience of Aboriginal people in the Kimberley and throughout Australia is shared with other First Nations people in countries colonized during the seventeenth and eighteenth centuries. The Norwegian peace philosopher, Johan Galtung, calls this legislated inequality 'structural violence'. Its effects are measured in high mortality rates for children and adults alike; in endemic alcohol and drug abuse; in 'socially transmitted disease' and in trans-generational psychic trauma. The effects of life on the frontier are recapitulated down through the generations and made worse by the violence of child removal from families, a state policy known today as the Stolen Generation.

In 2017, we recall a decade since our government ratified the United Nations Declaration on the Rights of Indigenous People. However, we have not seen these principles incorporated in Australian law. Many of us who have our customary law recognized in Native Title now work together to take charge of our own destiny and partner with like-minded people to deliver justice based on the First Water Law of the Mardoowarra.

This is a story of hope, innovation and cultural creativity as we explore our rights and responsibilities to create our own systems by going back to the principles of First Law, the law of 'country'. This First Law encompasses our relationship with each other, our neighbours, and most importantly our family of non-human beings – animals and plants. These relations are the key to our personal, community, cultural, economic and ecosystemic well-being.

As our Centre[1] demonstrates, our culture, science, heritage, and conservation economy is blossoming, founded upon our connectivity and cultural identities. Guided by First Water Law, our 'living water' systems are our life force, joining surface to ground water, uniting the diverse cultural landscape of the Kimberley. At the same time we are building collaborative knowledge systems, combining Western sciences, traditional knowledge and industry practice in sharing our most precious resources – water and biodiversity. We are reframing 'sustainable life' in our country.

Note
[1] See www.majala.com.au.

Further Resources

Fitzroy River Declaration, http://www.klc.org.au/news-media/newsroom/news-detail/2016/11/15/kimberley-traditional-owners-unite-for-the-fitzroy-river.

Galtung, Johan (1996), *Peace by Peaceful Means: Peace and Conflict, Development and Civilisation*. London: SAGE.

McInery, Marie (2017), 'Climate Justice: A Call to Broaden Science with Indigenous Knowledge', https://croakey.org/climate-justice-a-call-to-broaden-science-with-indigenous-knowledge/.

United Nations Declaration on the Rights of Indigenous Peoples, http://www.un.org/esa/socdev/unpfii/documents/DRIPS_en.pdf.

United Nations Permanent Forum on Indigenous Issues Background Guide 2017, https://www.humanrights.gov.au/united-nations-permanent-forum-indigenous-issues.

Western Australian Constitutional Act 1889, https://www.slp.wa.gov.au/legislation/statutes.nsf/main_mrtitle_185_homepage.html.

Dr Anne Poelina, MSc (Public Health and Tropical Medicine), MEd, MA (Indigenous Social Policy) is a Nyikina Warrwa Traditional Custodian from the Mardoowarra, Lower Fitzroy River, and Director, Walalakoo Native Title Body Corporate. She is an Adjunct Senior Research Fellow and DHS Scholar, Australian Government Research Training

Program, at The University of Notre Dame/Nulungu Research Institute, and Research Fellow at Charles Darwin University. Her current work focuses on legal protection of the Mardoowarra through customary law, conservation, culture, and science.

———•••••———

DEEP ECOLOGY

John Seed

Keywords: deep ecology, anthropocentrism, environmental ethics

Deep Ecology is a fundamental challenge to anthropocentrism or human chauvinism. This idea – that humans are the crown of creation, the source of all value, the measure of all things – is deeply embedded in the dominant global culture and consciousness. When we as humans investigate and see through our layers of anthropocentric self-cherishing, a most profound change in consciousness begins to take place. Alienation subsides. The human is no longer an outsider, existing apart. Humanness is then recognized as being merely the most recent stage of our existence, and as we stop identifying exclusively with this chapter of evolution, we can start to get in touch with ourselves as mammals, as vertebrates, as a species that has only recently emerged from the rainforest. As the fog of amnesia disperses, there is a transformation in relationships with other species, and in our commitment to them. What is described here should not be seen as merely intellectual. The intellect is merely one entry point to the process outlined, and the easiest one to communicate.

For some people, this change of perspective follows from actions on behalf of Mother Earth. 'I am protecting the rainforest' gradually becomes 'I am part of the rainforest protecting myself. I am that part of the rainforest recently emerged into thinking.' What a relief then! The thousands of years of imagined separation are over and we can begin to recall our 'true nature'. That is, the change is a spiritual one, 'thinking like a mountain' to borrow the words of Aldo Leopold. Sometimes this consciousness shift is referred to as 'deep ecology'.

As deep memory improves, as the implications of evolution and ecology are internalized, and as they replace the outmoded anthropocentric structures in our minds, there is an identification with all life. Then follows the realization that the distinction between 'life' and 'lifeless' is a human construct. Every atom in this body existed before organic life emerged 4,000 million years ago. Remember our childhood as minerals, as lava, as rocks? Rocks contain the potentiality to weave themselves into such stuff as this. We are the rocks dancing. Why do we look down on them with such a condescending air? They are an immortal part of us.

If we embark upon such an inner voyage, we may find, upon returning to present-day consensus reality, that our actions on behalf of the environment are purified and strengthened by the experience. We have found here a level of our being that moth, rust, nuclear holocaust or destruction of the rainforest genepool do not corrupt. The commitment to save the world is not decreased by the new perspective, although the fear and anxiety that were part of our motivation start to dissipate and are replaced by a certain disinterestedness. We act because life is the only game in town, but actions from a disinterested, less attached consciousness may be more effective.

Of all the species that have existed, it is estimated that less than one in a thousand exist today. The rest are extinct. As the environment changes, any species that is unable to adapt, to change, to evolve is extinguished. All evolution takes place in this fashion. In this way an oxygen-starved fish, an ancestor of ours, commenced to colonize the land. Threat of extinction is the potter's hand that moulds all forms of life. The human species is one of millions threatened by imminent extinction through climate disaster and other environmental changes. And while it is true that the 'human nature' revealed by 12,000 years of written history does not offer much hope that we can change our warlike, greedy, ignorant ways, the vastly longer fossil history assures us that we 'can' change. We 'are' the fish, and the myriad other death-defying feats of flexibility which a study of evolution reveals to us. A certain confidence, in spite of our recent 'humanity', is warranted. From this point of view, the threat of extinction appears as the invitation to change, to evolve. After a brief respite from the potter's hand, here we are back on the wheel again. The change that is required of us is a change in consciousness.

Deep ecology is the search for a viable consciousness. Surely consciousness emerged and evolved according to the same laws as everything else. Moulded by environmental pressures, the mind of our ancestors, time and again, must have been forced to transcend itself. To survive our current environmental pressures, we must consciously remember our evolutionary and ecological inheritance. We must learn to think like a mountain. If we are to be open to evolving a new consciousness, we must fully face up to our impending extinction – the ultimate environmental pressure – steep ourselves in the awful prognosis. This means recognizing that part of us which shies away from this truth, hides in intoxication or busyness from the despair of the human, whose 4,000-million-year race is run, whose organic life is a mere hair's breadth away from being finished. A biocentric perspective, the realization that rocks 'will' dance, and that roots go deeper than 4,000 million years, may give us the courage to face despair and break through to a more viable consciousness, one that is sustainable and in harmony with life again.

The term 'deep ecology' was coined by the Norwegian sceptical philosopher and eco-activist, Arne Naess (1912–2009). The deep ecology movement was carried forward by the US sociologist, Bill Devall (1938–2009), and philosopher George Sessions, (1938–2016). It was then picked up by Esprit founder and wilderness advocate, Doug Tompkins (1943–2015), whose Deep Ecology Foundation published an eleven-volume collection of Naess's work in 2005. From the 1980s on, deep ecology has become a major strand among ongoing debates in the academic field of environmental ethics.

Further Resources

Deep Ecology Foundation, http://www.deepecology.org/people.htm.
Devall, Bill and George Sessions, www.amazon.com/Deep-Ecology-Living-Nature-Mattered/dp/0879052473.
John Seed interviews and essays, www.rainforestinfo.org.au/johnseed.htm.
Katz, Eric, Andrew Light and David Rothenberg (eds) (2000), *Beneath the Surface: Critical Essays in the Philosophy of Deep Ecology*. Cambridge, MA: MIT Press.
Leopold, Aldo (1970), *A Sand Country Almanac*. New York: Ballantine.
Naess, Arne (1973), 'The Shallow and the Deep Long Range Ecology Movement'. *Inquiry*. 16 (1): 95–100.

John Seed is founder of the Rainforest Information Centre in Lismore, Australia. Since 1979 he has participated in direct actions and campaigns protecting rainforests worldwide. He has written and lectured on deep ecology, conducted experiential deep ecology workshops for twenty-five years and received an Order of Australia Medal (OAM) for services to conservation.

DEGROWTH

Federico Demaria and Serge Latouche

Keywords: ecological economics, political ecology, post-growth, Europe, green growth

Generally the degrowth project challenges the hegemony of economic growth and calls for a democratically led redistributive downscaling of production and consumption in industrialized countries as a means to achieve environmental sustainability, social justice, and well-being (Demaria *et al.* 2013). Degrowth is usually associated with the idea that smaller can be beautiful. However, the emphasis should not only be on less, but also on different. In a degrowth society everything will be different: activities, forms and uses of energy, relations, gender roles, allocations of time between paid and non-paid work, relations with the non-human world.

The point of degrowth is to escape from a society that is absorbed by the fetishism of growth. Such a rupture is therefore related to both words and things, symbolic and material practices. It implies decolonization of the imaginary and the implementation of other possible worlds. Degrowth politics does not aim for 'another growth', nor for another kind of development – sustainable, social, fair – but for the construction of another society, a society of frugal abundance (Serge Latouche), a post-growth society (Niko Paech) or one of prosperity without growth (Tim Jackson). In other words, it is not an economic project from the outset, not even a project for another economy, but a societal project that implies escaping from the existing economy as a material reality and imperialist discourse. 'Sharing', 'simplicity', 'conviviality', 'care' and the 'commons' are primary significations of what this society might look like (D'Alisa *et al.* 2015).

Although degrowth integrates bio-economics and ecological macroeconomics, it is actually a non-economic concept. On the one hand, degrowth certainly implies a reduction of social metabolism – the energy and material throughput of the economy – needed to meet existing biophysical constraints imposed by natural resource limits and ecosystem assimilative capacity. On the other hand, degrowth is an attempt to challenge omnipresent market-based relations and the growth-based roots of the social imaginary, replacing them by the idea of frugal abundance. It is also a call for a deeper democracy outside the mainstream democratic domain, including problems generated by technology. Finally, degrowth implies an equitable redistribution of wealth within and across the global North and South, as well as between present and future generations.

Over the last couple of decades, the single-minded ideology of growth has given rise to a seemingly consensual 'sustainable development' slogan – such a nice oxymoron! Its aim was to save 'the religion' of economic growth and to deny ecological breakdown. In this context, it became urgent to oppose the globalized capitalist market with 'another civilizational project' or, more specifically, to give visibility to a plan – as yet underground, but in formation for a long time. The rupture with development as a form of productivism for so-called developing countries, was foundational to this alternative project.

The term 'degrowth' was proposed initially by political ecologist André Gorz in 1972, and was used in the title of the the French translation of Nicholas Georgescu Roegen's essays in 1979. Degrowth was then launched by French environmental activists in 2001 as a provocative slogan to repoliticize environmentalism. In this, it met a need felt by both political ecology scholars and development critics. Thus, the phrase is not a concept symmetrical to economic growth, but rather a defiant political slogan with the objective of reminding people of the meaning of 'limits'. More specifically, degrowth is neither equivalent to economic recession nor to negative growth. The word should not be interpreted literally.

A degrowth transition is not a sustained trajectory of descent, but a transition to convivial societies living simply and in common. There are several ideas about the practices and institutions that facilitate such a transition and about the processes that bring them together and allow them to flourish. The attractiveness of degrowth emerges from its power to draw from and articulate principles of environmental justice and democracy and to formulate strategies including oppositional activism, grassroots alternatives, and institutional politics. The degrowth project brings together a heterogeneous group of actors who focus on issues from agroecology to climate justice. Degrowth complements and reinforces these topics, functioning as a connecting thread – the platform for a network of networks beyond single issue politics.

In fact, degrowth is not just an alternative, but rather a matrix of alternatives that reopens the human adventure to a plurality of destinies and spaces of creativity by throwing off the blanket of economic totalitarianism. It is about exiting the paradigm of *homo œconomicus* or Marcuse's 'one-dimensional man', the main source of planetary homogenization and the murder of cultures. Consequently, a degrowth society will not be established the same way in Europe, in sub-Saharan Africa or Latin America, in Texas or

Chiapas, in Senegal or Portugal. Instead, it is crucial to rediscover diversity and pluralism. That said: it is not possible to formulate 'turnkey' solutions for degrowth, but only to outline the fundamentals of a non-productivist sustainable society and to share concrete examples of transitional programmes. The degrowth design can take the form of a 'virtuous circle' of sobriety in the form of 8 'R's: re-evaluate, reconceptualize, restructure, relocate, redistribute, reduce, reuse and recycle (Latouche 2009). These eight interdependent objectives constitute a revolutionary rupture; one which will trigger the shift towards an autonomous society of sustainable and convivial sobriety.

The politics of this historical dynamic - the actors, alliances, institutions and social processes creating degrowth transitions remains the subject of lively debate in Europe and beyond. For instance, in September 2018 over 200 hundred scientists wrote an open letter to the major European institutions entitled 'Europe, It's Time to End the Growth Dependency'; it was then signed by almost 100,000 citizens.[1] The degrowth network includes over 100 organizations with 3,000 active members, mostly located in Europe but also in North and South America, the Philippines, India, Tunisia and Turkey.[2]

Notes

[1] The letter translated into 20 languages can be found here: https://degrowth. org/2018/09/06/post-growth-open-letter. The petition is available here: https:// you.wemove.eu/campaigns/europe-it-s-time-to-end-the-growth-dependency

[2] Find the map of the growing international degrowth network here: https://www. degrowth.info/en/map

Further Resources

D'Alisa, G., F. Demaria, and G. Kallis (2015), *Degrowth: A Vocabulary for a New Era*. London: Routledge, www.vocabulary.degrowth.org.

Degrowth blog: aspects of the degrowth discourses and movement, https://www. degrowth.info/en/blog/.

Demaria, F., F. Schneider, F. Sekulova, and J. Martinez-Alier (2013), 'What Is Degrowth? From an Activist Slogan to a Social Movement. *Environmental Values*. 22: 191–215.

Latouche, S. (2009), *Farewell to Growth*. Cambridge: Polity.

Media library: content on the topic of degrowth, https://www.degrowth.info/en/ media-library/.

University Masters Degree in Political Ecology, Degrowth and Environmental Justice, https://master.degrowth.org/-

World mailing list, http://www.criticalmanagement.org/node/3220.

Federico Demaria is an interdisciplinary socio-environmental scientist working on

political ecology and ecological economics at the Environmental Science and Technology Institute of the Autonomous University of Barcelona (ICTA-UAB). He is a visiting scholar at the International Institute of Social Studies in The Hague, Netherlands; as well as a member of the Research & Degrowth collective, and of EnvJustice, a research project that aims to study and contribute to the global environmental justice movement. He is also an organic olive farmer.

Serge Latouche is an Emeritus Professor of the University of Paris-Sud. He is a specialist in North–South economic and cultural relations, and in epistemology of the social sciences.

❖

DEMOCRATIC ECONOMY IN KURDISTAN

Azize Aslan and Bengi Akbulut

Keywords: democracy, ecology, gender emancipation, needs

The thread of the Kurdish Movement that follows the ideology developed by Abdullah Öcalan, which can be traced back to the founding of PKK (Kurdistan Workers' Party) in 1978, has moved away from its initially declared aim of an independent Kurdish state to defending Democratic Confederalism and Democratic Autonomy as primary organizational models. The project of Democratic Autonomy implies a process of organization within multiple aspects such as law, self-defence, diplomacy, culture, and ecology. It includes the construction of a communal and 'democratic economy' based on principles of gender emancipation and ecology. The Project re-embeds the economy within social processes, ensuring the access of all to the means of social reproduction – a reconfiguration defined by needs.

The main intellectual ground for the Kurdish project is provided by Abdullah Öcalan's critical writing on capitalist modernity. Öcalan deepens the Marxist analysis of capitalism by probing the universality of industrial/accumulationist capitalism. The project is also inspired by Murray Bookchin's ideas of social ecology and libertarian municipalism. Building on these intellectual roots, democracy, gender emancipation, and ecology are defined as the principles along which all economic relations are to be organized. Democracy implies that decision-making regarding what to produce and share, how to manage resources, and how distribution should be participatory and egalitarian. Among the means of societal decision-making are communes and councils at different scales and themes – neighbourhood, town, city; youth, women, education, economy, ecology, etc. – as well as venues such as energy cooperatives and water councils. Gender emancipation denotes the discursive and material subversion of the invisibility and devaluation of women's labour and knowledge, and the reconstruction of economic relations in ways to ensure women's participation in all decision-making processes. Ecology signifies the recognition that all nature is the common heritage of humans and non-humans and that all economic activities should be constrained by ecology as well as by society.

A democratic economy is a non-accumulationist economy wherein activities are not oriented towards an unquestioned imperative of economic development, but to fulfilment of the needs of all. This means the prioritization of use-value over exchange value, ensuring collective and equal access to land, water, and other local resources, and the positing of non-human nature as the non-commodifiable common heritage of all living creatures. Collective and equal rights over the means of social reproduction are upheld over efficiency and profit-orientation. The concrete proposals associated with this vision include realizing justice in land tenure, the reorganization of agricultural production on the basis of need, socialization of women's unpaid labour responsibilities by day-care centres and communal kitchens and local self-management of resources through energy cooperatives and water councils.

A notable example among the steps taken to operationalize this project is the municipal initiatives of providing land access to landless families. Plots in urban peripheries have been opened to collective cultivation by landless families, some 10 to 40 per plot, with technical and equipment support. The plots are linked to seedling camps, where indigenously developed seeds are being conserved. Production in these units is mainly subsistence-oriented, but they are also connected to direct producer–consumer hubs in urban centres for marketing surplus production. Another example is the network of women's cooperatives that the Kurdish women's movement spearheaded. Interlinked with production and distribution, these cooperatives are engaged primarily in agro-processing and textile manufacturing, and they market their products directly to consumers via the cooperative distribution hubs, *Eko-Jin's*. Most of the agro-processing cooperatives stem from and are linked to existing urban farming collectives. The cooperatives are further networked with municipal officials, activists, academics, and civil society groups under the broader umbrella of the women's movement, the Free Women's Congress (KJA) – a venue of debate and decision-making.

The Democratic Autonomy project foresees the organization of a self-sustaining autonomous economy as an indispensable aspect of political autonomy. It aims to organize the production of goods and services communally in order to pre-empt the functionality of the state within this field. In that sense, the project carries a close parallel with other autonomous movements such as the Zapatistas. It also resonates with the larger solidarity economy movement worldwide, as it deconstructs the imperative of capitalist development and prioritizes self-management, social justice, and ecological integrity.

Although much remains to be seen in how the project of democratic economy will further concretize a number of challenges can be discerned. Existing inequalities, such as those in landownership, are likely to imply challenges in organizing the economy along collective and egalitarian fulfilment of needs. The tension between the fulfilment of the needs of all as an organizing principle and the non-accumulationist stance of democratic autonomy is another node of challenge. While what constitutes needs is to be deliberated democratically, needs beyond self-production will inevitably pose the issue of how much surplus is to be 'accumulated' to meet them and if, collectively, such needs are seen as legitimate. More importantly, the escalation of armed and political violence from the Turkish state as well as the intensified diffusion of capitalist relations within the region, pose significant difficulties. Yet, what has enabled and continues to sustain this project are the solidarity networks found within the Kurdish people. While collectivism, sharing, and solidarity have always been strong cultural codes, the collective history of struggle has strengthened these networks most significantly. They have, in turn, served as the ground on which an autonomous democratic economy could be organized. In this sense, the commitment and the solidarity-based organization of Kurdish people is an invaluable opportunity.

Further Resources

Akbulut, Bengi (2017), 'Commons against the Tide: The Project of Democratic Economy', in Fikret Adaman, Bengi Akbulut and Murat Arsel (eds), *Neoliberal Turkey and Its Discontents: Economic Policy and the Environment under Erdoğan*. London: IBTauris.

Aslan, Azize (2016), 'Demokratiközerklikteekonomiközyönetim: Bakûrörneği', *Birikim*. 325 (May): 93–8; [Economic self-governance in democratic autonomy: The example of Bakûr (Turkish Kurdistan)], https://cooperativeeconomy.info/economic-self-governance-in-democratic-autonomy-the-example-of-bakur/.

Cooperative Economy, https://cooperativeeconomy.info.

Madra, Yahya M. (2016). 'Democratic Economy Conference: An Introductory Note'. *South Atlantic Quarterly*. 115: 211–22.

Öcalan, Abdullah (2015), *Manifesto for a Democratic Civilization*, volume I: *Civilization: The Age of Masked Gods and Disguised Kings*. Porsgrunn: New Compass Press.

Azize Aslan is a PhD student at Benemérita Universidad Autónoma de Puebla, Instituto De Ciencias Sociales y Humanidades Alfonso Vélez Pliego. She is working on a comparative study of Zapatista and Kurdish Movements; has taken an active part in the Democratic Economy project and written extensively on the issue.

Bengi Akbulut has a PhD in Economics, and is an Assistant Professor in the Department of Geography, Environment and Planning, Concordia University, Canada. She works on the political economy of development, political ecology, the commons, and alternative economies.

DIRECT DEMOCRACY

Christos Zografos

Keywords: direct democracy, self-governance, autonomy, socio-ecological transformation

Direct democracy is a form of popular self-rule where citizens participate directly, continuously, and without mediation in the tasks of government. It is a radical form of democracy that favours decentralization and the widest possible dispersal of power, eliminating the distinction between rulers and those governed. It is premised upon the principle of political equality, understood as the requirement that all voices in society are equally audible. Its key institution is the deliberative assembly. These assemblies involve meetings where citizens make decisions by listening to, and discussing, different views on a matter, reflecting on each view, and trying to arrive at a common decision without coercion. Direct democracy is distinguished from representative democracy, which involves electing representatives who decide on public policies. Nevertheless, elements of direct democracy, such as the referendum, are also present in existing representative democracies.

The practice of direct democracy is very old, indeed ancient. Ancient,

5th century BCE Athens is the oft-quoted example of direct democracy where adult, male citizens participated directly in public decision-making. The exclusive character of Athenian democracy that barred slaves, women and foreigners from participation in decision-making suggests that it was a very limited form of democracy, although relevant in terms of its direct democracy institutions and forms of participation. If we think of democracy as 'government by discussion', one can also trace its roots in a long, non-Western tradition outside and roughly contemporary to Athens, such as the northern Indian experiences of the city of Vesali and the Sabarcae/Sambastai people recorded by both ancient Indian and Greek sources. In terms of intellectual origins, a key modern influence is Jeanne-Jacques Rousseau and his ideas on representation and government. For Rousseau, handing over one's right of self-rule to another person was a form of slavery; he thus rejected binding legislation on issues that citizens had not previously deliberated and agreed upon. A key related concept is autonomy. According to Castoriadis, autonomy involves the capacity of society to collectively and continuously question and change its norms and institutions, which is based on the belief that society itself is the only legitimate source for doing so. Castoriadis criticized dogmas that pose external rules which limit autonomy or that justify and determine collective decisions by ascribing them to some authority outside society (e.g. God, historical necessity, etc.), a condition he termed as heteronomy. In sum, direct democracy allows citizens to control decisions over their own destinies, educates them in participatory decision-making instead of relying on self-serving politicians, and produces highly legitimate decisions (Heywood 2002).

With regard to post-development (Rahnema and Bawtree 1997), the transformative potential of direct democracy can be considered in two ways: on the one hand, it helps challenge the hegemony of single ways of thinking and the colonization of minds by a heteronomous imaginary; on the other hand, it helps to build alternatives to development in the practice. This potential can be evidenced in the way in which contemporary social movements but also non-state polities around the world bring direct democracy into play.

In Spain, assembly-based decision-making processes popularized during the *indignados* movement have empowered 'right-to-housing' social movements to disrupt the urban capital accumulation dynamics of Spanish capitalism (García 2017), as well as municipal governments to pursue more equitable and environmentally sustainable models of the city through binding citizen consultations. In India, radical ecological democracy initiatives, such as the Arvari River Parliament of 72 riverine villages in

Rajasthan, signal attempts to achieve transitions towards a bioregional vision of ecological units governed democratically by local communities, at the core of which lies a pledge to cultural diversity, human well-being, and ecological resilience. The governance model of the Kurdish autonomous canton of Rojava, which emphasizes gender equality in political office and participation, incorporates direct democracy into its decision-making; its aim is to transform society on the basis of the principles of Murray Bookchin's social ecology, thus becoming an exemplary organization for future democratic confederalist systems of regional governance. And throughout the American continent, many indigenous, campesino, and Afro-descendant communities practise self-governance and assembly-based decision-making in their efforts to materialize into life projects the principles of autonomy, communality and respect for diverse forms of life that stem from their cosmovisions.

Conversely, a 'darker' side of direct democracy lies precisely in its capacity to prevent transformation. The Appenzell-Innerrhoden Swiss canton that is celebrated as an example of direct democracy conceded voting rights to women only in 1991 when forced by the Swiss Federal Supreme Court. Moreover, the canton has registered the highest Swiss canton vote in favour of banning minarets. Another criticism against direct democracy is that supporters romanticize it, overlooking that states may be better vehicles for achieving radical transformations due to their capacity to coordinate and mobilize resources across larger areas, something crucial in a globalized world. Critics also question the willingness of citizens to be constantly engaged in the governance of everyday life, and criticize direct democracy as a romantic nostalgia of a 'liberal left' by pointing to historical examples like the 1871 Paris Commune to argue for its inability to sustain itself and as proof of its limitations.

Other critics draw attention to limitations related to central traits of the deliberative process that characterizes direct democracy, pointing out that those traits limit its capacity to pursue radical socio-ecological transformations. For instance, critics contend that the emphasis on decisions taken on the basis of consent downplays the importance of conflict, dissent and difference for effecting such transformations; that the role of reason and rational argumentation used to arrive at consensual decisions underplays the crucial role that emotions, imagination, narrative, socialization, and bodily activity play in producing transformation; and that past evidence suggests that strong leadership might be more crucial than horizontality, a central tenet of direct democracy, for achieving transformations.

Despite these criticisms, it seems certain that the ideal and practice of direct democracy has historically inspired and still motivates individuals and communities to boldly attempt to create worlds which are different and better to those they inhabit. In that sense, direct democracy holds promise for helping in decolonizing minds and challenging hegemonic ways of thinking, acting, and being. At its best, direct democracy becomes a different way of being.

Further Resources

A Kurdish Response to Climate Change, https://www.opendemocracy.net/uk/anna-lau-erdelan-baran-melanie-sirinathsingh/kurdish-response-to-climate-change.

Centre for Indigenous Conservation and Development Alternatives (CICADA), http://cicada.world.

García, Lamarca M. (2017), 'From Occupying Plazas to Recuperating Housing: Insurgent Practices in Spain', *International Journal of Urban and Regional Research*. doi:10.1111/1468-2427.12386.

Heywood, Andrew (2002), *Politics*. New York: Palgrave Macmillan.

O'Connor, Kieran (2015), 'They Don't Represent Us', Summary of a discussion about democracy and representation between Jacques Rancière and Ernesto Laclau, http://www.versobooks.com/blogs/2008-don-t-they-represent-us-a-discussion-between-jacques-ranciere-and-ernesto-laclau.

Rahnema, Majid and Victoria Bawtree (eds) (1997), *The Post-development Reader*. London: Zed Books.

Christos Zografos is Ramón y Cajal Senior Research Fellow with Pompeu Fabra University. His research in political ecology and ecological economics focuses on political conflict and environmental transformation. He is a member of the Research & Degrowth collective in Barcelona, and a Visiting Professor at Masaryk University, Czech Republic.

———◦◦◦◦———

EARTH SPIRITUALITY

Charles Eisenstein

Keywords: science, spirituality, development

Earth spirituality refers to a belief system that embraces the sentience, sacredness, and conscious agency of nature and its non-human beings. It takes diverse names, corresponding to somewhat different orientations such as neo-paganism, wicca, animism, or pantheism. It is closely aligned with, and often consciously borrows from, ancient and indigenous conceptions of Nature. As anthropologist Frederique Apffel-Marglin puts it:

> As the ethnographic record for indigenous and traditional societies abundantly attests . . . sentiments of gratitude, reciprocity, responsibility, and the like are addressed to the non-human world through the spirits of the earth, the animals, the seeds, the mountains, the rains, the waters, and so on. (Apffel-Marglin 2012: 39)

Earth spirituality stands in contrast to the scientific materialist view of nature that is governed by impersonal forces and populated by animals and plants lacking full beingness. It therefore delegitimizes technologies that subjugate, and often destroy, nature. If nature possesses some kind of intelligence, then we can no longer impose human design upon it with impunity. If animals, plants, soil, water, mountains, rivers, and so on are sentient subjects, we can no longer in good conscience treat them as instruments of human utility. We must take into account the well-being, integrity, and even the dignity, of all beings, and not treat them as mere 'resources'.

The incompatibility of earth spirituality with normative scientific belief opens it to charges of anthropomorphic projection. Just as a child imagines that her teddy bear is hungry, so does the romantic hippie environmentalist imagine that the land is angry or the river feels insulted or the mountain wants to keep its gold ore. However, this association of earth spirituality with puerile fantasy or New Age woo is also a colonialist narrative tarring indigenous cultures with the same disdainful perspective. It suggests that indigenous peoples – who are nearly unanimous in their personification of nature – are like children, superstitious and epistemically primitive, and therefore in need of education into modern knowledge systems. This is a key legitimizing assumption of development.

A related critique accuses earth spirituality of cultural appropriation. It sees indigenous perceptions of the sentience and personhood of nature as mere cultural 'beliefs' and not as revelations of something real. While it is certainly common for seekers from the dominant culture to array themselves in the intellectual costume of borrowed indigenous beliefs in an attempt to compensate for their own crumbling structures of meaning, indigenous world views may contain knowledge essential to material and psychic well-being. The transmission of these world views and their associated practices is a reversal of traditional development, which said, 'We know how to live better than you; we know how to "know" better than you.' Today, seekers of earth spirituality sense that other cultures have preserved the knowledge they seek.

A third critique is that environmentalists adopting such beliefs make themselves vulnerable to accusations of flakiness and exclusion from policy circles, to which language such as 'the consciousness of the forest' is foreign. While earth spirituality generally concords with the goals of the environmental movement, and while many environmentalists privately sympathize with it, the public narrative around environmental protection, and particularly climate change, mostly invokes utilitarian arguments – we should stop destroying nature because otherwise bad things will happen to us. These utilitarian arguments are easily framed in the language of public policy. Unfortunately, they are the first step down a slippery slope. The next step is the quantification and monetization of ecosystem services, putting a number on nature's utility so that it can be optimally allocated via market mechanisms. It is fine to drain a wetland here if another one is restored there; to cut down a forest here and replant one there; even, in the extreme, to replace all of nature with a technological substitute were that to become possible.

Because development has in large part meant the conversion of nature into commodities, earth spirituality is indeed an anti-development or post-development belief system. By insisting that nature has an inherent value beyond its utility, earth spirituality precludes the normal capitalist disposition of the biosphere. Just as it is immoral to kill human beings in order to harvest their organs, so also it is immoral to want only to destroy non-human beings for their use-value. A key political corollary of earth spirituality is 'rights of nature', which has recently entered the legal codes of several countries including Bolivia, Ecuador, and New Zealand.[1] All of these countries have a strong indigenous presence, but this alone would not be enough to legalize the rights of nature were it not for the resurgence of earth spirituality in the dominant culture.

While in 'less developed' places earth spirituality is an unbroken continuation of an ancient earth-centred world view, in the West the line of continuity with a pagan or pantheistic past is much more tenuous, having endured centuries of repression under the Church, market capitalism and the Scientific Revolution. Today, as these institutions unravel, earth spirituality is resurgent as both theory and practice. It is, for example, a foundational principle for the western ecovillage movement, most famously Findhorn with its ritual communication with nature spirits and Tamera with its stone circle, but also Earth Haven, The Farm, the Sirius Community, and many more. Talk to the most nuts-and-bolts permaculturist and you will usually find that he, and especially she, privately embraces some form of earth spirituality.

Finally, the seeming opposition between science and earth spirituality is crumbling in light of recent discoveries. Plant intelligence, mycelial intelligence, the ability of water to carry information and the complexity of animal communication lend scientific credence to the idea that non-humans have subjective agency and inner experience. And, of course, Gaia theory suggests that the entire planet is alive and (though few ecologists would say so publicly) conscious.

Do these developments point to a future world that once again embraces earth spirituality as a basis for technology and the relationship between humans and other beings? Is it the West now that will pursue a course of 'development', perhaps a different civilizational path, inspired by the remnants of the world view it nearly extinguished?

Note

[1] Harmony with Nature (UN), http://www.harmonywithnatureun.org/rightsOfNature/.

Further Resources

Apffel-Marglin, Frederique (2012), *Subversive Spiritualities*. USA: Oxford University Press.
Charles Eisenstein's website, http://charleseisenstein.net/.

Charles Eisenstein is a writer and speaker. His most recent books are *Sacred Economics* (North Atlantic Books, 2011) and *The More Beautiful World Our Hearts Know Is Possible* (North Atlantic Books, 2013).

ECO-ANARCHISM

Ted Trainer

Keywords: limits to growth, simplicity, community sustainability, eco-anarchism

The conventional definition of 'development' in terms of striving endlessly for growth and affluence inevitably creates and accelerates ecological destruction, inequality and poverty, social breakdown, and armed conflict over resources and markets. As argued in *Abandon Affluence!* (Trainer 1985), the key to eliminating this potentially fatal predicament is to recognize how mistaken and vicious the dominant conception of development is, and to replace it with The Simpler Way vision, as we call it in Australia. This perspective on the global situation focuses on the largely unrecognized fact that in a sustainable and just society, per capita resource consumption would have to be cut to around 10 per cent of present rich world levels. This basic 'limits to growth' case is outlined on the website.[1] The analysis is now overwhelmingly strong and has huge and inescapable implications for development ends and means. In rich and poor countries, the goal has to be mostly small-scale settlements, highly self-sufficient, and self-governing, informed and driven by a culture of simplicity, frugality, and non-material sources of life satisfaction. Only communities of this kind can get per capita resource consumption down sufficiently while enabling a good quality life for all the world's people.

What is easily overlooked is the fact that these arrangements must be eco-anarchist. Only thoroughly participatory self-governing communities can run small-scale local economies well. There would still need to be some centralized and state-level systems, but the national economy would be reduced to a fraction of current production, trade, and GDP. The economy would be zero-growth and geared to provisioning towns and regions with small quantities of basic inputs such as cement, irrigation pipe and light machinery. Communities would be in control of their own affairs through citizen initiatives, with minimal dependence on officials or bureaucracies. There would be voluntary citizen committees, working bees, informal discussion, commons, spontaneous action, and town meetings. There would not be the surplus resources for centralized states to run local systems. More importantly, bureaucracies possess neither local knowledge nor the capacity to support the grassroots energy and cohesion needed for change across the board. Unless political procedures are thoroughly participatory empowerment, solidarity, and right decisions, actions will not result.

Above all, genuine development cannot be driven by the quest for wealth; non-material sources of life satisfaction must replace individualistic, competitive acquisitiveness. International acknowledgement of this is essential to help peasant and tribal societies avoid identifying affluence with progress. And that, in turn, enables the preservation and celebration of traditional cultures as a further bulwark against the onslaught of Western consumerism. Despite the progress being made by rich world movements such as Voluntary Simplicity, Eco-Village, Downshifting, and Transition Towns, the eco-anarchist revolution will probably be led by peasant and tribal peoples. It is critical not to regard this alternative as inferior or a consolation alongside the supposedly superior consumer-capitalist path. Large numbers of people around the world are more or less already on this path, for instance, within La Via Campesina, Chikukwa, and Zapatista movements.

In addition to strongly asserting a particular vision regarding social goals, The Simpler Way has direct implications for means. It becomes very clear that strategy must be eco-anarchist, once the standard 'eco-socialist' theory of transition is examined. Eco-socialists seek to take state power in order to implement post-capitalist arrangements from the political centre. They do not try to enable the kind of alternatives discussed above; they are strongly inclined to work to 'release the industrial system from the contradictions of capitalism in order to raise everyone to high living standards'. Eco-anarchist strategy gives priority to grassroots agency and the cultural revolution whereby ordinary people come to embrace alternative ideas, systems, and values. As Kropotkin and Tolstoy realized, taking state power is a waste of time unless and until people come to appreciate the need for self-governing participatory communities. The emergence of that vision and commitment is in effect the revolution, and it makes possible the subsequent change of structures. Taking or eliminating state power are consequential to that.

A major concern of The Simpler Way project[2] is to provide practical detail showing how this vision of eco-anarchist development can be realized in both rich and poor countries. In addition to this, a 53-page report[3] explains how a suburb in an affluent city such as Sydney might be remade to cut resource, dollar and ecological costs by 90 per cent. City suburbs, rural towns and Third Word villages can easily meet most of their basic needs via local resources and cooperative arrangements. Instead of feeling compelled to compete in or purchase from the global economy, the concern is to maximize independence by collective living. The Simpler Way transition strategy focuses primarily on working within Transition Towns, De-Growth, Permaculture, and Eco-village movements, and especially Third World villages, where many already model localist solutions, community self-government, and 'prefiguring'.

Notes

[1] The Limits To Growth Analysis Of Our Global Situation, www.thesimplerway. info/LIMITS.htm.

[2] The general vision is found at http://thesimplerway.info/THEALTERNTIVELong. htm.

[3] http://thesimplerway.info/RemakingSettlements.htm.

Further Resources

Kropotkin, Peter (1912), *Fields, Factories and Workshops.* London: Nelson.

Sarkar, Saral (1999), *Eco-Socialism or Eco-Capitalism? A Critical Analysis of Humanity's Fundamental Choices.* London: Zed Books.

The Simplicity Institute, www.SimplicityInstitute.org.

Trainer, Ted (F.E) (1985), *Abandon Affluence!* London: Zed Books.

Transition Towns, www.TransitionNetwork.org.

Via Campesina, https://viacampesina.org/en/.

Ted (E.F.) Trainer is a retired lecturer from the School of Social Work, University of New South Wales in Sydney. He has written numerous books and articles on global problems, sustainability issues, radical critiques of the economy, alternative social forms, and the transition to them. He is developing an alternative lifestyle education site known as Pigface Point, and a website for use by critical global educators.

———•••———

ECOFEMINISM

Christelle Terreblanche

Keywords: capitalist patriarchal economics, ideological dualism, meta-industrial labour, subsistence, eco-sufficiency

Ecofeminists spell out historical, material, and ideological connections between the subjugation of women and the domination of nature. As an evolving movement, they speak to a diverse body of political theory including feminist, decolonial, and environmental ethics, urging examination of how foundational concepts are embedded in and corrupted by traditional sex-gendered assumptions. From its beginnings in the 1960s, ecofeminist theory was inspired by grassroots direct action. Ecofeminism grew rapidly alongside the anti-nuclear and peace movements of the 1970s and '80s, and amidst rising public concern over environmental degradation. Women activists are found wherever the social and ecological reproduction of life is threatened: whether by toxic waste, race violence, exploitation of care workers, biodiversity loss, deforestation, commoditized seeds, or dispossession of ancestral lands for 'development'.

Ecofeminists assert that human emancipation from historically patriarchal attitudes cannot be achieved without the liberation of all 'othered' beings. They see how women in the global North and peasants and indigenous peoples in the South can combine in a single authentic political voice. The reason is that these social groupings are skilled in caring for human and non-human life. As a politics, ecofeminism, therefore, is *sui generis* and not simply an offshoot of feminism, Marxism, or ecology. Notwithstanding some cross-fertilization of ideas, ecofeminism rearticulates feminist concerns about social equality by linking it to environmental justice and integrity.

Ecofeminism is sometimes regarded as a revival of ancient wisdom on the interconnectedness of 'all life'. An example would be India's Chipko women centuries ago, who protected forests from logging, with their arms wrapped around the trees. However, the actual term *ecofeminisme* is attributed to French feminist, Francoise D'Eaubonne's 1974 appeal for a revolution to save the ecosphere; a total reconstruction of relations between both humans and nature, and men and women. Pioneering theorist Carolyn Merchant's historical analysis of the European scientific revolution, *The Death of Nature*, exposed the determination of modernity's fathers to master women's reproductive sovereignty through institutionalized witch-hunts. The specialized knowledge of herbalists and midwives was replaced by a 'profession of medicine' positing nature and the body as 'machines'. This abolished the precautionary principle inherent in women's care-giving labours, while at the same time reinforced an ancient dualist ontology of men's rational superiority and control over 'others' such as 'unruly' women and 'chaotic' nature.

Mainstream liberal modernists have often read the ecofeminist critique upside down as entrenching the patriarchal idea that women or natives are 'essentially closer to nature' and thereby inferior. In fact, ecofeminists deconstruct old hegemonic binaries stemming from the 'man over nature' dualism, revealing just how these are used by people of sex-gender, ethnic, and class privilege to maintain their social domination by 'othering'. Understood in this way, an ecofeminist standpoint can help deepen a person's reflexive self-awareness as to how they themselves are served by existing power relations.

Internationally, women do 65 per cent of all work for 10 per cent of wages while in the global South, women produce 60 to 80 per cent of all food consumed. Following research in colonial Africa and South America, Maria Mies and her German Bielefeld School associates proposed a 'subsistence perspective' validating the ecological knowledge of women and peasants as producers and provisioners of life. Since the 1980s, this economic argument has mobilized ecofeminism as a post-development politics, anticipating contemporary alternatives such as the Latin American indigenous *buen vivir* or 'good life' worldview, and recent European attention to de-growth and solidarity economies. Another exposé of 'maldevelopment' has been Vandana Shiva's account of how the communal food sovereignty achieved by Indian women farmers was lost after the introduction of twentieth-century Green Revolution technologies.

As financial and technological fixes deepen the ecological crisis, ecofeminists uncover the complex class, ethnic, and sex-gendered character of capitalist appropriation. Being a materialist politic grounded in labour, it is by definition non-essentialist, connecting the dots between overconsumption in the affluent industrialized global North and its 'taps and sinks' in the South. For it is the peripheries of capitalist patriarchal productivism that carry its polluting fallout – as ecological debt on indigenous communities, and as an embodied debt on living women and future generations. Materialist ecofeminists such as Ariel Salleh, Mary Mellor, Eva Charkiewicz, Ana Isla, and others link subsistence and eco-sufficiency. Their structural critiques of reductionist economics point to its blindsiding of reproductive work in homes and in fields – and of the natural cycles on which capitalism depends.

Ecofeminists argue that such reproductive labour stands *a priori* to capitalist and Marxist valorization of production and exchange value as the driver of accumulation. Salleh conceptualizes unspoken reproductive workers – women, peasants, and indigenes – as a worldwide majority 'meta-industrial class' whose skills express an 'embodied materialist' epistemology and ethic. Their regenerative modes of provisioning at the cusp of nature are a readymade political and material response to the environmental crisis. Such workers exist around the world in a vast, yet seemingly invisible, patchwork of non-alienated work, maintaining life in a complex web of humanity-nature relationships. Meta-industrial labour infuses ecological cycles with a net positive 'metabolic value'. Clearly, ecofeminism expands the focus of traditional Marxist class analysis. And indeed, its theorization of the 'naturalized' underpinnings of capitalist appropriation through reproductive labour is being taken up on the academic left. There is always a risk, however, that women's theorizations will be repackaged in existing patriarchal meta-narratives.

An ecofeminist politics aims to foster human emancipation through regenerative solidarity economies based on sharing. It puts complexity before homogeneity, cooperation before competition, commons before property, and use value before exchange value. This emancipatory politics is gaining recognition for its capacity to elucidate convergences between the concerns of ecology, feminism, Marxism, and life-centered indigenous ethics such as *swaraj* in India and the African ethic of *ubuntu*. The analysis provides a systemic sociological foundation for all post-development alternatives, which seek both equality and sustainable ways of living. Ecofeminists argue for a world-view based on care for the diversity of all life forms.

Further Resources

Merchant, Carolyn (1980), *The Death of Nature: Women, Ecology and the Scientific Revolution*. San Francisco: Harper.

Mies, Maria and Vandana Shiva (1993), *Ecofeminism*. London: Zed Books.

Salleh, Ariel (ed.) (2009), *Eco-Sufficiency and Global Justice: Women Write Political Ecology*. London: Pluto Press.

Women in Diversity, http://www.navdanya.org.

WoMin – African Women Unite Against Mining, https://womin.org.za.

World March of Women, www.marchemondiale.org.

Christelle Terreblanche is a PhD candidate in Development Studies at the Centre for Civil Society, University of KwaZulu-Natal, South Africa. Her research interests include ecofeminism, political ecology and ecological justice. She is a veteran political correspondent and former spokesperson for the South African Truth and Reconciliation Commission.

ECOLOGY OF CULTURE

Ekaterina Chertkovskaya

Keywords: ecology, culture, social movements, Russia, environmental justice

The 'ecology of culture' may be defined as a conceptual framework and space of struggle that brings together the ecological and cultural sustainability of places and spaces in which people live. The concept was introduced in 1979 by Dmitry Sergeyevich Likhachyov – a Soviet scholar and public intellectual who survived the Solovki prison camp and the Leningrad Blockade. Referred to as the 'conscience of the nation', Likhachyov was notably outspoken and voiced his opposition to issues such as the Siberian river reversal – a massive project planned by the Soviet government since the 1930s, but finally abandoned in 1986.

Likhachyov saw ecological sustainability as crucial for life on Earth and critiqued the understanding of progress in terms of industrial expansion. Equally important for him was the continuity of culture and cultural heritage, which was part of his broader understanding of ecology. The simultaneous emphasis on environment and culture is not surprising. The productivist vision of progress in the Soviet Union and the ideologies surrounding it resulted not only in dramatic interventions into the natural environment, but also into what Likhachyov called cultural environment, such as the architecture of urban spaces, reminiscent of pre-revolutionary times and was to be reorganized in line with Soviet visions. Though the Soviet era undoubtedly brought remarkable architectural styles, planning, public transport, and monuments into urban landscapes, a lot of landmarks and symbols of the past were mercilessly destroyed. Ironically, today the Soviet cultural legacy itself is often under threat, with the urge for quick gain being a key feature of post-Soviet capitalism.

The ecology of culture, according to Likhachyov, is as important for people's moral life as it is for the environment in a biological sense for the sustainability of physical life – with the two being inseparable: 'There is no gap between the two, as there is no clear border between nature and culture' (Likhachyov 2014/1984: 90). The ecology of culture does not imply some high-brow or exclusive understanding of culture, but the entire range of cultural practices through which people give meaning to their lives. It may come from remarkable architectural constructions, such as the Russian avant-garde, which reminds one of the ambitions and potentialities expressed in the early period of the Soviet era, or the folk art that is used to decorate village houses. This may also be a harmoniously organized landscape, which opens a view over the river or creates a sense of space. The ecology of culture, however, is not limited to material or physical phenomena and may also include, for example, music,[1] dance, and literature.[2]

While the ecology of culture is not used explicitly by social movements today, it offers a helpful articulation of urban struggles in post-Soviet Russia, which often resist the destruction of material and immaterial heritage and infill development. Across the country, architectural destruction takes place with astonishing frequency and is actively resisted by social movements such as Archnadzor[3] ('Architectural watch') in Moscow, Zhivoi Gorod[4] ('Living city') in St Petersburg and similar movements in Ufa, Tver, Vologda, and other cities. They bring together city residents, ethnographers, ecologists, and other people.

If a building is at risk, highlighting by whom it was built, what is special about it and which figures are associated with the place becomes part of the resistance campaign, which shows the centrality of the ecology of culture to urban struggles. One such action was undertaken for the protection of a building that was connected to Tolstoy's family. The figure of Tolstoy, is accompanied by a quote from *War and Peace* 'it is much more honourable to admit one's mistake than to let matters become irreparable', became central to the activist campaign. Struggles with a strong element of the ecology of culture are often in tune with ecological sustainability, with both cultural and biophysical environment being under threat from capital accumulation-centred development. This is the case in the campaign against building a judicial and residential quarter on the embankment of the Neva river next to the Peter and Paul fortress in St Petersburg and for creating a park instead, suggested to be named after Likhachyov. Here, activists argue for a green space in the city centre, which would also fit into the current landscape.

The movements that may be articulated in terms of the ecology of culture have a strong emphasis on the importance of participatory justice, which brings the ecology of culture close to environmental justice frameworks and struggles. Some of these movements are building alliances and exchange information and experiences, fostering new practices and processes of self-organizing. In the city of Zhukovskiy the mobilization of residents for forest protection ultimately contributed to electing Zhukovskiy People's Council – an organization for coordinating residents' control over the official powers of the city. In response to continuous threats to the architectural heritage of Moscow, numerous initiatives that aim to share stories and knowledge about the city have emerged, involving walks and lectures run by ethnographers and city enthusiasts. These initiatives are usually not commercially driven but rather convivial and collective in nature, contributing to the ecology of culture and raising awareness of constant threats to the city.

The ecology of culture has the potential to unite different and somewhat dispersed struggles, framing them as addressing a common problem. Due to its connection to ecological sustainability, the ecology of culture may also help to pave the way to increased public awareness about ecological issues. For this transformative potential to be realized, however, the ecology of culture needs to be treated as a living concept, which is open to being enriched with new meanings and new people who can contribute to its development.

Notes
[1] See Sonevytsky and Ivakhiv (2015).
[2] See also Likhachyov (2000/1992–93).

[3] Archnadzor ('Architectural watch') – social movement aimed at protecting the architectural heritage of Moscow. See www.archnadzor.ru (in Russian).

[4] Zhivoi Gorod ('Living city') – social movement that unites people who love and care about St. Petersburg. See www.save-spb.ru (in Russian, some information in English).

Further Resources

Likhachyov, Dmitry Sergeyevich (2000/1992–93), 'Ekologiya kul'turi', in *Russkaya kul'tura*. Moskva: Iskusstvo.

————— (2014/1984), 'Ekologiyakul'turi', in *Zametki o russkom*. Moskva: Azbuka-Attikus/Kolibri.

Sonevytsky, Maria and Adrian Ivakhiv (2015), 'Late Soviet Discourses of Nature and the Natural: Musical *Avtentyka*, Native Faith, and "Cultural Ecology" after Chornobyl', in Aaron S. Allen and Kevin Dawe (eds), *Current Directions in Ecomusicology: Music, Nature, Environment*. London: Routledge.

Ekaterina Chertkovskaya is a researcher in degrowth and critical organization studies and a member of the editorial collective of *ephemera* journal. She is presently based at Lund University, Sweden, and was a co-coordinator of the degrowth theme for the Pufendorf Institute for Advanced Studies at Lund University (2015–16).

ECO-POSITIVE DESIGN

Janis Birkeland

Keywords: sustainable cities, social justice, design for nature, open-system, net-positive design

It is possible to have green buildings with zero carbon or energy impact, but they do not count embodied energy use or the ecological destruction that happens during construction. That is, they do not give back to nature more than they take, so they are not sustainable. Seeing 'buildings as landscapes', instead of a set of component products, reveals opportunities to use structure and surface to create new ecological spaces. 'Design for eco-services' increases the multiple 'free services' that nature provides. This may include play-gardens, passive solar retrofit modules and 'green scaffolding', diverse space-frame structures around, between or even on the inside of buildings. Buildings can contribute to sustainability, if they provide more benefits than no building at all, and if existing cities are retrofitted to turn ecological and social deficits into gains. Eco-positive design applies open-system thinking to built environment design as well as to development control and assessment in urban planning.

Typically, physical construction concentrates wealth, privatizes resources, extinguishes biological and cultural diversity, and reinforces wasteful and inequitable living conditions. In short, it closes off future options. Conventional approaches to sustainable design do little to reverse this. Cities could increase life quality and regenerate the environment, yet this is not enough. Advocates of eco-positive or net-positive design insist that the 'positive ecological footprint of nature' must exceed humanity's negative footprint. To increase future options and compensate for unavoidable cumulative ecological impacts, cities as a whole must create net ecological and social gains. This is technically possible. For example, as demonstrated in Renger *et al.* (2015), buildings can sequester more carbon than emitted over their lifecycles and, as suggested in Pearson *et al.* (2014), design strategies such as green scaffolding can provide three dozen ecosystem services.

Based on a critique of conventional sustainable design approaches, an alternative eco-positive framework enables a form of construction that gives more to nature and to society than it takes. First, cities must not only integrate living nature with buildings, but also increase the total 'ecological base'. Whereas standard sustainable design paradigms, as summarized by Hes and du Plessis (2014), call for designing 'with' or 'like' nature, eco-positive design is 'for' nature. Secondly, a prerequisite of a sustainable democracy is equitable, eco-productive human environments that meet basic needs and more. Cities should increase direct, universal access to livelihood, social engagement, and individual well-being, in other words, the 'public estate'.

The failure of current sustainable design approaches to contribute in net-positive ways can be traced to the prevailing closed-system paradigm. This dates back to the 1960s 'spaceship earth' metaphor and the 1970s 'limits to growth' argument, which were supposed to counter the linear model of industrial progress. A 1980 *World Conservation Strategy* called for 'living within the carrying capacity of supporting ecosystems' – another 'limits to nature' argument. These negative messages failed, probably because 'limits' suggested lower living standards. The influential 1987 *Brundtland Report* perpetuated the closed-system model. It also omitted the built environment from consideration and relegated sustainability planning to decision-making institutions. Biophysical sustainability is a design problem, not simply a management one. It requires changing not only physical structures but the institutions and decision-making frameworks that shape them.

Design can create positive, synergistic relationships that multiply public benefits and ecological gains. By contrast, political decision-making relies on closed-system frameworks for comparing alternatives or allocating costs among priorities. Reductionist, computerized technocratic decision models are nevertheless regarded as 'higher-order thinking', while design is marginalized as a soft, subordinate subset. Yet eco-positive design introduces an open-system model to guide fundamental paradigm shifts in both the conceptual and material realms. First, this means replacing reductionist decision-making with design-based frameworks that can increase symbiotic human – nature relationships. Second, it means transforming ecologically terminal physical structures so as to increase vital ecological space and maximize public benefits.

As noted, sustainable design is now characterized by closed-system thinking. This originated with 'resource autonomous' buildings in the 1970s, and 'closed loop' manufacturing in the 1990s. In a closed-system, design solutions are essentially 'recycling'. Since modern buildings are composed of many manufactured products, recycling and/or 'upcycling' to higher economic values can reduce many impacts. However, it cannot in itself create eco-positive environments. Even a zero-impact building would not necessarily contribute positively to ecological sustainability. Since the aim of recycling is zero performance, the usual standards only require a reduction in negative impacts. Further, impact assessments rely on imaginary 'system boundaries', beyond which cumulative and embodied impacts become too amorphous and complex to measure. These remain uncounted. This limits the responsibility of design teams to measurable impacts within artificial boundaries, such as property lines, thereby excluding wider ethical issues.

Today, tick box commercial 'green building rating tools' remain dominant in the planning profession. These private-sector, voluntary certification schemes were introduced by green building councils, formed in many countries since 1990, to forestall the demands of sustainable building regulation. Such rating tools exemplify reductionist, bounded thinking. They only aim to reduce existing or predicted impacts relative to the norm. Therefore, a reduction in negative impacts is often called 'positive'; yet, from a wider perspective, a 40 percent less energy usage can turn out to be a 60 percent overall increase.

The application of conventional rating tools also leaves out fundamental sustainability issues such as social justice. It neglects potential public benefits, encourages trade-offs between incommensurable values, counts financial savings through worker productivity as 'ecological' gains, and so on. By contrast, upfront eco-positive design methods are affordable in disadvantaged regions. The social baseline requires equity and environmental justice on a region-wide basis, and this is assessable by flows analyses, as outlined in Byrne *et al.* (2014). These new standards prescribe an entirely new architecture.

Further Resources

Birkeland, Janis (2008), *Positive Development: From Vicious Circles to Virtuous Cycles through Built Environment Design.* London: Earthscan.

Byrne, Jason, Neil Sipe and Jago Dodson (eds) (2014), *Australian Environmental Planning.* London: Routledge.

Hes, Dominique and Chrisna du Plessis (2014), *Designing for Hope: Pathways to Regenerative Sustainability.* London: Taylor and Francis.

Pearson, Leonie, Peter W. Newton and Peters Roberts (eds) (2014), *Resilient Sustainable Cities.* London: Routledge.

Renger, Birte Christina, Janis L. Birkeland and David J. Midmore (2015), 'Net Positive Building Carbon Sequestration', *Building Research and Information.* 43 (1): 11–24.

Report of the World Commission on Environment and Development: Our Common Future (Brundtland Report) (1987),
http://www.un-documents.net/our-common-future.pdf.

World Conservation Strategy (1980),
https://portals.iucn.org/library/sites/library/files/documents/WCS-004.pdf.

Janis Birkeland is an Honorary Professorial Fellow in Building and Planning at the University of Melbourne and author of several books and articles. She was a lawyer, architect, and planner in the USA before embarking on a twenty-five-year academic career in Australia.

ECO-SOCIALISM

Michael Löwy

Keywords: capitalism, eco-socialism, Marxism, consumption, authentic needs, qualitative transformation

The capitalist system cannot exist without unlimited 'development', 'growth', and 'expansion'. A radical post-development alternative must therefore be a post-capitalist one. Eco-socialism is one such system alternative. This is a current of ecological thought and action that builds on the fundamentals of Marxism, while shaking off its productivist dross. Eco-socialists see both the logic of markets and the logic of bureaucratic authoritarianism as incompatible with the need to safeguard the environment. Thinkers such as Rachel Carson or James O'Connor (USA), André Gorz (France), Frieder-Otto Wolff (Germany), and Manuel Sacristan (Spain) are among the pioneers of eco-socialism. More recently, works by Joel Kovel, John Bellamy Foster, and Ian Angus have developed the eco-socialist argument.

The rationality of capitalist accumulation, expansion, and development – particularly in its contemporary neoliberal form – is driven by short-sighted calculation and stands in intrinsic contradiction to ecological rationality, and the long-term protection of natural cycles. Ruthless competition, the demands of profitability, a culture of commodity fetishism, and transformation of the economy into an autonomous sphere beyond the control of society or political powers – all destroy nature's balance.

A radical alternative economic policy would be founded on the non-monetary criteria of social needs and ecological equilibrium. The replacement of the micro-rationality of profit-making by a social and ecological macro-rationality demands a change of civilizational paradigm, concerning not only production, but also consumption, culture, values, and lifestyle.

In an eco-socialist society, entire sectors of the productive system would be restructured and new ones developed, so that full employment is assured. However, this is impossible without public control with democratic planning over the means of production. Decisions on investment and technological change must be taken away from banks and capitalist enterprises in order to serve the common good. An economy in transition to eco-socialism should be 're-embedded' as Karl Polanyi would say, in the social and natural environment. Democratic planning means productive investments are chosen by the population, not by 'laws of the market', or an omniscient politburo. Far from being 'despotic', such planning is the exercise of a society's freedom, its liberation from alienation and from reified 'economic laws'.

Planning and the reduction of labour time are the two most decisive human steps towards what Marx called 'the kingdom of freedom'. A significant increase of free time is, in fact, a condition for the participation of working people in the democratic discussion and management of both economy and society. The passage from capitalism's 'destructive progress to socialism, is a historical process, a permanent revolutionary transformation of society, culture, and subjectivity. This transition would lead not only to a new mode of production and an egalitarian society, but also to an alternative 'mode of life', a new eco-socialist civilization, beyond the reign of money.

Such a revolutionary transformation of social and political structures cannot begin without active support of an eco-socialist programme by a majority of the population. The development of socialist consciousness and ecological awareness is a process, whereby the decisive factor is people's own collective experience of struggle through local and partial confrontations.

Some ecologists believe that the only alternative to productivism is to 'stop growth' altogether, or to replace it by negative growth – what the French call *décroissance*. This is based on a drastic reduction of consumption, cutting by half the expenditure of energy as individuals renounce central heating, washing machines, and so on. Eco-socialists emphasize instead a 'qualitative transformation' of production and consumption. This means putting an end to the monstrous waste of resources by capitalism, based on large-scale production of useless and/or harmful products such as the armaments industry. Many of the 'goods' produced by capitalism have inbuilt obsolescence; they are designed wastefully for rapid replacement to generate profit. From an ecosocialist perspective, the issue is not so much one of 'excessive consumption' therefore, but the 'type' of consumption. An economy based on mercantile alienation and compulsive acquisition of pseudo-novelties imposed by 'fashion' is simply incompatable with an ecological rationality.

A new society would orient production towards the satisfaction of authentic needs, beginning with those described as 'biblical' – water, food, clothing, housing – and basic public services such as health, education, and transport. Authentic needs are clearly distinguished from artificial or ficticious needs induced by a manipulative advertising industry. Advertising is an indispensable dimension of the capitalist market economy but it has no place in a society transitioning to socialism. Here, people's information on goods and services would be provided by consumer associations. The test for distinguishing authentic from artificial needs, is to see whether they persist after the suppression of advertising.

Eco-socialists work to build a broad international alliance between the labour movement, the ecological, indigenous, peasant, feminist, and other popular movements in the global North and South. These struggles may lead to a socialist and ecological alternative, but not as an inevitable result of contradictions of capitalism or 'iron laws of history'. One cannot predict the future, except in conditional terms. What is clear however is that in the absence of ecosocialist transformation, that is to say, a radical change in civilizational paradigm, the logic of capitalism can only lead the planet into dramatic ecological disasters, threatening the health and even lives of billions of human beings, perhaps even the survival of our species.

Further Resources

Angus, Ian (2016), *Facing the Anthropocene*. New York: Monthly Review Press.

Bellamy Foster, John (2009), *The Ecological Revolution*. New York: Monthly Review Press.

Capitalism Nature Socialism, http://www.tandfonline.com/loi/rcns20.

Climate and Capitalism, http://climateandcapitalism.com.

Kovel, Joel (2007), *The Enemy of Nature*. London: Zed Books.

Lowy, Michael (2015), *Ecosocialism: A Radical Alternative to Capitalist Catastrophe*. New York: Haymarket Books.

Philosopher **Michael Löwy** was born in 1938 in Brazil and has lived in Paris since 1969. He is currently Emeritus Research Director at the Centre National de la Recherche Scientifique (CNRS) and his books and articles are translated into twenty-nine languages. Lowy co-authored *The International Ecosocialist Manifesto* (2001) with the late Joel Kovel, editor of the US based journal *Capitalism Nature Socialism*.

ECOVILLAGES

Martha Chaves

Keywords: ecovillage, cosmo-centric paradigms, intentional community, spirituality

Ecovillages constitute a growing movement of practical alternatives that challenges the destructive human-centered worlds created by most modern societies. Defining an 'ecovillage' is a difficult task as the concept is constantly evolving in the way it is envisioned and practised. The most current definition used by the Global Ecovillage Network is 'an intentional or traditional community using local participatory processes to holistically integrate ecological, economic, social, and cultural dimensions of sustainability in order to re-generate social and natural environments'.[1]

The term 'ecovillage' was first defined by Robert Gilman as a 'human-scale full-featured settlement in which human activities are harmlessly integrated into the natural world in a way that is supportive of healthy human development, and can be successfully continued into the indefinite future'.[2] This definition is closely connected to community experiments into alternative living growing out of various strands such as the back-to-the-land movements of the '60s and '70s, predominantly from the global North. The roots of these movements can be found in a deep discontent with the dominant post-war narratives of industry-driven development and with the movement seeking alternatives to materialist values, reconnection to nature, and human-scale settlements.

Gilman's definition has been criticized for being too aspirational in nature, as it describes an end-state which very few settlements have reached. Dawson instead states five defining characteristics of ecovillages:

> [P]rivate citizens' initiatives; in which the communitarian impulse is of central importance; that are seeking to win back some measure of control over community resources; that have a strong shared values base, often referred to as 'spirituality'; and that act as centers of research, demonstration and in most cases training.[3]

Dawson admits, however, that this definition is more representative of the global North, with the most common feature in the global South being the use of the ecovillage concept to build alliances between very different actors such as traditional/indigenous communities with NGOs, local governments and/or urban dwellers seeking to build networks to alleviate poverty, address climate change, strive for socio-environmental justice, and find respect for cultural, territorial, and spiritual diversity.

In practice, there are many ecovillage expressions. There are educational centres such as Findhorn in Scotland (500 people), the intentional community of The Farm in Tennessee, USA (200 residents), and the permaculture design site of Crystal Waters in Australia (over 200 people). There are also organized networks such as Sarvodaya, which are 2,000 active sustainable villages in Sri Lanka; 100 traditional villages in Senegal equipped by the government with techniques learnt from ecovillages; and the eco-town of Auroville in southern India, inspired by Sri Aurobindo, with around 2,400 residents. There are also numerous small-scale ecovillages such as the family Anthakarana (9 residents) or Aldeafeliz (30 residents) in the Colombian Andes. Through this diversity we can see how Gilman's classic ecovillage definition has evolved to become more inclusive, encompassing both intentional and traditional communities.

Although the ecovillage movement is sometimes criticized for its perceived elitist and isolationist tendencies, significant shifts have taken place in recent decades, resulting in a transformation of its identity, role and potential. Increasingly connected through alliances and collaborations with more progressive elements of society, the ecovillage movement is broadening to include other types of community networks and initiatives while experimenting with intercultural dialogue and collective action. This can be seen in the Council of Sustainable Settlements of the Americas (CASA) network, which has expanded beyond the ecovillage realm to include indigenous and Afro communities, as well as the Hare Krishna movement and urban professionals. In countries such as Colombia and Mexico, members of the CASA network mix traditional knowledge, customs and rituals with urban ones to find new and stronger connections to nature, and improve their eco-living practices.[4]

Ecovillages are contributing to a new kind of social movement characterized by the diversity of actors, decentralized forms of leadership, and a genuine effort at living in synchronization with the Earth. Their activism is based on the power of social networks and communication technologies, where lived examples of low-carbon lifestyles are hybridized with modern technology and spiritual values. Ecovillages, therefore, represent a transition strategy for bringing together diverse progressive actors to visualize cosmo-centric paradigms, with the goal of demonstrating to the broader public the viability and, indeed, joy that can be derived from living a low-impact lifestyle closer to nature. In this way, ecovillages are closely connected to the degrowth movement through the downscaling of production and consumption, as well as to the transition town movement which presents a broader model of community engagement.

Apart from the numerous challenges of practising sustainability in a communal setting, one of the greatest limitations for the ecovillage movement is generating wider appeal and change at the institutional level. Given that its values are neither market-driven nor mainstream, it has limited appeal to most governments, hence the question of the extent to which it provides a viable alternative to the individualism and materialism of most modern worlds. Furthermore, at least in the global North, replicating the ecovillage model is proving difficult due to land prices and planning permissions. Having recognized the need for greater societal engagement, the transformative potential of the ecovillage movement is found in its evolution towards learning centres for transition where post-developmental ideas, skills, and alliances are increasingly being percolated into mainstream society.

Notes

[1] Global Ecovillage Network (GEN) is a growing network of sustainable communities and initiatives that bridge different cultures, countries, and continents. GEN serves as umbrella organization for ecovillages, transition town initiatives, intentional communities, and ecologically minded individuals worldwide, http://gen.ecovillage. org/.

[2] Gilman, 'The Ecovillage Challenge', p. 10.

[3] Dawson, 'From Islands to Networks', p. 219.
[4] Chaves, 'Answering the "Call of the Mountain"'.

Further Resources

Chaves, Martha Cecilia (2016), *Answering the 'Call of the Mountain': Co-creating Sustainability through Networks of Change in Colombia*. Doctoral thesis, Wageningen University.

Dawson, Jonathan (2013), 'From Islands to Networks: The History and Future of the Ecovillage Movement', in Joshua Lockyer and James R. Veteto (eds), *Environmental Anthropology Engaging Ecotopia: Bioregionalism, Permaculture, and Ecovillages*. New York: Berghahn Books.

Gilman, Robert (1991), 'The Ecovillage Challenge: The Challenge of Developing a Community Living in Balanced Harmony – with Itself as well as Nature – Is Tough, but Attainable', *In Context*. 29: 10–14.

The Council of Sustainable Settlements of Latin America (CASA), http://www.casacontinental.org/.

Dr Martha Chaves is an independent scientist working at the interface of natural and social sciences through the Colombian research group MINGAS enTransición. She is also active with the NGO Mentesen Transición (Transitional Minds Foundation), which promotes cultural and practical transitions to sustainable living. She contributes to the CASA network in Colombia and has close connections with the University of Wageningen, Netherlands, and the University of Quindío, Colombia.

ENERGY SOVEREIGNTY

Daniela Del Bene, Juan Pablo Soler, Tatiana Roa

Keywords: energy, resistance, alternatives, sovereignty, transformation

Energy sovereignty (ES) refers to political projects and visions towards a just generation, distribution, and control of energy sources by ecologically and culturally grounded mobilized communities, both urban and rural, in ways that do not affect others negatively, and with respect for ecological cycles. Energy sovereignty acts as a slogan for organizations and movements to reclaim the right to take decisions about energy, understood as a natural commons, and basis of life for all. It also refers to the plurality of systemic alternatives under way that challenge the dominant energy paradigm controlled by centralized powers.

The concept of ES has been used since the 1990s in Latin America to challenge the privatization of basic services by transnational corporations and the 'corporatization' of the state enterprises. Similar to the claim for food sovereignty by farmers' movements, ES has become popular among organizations and movements globally, especially after 2000, as a response to multiple forms of extractivism, energy poverty, corporate oligopoly, patriarchy, privatization and trade agreements, wars and crimes to secure provision of fossil fuels.

More recently, it has also become a response to climate change and to the fossil fuels industry. For instance, energy sovereignty has been included in the new constitutions of Ecuador and Bolivia. In Europe, the issue has been addressed in several campaigns questioning the energy oligopoly and seeking to create new public enterprises. Barcelona is a case in point. In the Germany where the transition to renewables is strong, *Energiewende*, meaning energy democracy is mostly used. There are moves to re-municipalize urban energy utilities and grids in Boulder, USA, Hamburg, Berlin, London.

Energy Sovereignty defends the right to decide what source of energy to exploit, how much to produce, how, by whom, where, and for whom. In line with ecofeminist perspectives, it calls for decolonizing the hegemonic structure of the energy model. Decolonizing energy requires questioning deeply rooted beliefs such as the universalizing understanding of Energy with a capital 'E' as the abstract and uniform commercial generation of energy, and as a function of capital accumulation. It also involves differentiating Energy from the incommensurable and contextually diverse uses of 'energy', with a small 'e' (Hildyard *et al.* 2012), which is able to adapt across time and space to different ecologies and human geographies.

The alliance between actors – environmental justice organizations, peoples affected by energy projects, trade unions and urban dwellers – across sectors shows not only the complexity but also the great potential of ES as a political project. In Colombia, the movement of dam-affected people, *Movimiento Ríos Vivos*, urges that any proposed dam must include both energy and water sovereignty, owing to the close connection between communities and their water cultures, as well as the direct relation between the historical domination over peoples and that over water resources.[1]

In Brazil, the national Movement of Dam Affected People (MAB) allied with trade unions in the *Plataforma Operária e Camponesa para Energia* or Workers and Farmers Platform for Energy,[2] met to discuss the historical debt that megaprojects and energy corporations owe to those affected, and to draw-up a proposal for an energy and mining policy for the country (*Proyecto Energetico Popular*).

In the US and internationally, the federations Trade Unions for Energy Democracy (TUED) and Public Service International (PSI) also understand energy as a crucial common issue across most social sectors. Energy is important for restructuring economic and productive relations, and to properly address public health and workers' safety.

Energy sovereignty challenges the opposition of 'urban' and 'rural' when it comes to socio-environmental impacts: the ones affected by an unjust energy model are not only those displaced by megaprojects, but all those on which the socialization of costs is imposed and from whom extra profits are mined. The urban energy poor should be considered impoverished or robbed, and democratic processes distorted by the 'revolving door' between politicians and energy entrepreneurs. In Spain, United Kingdom or Bulgaria, for instance, urban dwellers have organized to denounce skyrocketing electricity tariffs and the violation of laws designed to protect vulnerable families – for example, in Barcelona, by the Alliance against Energy Poverty.

Energy sovereignty also tackles the issue of technology and knowledge within energy transitions. It calls for decentralization, relocalization, and differentiation of energy generation, technology, and knowledge. It poses an epistemic challenge to reconsider our 'territory' not as a mere repository of natural resources, but as a socio-cultural whole, where one makes sense of existence and where one bases and roots conscious, responsible and joyful political *proyectos de vida* or life projects (Escobar 2008). Or, as other Latin American communities say, *planes de permanencia en los territorios* (plans for staying in the territory) or *proyectos de buen vivir* (plans of living well).

Proposals for ES inevitably meet with limitations and conflicts. As it shakes the basis of production relations, it challenges powerful sectors of our societies: energy companies, constructors, finance and political elites, the military establishment, and so on. What will be, for example, the implications for the structures of modern states and governments? Will ES require a restructuring of the administrative context in order to manage a new energy model? How can it avoid closed and exclusionary groups, and instead promote collaboration of open communities, perhaps based on a subsidiary principle? Will initiatives for ES ultimately help redefine limits to consumption and establish patterns of energy-usage that are truly sustainable for a given territory?

Despite the depth of these challenges, a closer look reveals that different models are already implemented and functioning, for example, the rural electrification cooperatives in Costa Rica (among which, COOPELESCA), the cooperatives SOM ENERGIA and GoiEner in Spain, and RETENERGIE in Italy, along with urban re-municipalization initiatives. They need to be valued and defended as powerful potential multipliers.

Notes
[1] See more details on the work in Colombia of the Movimiento Rios Vivos at https://defensaterritorios.wordpress.com/2016/08/25/politica-energetica-colombiana-y-propuestas-del-movimiento-rios-vivos-para-su-transformacion/.
[2] See Movimento dos Atingidos por Barragens (MAB), http://www.mabnacional.org.br/category/tema/plataforma-oper-ria-e-camponesa-para-energia.

Further Resources

Angel, James (2016), *Towards Energy Democracy*.Transnational Insitute. https://www.tni.org/en/publication/towards-energy-democracy.

Catalan Network for Energy Sovereignty (XSE) 2012, Energy Sovereingty, http://www.odg.cat/sites/default/files/energy_sovereignty_0.pdf.

Escobar, Arturo (2008), *Territories of Difference: Place, Movements, Life, Redes*. Durham, NC: Duke University Press.

Hildyard, Nicholas, Larry Lohman and Sarah Sexton (2012), *Energy Security for What? For Whom?* Sturminster Newton: The Corner House.

Kunze Conrad and Soren Becker (2014), *Energy Democracy in Europe*. https://www.rosalux.de/fileadmin/rls_uploads/pdfs/sonst_publikationen/Energy-democracy-in-Europe.pdf.

Daniela Del Bene is coordinator of the EJAtlas.org at the Autonomous University of Barcelona, member of the Research & Degrowth collective and the Catalan Network for Energy Sovereignty (XSE).

Juan Pablo Soler is a member of Colombian Movement of Peoples Affected by Dams and in Defense of Territories – Movimiento Ríos Vivos – and the Latino–American Movement of People Affected by Dams – MAR.

Tatiana Roa is the General Coordinator of Censat Agua Viva – Amigos de la Tierra, Colombia.

ENVIRONMENTAL JUSTICE

Joan Martinez-Alier

Keywords: social metabolism, political ecology, ecological distribution conflicts, EJ Atlas

A global movement for environmental justice is helping to push society and economy towards environmental sustainability. It is born from 'ecological distribution conflicts' (Martinez-Alier 2002), a term for collective complaints against environmental injustices. For instance, a factory may be polluting a river which belongs to nobody or belongs to a community that manages the river – as studied by Elinor Ostrom and her school on the commons. This is not a damage valued in the market. The same happens as climate change makes glaciers recede in the Andes or in the Himalayas, depriving communities of water. More than market failure – implying that such externalities could be valued in money terms and internalized by the price system – these are 'cost-shifting successes', as Karl William Kapp says, and they lead to complaints from those bearing them. If the complaints are effective, which is not the general rule, the activities may be banned.

In the United States, such ecological distribution conflicts perceived in terms of persistent injustices towards 'people of colour' gave rise to a social movement in the 1980s (Bullard 1993). The words 'environmental justice' (EJ) then began to be used in struggles against the disproportionate dumping of toxic waste in poor black communities. As early as 1991, at the Washington, DC 'People of Color Environmental Leadership Summit' ties were forged 'to begin to build a national and international movement of all peoples of color to fight the destruction and taking of our lands and communities'.

The number of ecological distribution conflicts around the world focusing on resource extraction, transport, and waste disposal is on the rise. There are several local complaints but there are also successful examples where projects are stopped and alternatives developed. Environmental justice is a powerful lens for making sense of struggles over negative impacts. Economic growth is changing the global 'social metabolism', in other words, flows of energy and materials impacting human livelihoods and nature conservation worldwide. Today's industrial economy has a colossal appetite for materials and energy. Even a non-growing industrial economy needs 'fresh' supplies of fossil fuel because energy is not recycled, and it also needs new supplies of materials which are recycled only in part. Such requirements increase with economic growth.

With industrialization, increased amounts of carbon dioxide are deposited in the atmosphere increasing the greenhouse effect or acidifying oceans. This kind of economy is not circular; it is entropic. Aquifers, timber and fisheries are overexploited, soil fertility is jeopardized and biodiversity is depleted. This changing social metabolism gives rise to ecological distribution conflicts that sometimes overlap with other social conflicts over class, ethnicity or indigenous identity, gender, caste, or territorial rights.

A global movement for environmental justice is slowly asserting itself as shown in the Environmental Justice Atlas (Martinez-Alier *et al.* 2016). Other kinds of depletion occur as metabolic growth is required for mining, dams, gas fracking, plantations, and new transport networks. Little by little these developments reach every remaining corner of the planet, undermining the environment, as well as the conditions of existence of local populations, and they complain accordingly. There is potentially a difficult alliance between the EJ movement and the conservation movement, though a call for an easier convergence between the Degrowth movement and EJ was first made in 2012 (Martinez-Alier 2012).

Ecological distribution conflicts are different from economic distribution conflicts over salaries, prices, and rents. They are conflicts over conditions of livelihood, access to natural resources, and the distribution of pollution. Their protagonists are less likely to be industrial workers than indigenous women struggling against open cast mining, peasants against invading oil palm plantations or urban citizens and waste recyclers against incineration (as in so many cases in the EJ Atlas). Such conflicts are different from the classic struggles between capital and labour although they sometimes overlap.

These conflicts are struggles over valuation in two senses. The first sense concerns which values should be applied while taking decisions regarding the use of nature in particular projects, for example market values including fictitious money values through contingent valuation or other methods; livelihood values; sacredness; indigenous territorial rights; ecological values in their own units of account. Second, and more importantly, which social grouping should have the power to include or exclude the relevant values, to weigh them and to allow trade-offs. For instance, do sacred indigenous territorial rights carry a veto power (Martinez-Alier 2002)?

Since the mid-1990s, a connection was established between the EJ movement in the United States and the environmentalism of the poor in Latin America, Africa, and Asia. This followed the death of Chico Mendes in 1988 fighting deforestation in Brazil and the 1995 death of Ken Saro-Wiwa and Ogoni comrades in the Niger Delta fighting oil extraction and gas flaring by Shell. Also in the mid-1990s, 'liberation theologist' Leonardo Boff's book *Cry of the Earth, Cry of the Poor* (1995) made the connections between poverty and environmental complaints. His work is vindicated in the Papal encyclical *Laudato Si* (2015) which is itself a call for environmental justice.

Since the 1980s, the EJ movement has initiated a set of concepts and a campaign. Proposals for leaving fossil fuels in the soil were developed by *Acción Ecológica* in Ecuador, ERA in Nigeria and the network Oilwatch since 1997. Resistance to socio-environmental injustice has given birth to many Environmental Justice Organizations (EJOs) pushing for alternative social transformations and deploying a new vocabulary of environmental justice, which includes terms and phrases such as popular epidemiology, sacrifice zones, climate justice, water justice, food sovereignty, biopiracy; China's 'cancer villages', Brazil's 'green deserts', Argentina's *pueblos fumigados* (fumigated people), India's 'sand mafias'; environmental racism, 'tree plantations are not forests', the ecological debt, land grabbing, and ocean grabbing. Several networks are using such concepts in various languages, inventing their own songs, displaying their own banners and making their own documentaries.

Further Resources
Boff, Leonardo (1995), *Cry of the Earth, Cry of the Poor.* New York: Orbis Books.
Bullard, Robert D. (ed.) (1993), *Confronting Environmental Racism: Voices from the Grassroots.* Boston: South End Press.
His Holiness, Pope Francis (2015), *Laudato Si'.* http://w2.vatican.va/content/dam/francesco/pdf/encyclicals/documents/papa-francesco_20150524_enciclica-laudato-si_en.pdf.
Martinez-Alier, Joan (2002), *The Environmentalism of the Poor: A Study of Ecological Conflicts and Valuation.* Cheltenham: Edward Elgar.
———— (2012), 'Environmental Justice and Economic Degrowth: An alliance between Two Movements', *Capitalism Nature Socialism.* 23 (1): 51–73.
Martinez-Alier, Joan, Leah Temper, Daniela Del Bene and Arnim Scheidel (2016), 'Is There a Global Environmental Justice Movement?', *Journal of Peasant Studies.* 43 (3): 731–55.
The EJOLT glossary, http://www.ejolt.org/section/resources/glossary/.
The Environmental Justice Atlas (EJ Atlas), http://ejatlas.org.

Joan Martinez-Alier is a senior researcher on ecological economics and political ecology in the Environmental Science and Technology Institute of the Autonomous University of Barcelona (ICTA-UAB). He is the author, among other books, of *Ecological Economics: Energy, Environment and Society* (Blackwell, 1987) and *The*

Environmentalism of the Poor: A Study of Ecological Conflicts and Valuation (Edward Elgar, 2002).

————•◦◦•————

FOOD SOVEREIGNTY

Laura Gutiérrez Escobar

Keywords: La Vía Campesina, agroecology, agri-food systems

As defined by the transnational peasant movement La Via Campesina (The Peasant Way), food sovereignty refers to 'the right of peoples to healthy and culturally appropriate food produced through ecologically sound and sustainable methods, and their right to define their own food and agriculture systems' (La Via Campesina 2007). This group first articulated the concept at the 1996 World Food Summit (WFS) in Rome, convened by the United Nations' Food and Agriculture Organization (FAO), in response to 'food security'. Food security is a guiding principle for government and multilateral agency policies to combat world hunger and rural poverty. By proposing Food Sovereignty, La Via Campesina rejected the increasing attempts by global elites to define food security under a neoliberal market framework.

Governments, multilateral organizations, and food corporations legitimize 'Free' Trade Agreements and policies in the name of 'food security'. As a result, peoples and nations, particularly in the global South, have become increasingly dependent on international markets to acquire 'cheap' food, and they are vulnerable to speculation, land grabbing, dumping, and other unequal practices that undermine their capacity to feed themselves. Peasant farmers become landless and dispossessed migrants in the cities, as they cannot compete against subsidized food imports, or semi-servants of agri-business projects that supply raw materials for various industries – from the fast and highly processed food industry to the so-called clean energy sector for the production of ethanol and other agro-fuels.

Food security reinforces the core principles of modern industrial agriculture and the Green Revolution, including the use of capital-intensive and chemical-based inputs, mono-cropping, and 'improved' seeds such as hybrids and, more recently, genetically modified varieties. The industrialization of agriculture and food has resulted in the expansion of 1) 'green deserts', or plantations where only plants considered profitable can grow and reproduce, thus threatening peasant agriculture and (agro) biodiversity; and 2) highly processed food and fast food and feedstuff that poison humans and animals. In other words, food security aligns with the Western paradigm of 'development' as the exploitation and manipulation of plants and animals in laboratories, feedlots, plantations, factories, and markets for the exclusive benefit of (some) humans (Shiva 2000).

The concept of food sovereignty has changed since 1996, as multiple organizations, communities, and NGOs within and outside La Via Campesina have adapted and debated it in response to their diverse life and political conditions. One such debate is around the issue of how to scale up the food sovereignty movement, so that it becomes environmentally, socio-economically and politically resilient in the face of climate change and corporate power, without losing the diversity and autonomy of localized food sovereignty initiatives (McMichael 2013).

Another topic of discussion relates to the term 'sovereignty'. To be sure, La Via Campesina has framed food sovereignty within an autonomy paradigm, which involves the right of all peoples and nations to autonomy or self-government in order to define their own food systems, rather than being subjected to the demands and interests of distant and unaccountable markets and corporations. Food sovereignty then prioritizes local and national economies and markets and empowers peasant and small-scale agriculture. However, some agrarian movements have proposed 'food autonomy' – as the prerequisite for food sovereignty – to emphasize the place-based character of food production, non-liberal forms of democratic decision-making, and autonomy from state institutions. To reflect such semantic and political complexity, we propose the dual concept of food sovereignty and autonomy.

Food sovereignty and autonomy have become a rallying cry for a wide-range of struggles concerning land, water, and seed grabbing; agro-toxics; the corporatization of agri-food systems; biodiversity conservation and nature's rights; agricultural biotechnology and the patenting of life forms; farm workers' labour and human rights; malnutrition and hunger; and food provision in urban centres, among others (Desmarais 2007).

Food sovereignty and autonomy originated in – and continually emerge from – the knowledge, histories, and experiences of people and communities in struggle around the world. As a result, food sovereignty and autonomy is an analytical framework, a social movement, and a political project

(McMichael 2013). It entails a radical alternative that seeks to transform the structural inequalities embedded in agri-food systems – including the aid and development discourses and institutions – and it also explores transition to different models based on life-affirming principles. Food sovereignty and autonomy imply the defence of the knowledge, practices, and territories of food producing peoples – including peasants, fishermen, pastoralists, and urban farmers – as spaces for the reproduction and thriving of life and multispecies communities. This contrasts with the rationalized management of plant and animal life and death for profit and economic growth under the corporate food system. The transformative actuality and potential of food sovereignty and autonomy resides in the defence of three of such life-affirming principles: commons, diversity, and solidarity.

First, under a food sovereignty and autonomy paradigm, seeds, land, water, knowledge, biodiversity – and anything else that sustains materially and symbolically or spiritually, a people in a territory – are considered a commons. Rather than 'resources' to be exploited and privately appropriated, a commons perspective recognizes their collective and inalienable condition.

Second, food sovereignty and autonomy recognize that (agro) biodiversity and cultural diversity are intrinsically interdependent. Against the homogenizing tendencies of modern agri-food systems, food sovereignty protects and encourages the multiplicity of food producing systems around the world which thrive on the diversity of seeds, animals, foods, knowledge, labour practices, kinds of markets, landscapes, and ecosystems.

Third, solidarity among food producers and consumers around the globe, and with coming generations is fundamental for food sovereignty and autonomy. Strategies to encourage solidarity include 'transparent trade that guarantees just income to all peoples and the rights of consumers to control their food and nutrition' and environmentally responsible food systems, such as those based on agroecology, so that future generations can sustain themselves in 'the territories'. At its heart, food sovereignty and autonomy imply food systems based on 'new social relations free of oppression and inequality' (La Via Campesina 2007).

Further Resources

African Centre for Biodiversity, http://acbio.org.za/.

Desmarais, Annette Aurelie (2007), *La Via Campesina: Globalization and the Power of Peasants*. London, UK: Pluto Press.

ETC Group, http://www.etcgroup.org/.

La Via Campesina (2007), *Food Sovereignty and Trade Nyéléni Declaration*. https:// viacampesina.org/en/index.php/main-issues-mainmenu-27/food-sovereignty-and-trade-mainmenu-38/262-declaration-of-nyi.

McMichael, Philip (2013), 'Historicizing Food Sovereignty: A Food Regime
    Perspective', *Conference Paper, No. 13. International Conference. Food Sovereignty: A
    Critical Dialogue*. Yale: Yale University and the Journal of Peasant Studies.
Network for a Latin America Free of Transgenics (RALLT by its Spanish acronym),
    http://www.rallt.org/.
Shiva, Vandana (2000), *Stolen Harvest: The Hijacking of the Global Food Supply*.
    Cambridge: South End Press.

Laura Gutiérrez Escobar was born in Bogotá, Colombia, holds an undergraduate
degree in history from the *Universidad Nacional de Colombia* at Bogotá, a masters
in Latin American Studies from the University of Texas at Austin, and a PhD in
Anthropology from the University of North Carolina at Chapel Hill.

FREE SOFTWARE

Harry Halpin

Keywords: computers, software, intellectual property, commons

As the world is increasingly woven together by a dense interconnected
network of computers, the central question of our era is becoming
technological: How can freedom be maintained in a world that is increasingly
run not by humans but by software? Free software provides an answer to
this question by updating the traditional pre-digital notion of freedom to
encompass the dependency of humanity on software. The question that any
future form of politics must answer is not just the preservation of human
freedom, but its expansion via increased popular autonomous control of the
computational infrastructure by the users themselves.

At present, internet-driven capabilities are monopolized by a few large
Silicon Valley firms such as Google, Apple, Facebook, and Microsoft. Our
extended cognitive capabilities are mediated via software that are effectively
privatized. This signals a distinct turn in capitalism, with digital labour in
the form of 'programming' becoming the new hegemonic form of labour.
This does not mean that traditional factory jobs and resource extraction
have been rendered obsolete – far from it! However, this kind of labour
is being pushed under increasingly brutal and precarious conditions in

countries on the 'periphery' and excluded zones in core countries. Failure to invest in software in favour of industrial production and resource extraction makes 'developing countries' mere cogs providing low-margin goods and cheap labour while capitalism re-organizes around software.

Software drives automation, the replacement of human labour by machines. The language that coordinates these machines globally is code. The computer is defined as a universal Turing machine, a machine that is infinitely flexible in comparison with other specialized tools, as the same machine can be re-organized to be more efficient or re-programmed for new capabilities. The core of capitalism is no longer the factory, but code.

What if people could control the code themselves? 'Free software' inscribes four fundamental freedoms in the code itself:

1. the freedom to run the program as you wish, for any purpose
2. the freedom to study how the program works, and change it so it does your computing as you wish
3. the freedom to redistribute copies so you can help your neighbour
4. the freedom to distribute copies of your modified versions to others.

These freedoms mean that people can control the software for their own purposes by virtue of having access to the source code – as put by the Free Software Foundation, 'free software is a matter of liberty, not price' (Stallman 2017). Free software is a political programme that goes beyond just 'open source' and 'open access' to code, although it provides open access to code as it is necessary for freedom.

Free software was invented as a hack in American copyright law by the hacker Richard Stallman at Massachussets Institute of Technology (MIT), who saw that the culture of sharing software developed by hackers was being enclosed by commercial ventures such as Microsoft. In order to create a legally binding resistance to these new cognitive enclosures, Stallman created the General Public License (GPL). As the copyright of software is given by default to the developer, the developer can license their software to an unlimited number of people, thus preserving the four fundamental freedoms for posterity. The GPL license requires that all derivative works also use the GPL, so copyright's traditional required restrictions transforms into a 'copyleft' that instead requires that the four freedoms be granted. Other 'open source' licences such as the MIT licence or most Creative Commons licences – that directly assign copyright to the public domain – do not prevent derivative works from being enclosed in a proprietary manner. With the GPL licence, not only can a particular piece of software be guaranteed to preserve human capabilities, the software commons can virally grow. The

GPL has been a remarkably successful licence and software methodology. For instance, GNU/Linux runs most of the internet architecture today and even Google's Android is based on a free software kernel, although Google outsources vital components to its proprietary cloud.

Free software solves previously insurmountable problems for those seeking technological sovereignty on both an individual and collective scale. First, it allows programmers to form new kinds of social solidarity via the collective programming of code, as opposed to proprietary software development that is kept in the silo of a single corporation. Second, free software users are enabled by design to become programmers themselves, as they have the capability to learn how to programme and make modifications to the code. Third, open source is the only guarantee of security as it allows experts to audit the code. There are no licence fees and security updates are free, preventing many cyber-attacks. Lastly, although code may be kept 'in the cloud', that is hosted on other people's computers, versions of the GPL such as Affero GPL can guarantee that source code of software that runs on servers would be available as part of the commons. The GPL is a prerequisite for decentralizing the Internet and challenging the power of informational capitalism.

Free software is vital for the future of social movements. One cannot go backwards to a pre-industrial life without software and computers. Computers are a mathematical formalization of an abstract philosophical theory of causation in the material world, and so they can take many forms, from quantum to biological computing, and in the future hopefully forms that integrate into ecology. To reject computers entirely would, if taken to its extreme, reject all machinery and reduce humanity to perpetual drudgery and provincialism – hardly a promising future. It is equally naïve to imagine that capitalism will accelerate via the spread of computers into some socialist utopia without work. From the spread of Indymedia to Rhizomatica's community-run mobile telephony in Oaxaca, to the plan for autonomous ecological infrastructure that uses free software in Rojava, free software quietly has been aiding social movements for decades by providing the underlying software needed by struggles. In terms of post-development, what is needed is a strategy that increases both individual and collective freedom via the technological extension of human capabilities. As these capabilities increasingly depend on computers, free software provides a necessary tactic in the struggle to take software back from the enclosures – and give power to the people.

Further Resources

Gandy, Robin (1995), 'The confluence of ideas in 1936', in Rolf Herken (ed.), *The Universal Turing Machine a Half-Century Survey*. Berlin: Springer.

GNU Operating System, https://www.gnu.org/licenses/.

Levy, Stephen (2001), *Hackers: Heroes of the Computer Revolution*. New York: Penguin Books.

Moglen, Eben (1999), 'Anarchism Triumphant: Free Software and the Death of Copyright', *First Monday*, 4 (8).

Stallman, Richard (2017), 'What Is Free Software?', *The Free Software Foundation*. http://www.gnu.org/philosophy/free-sw.html.

Harry Halpin is a scientist at INRIA, the French national research institute for digital sciences in Paris, and visiting researcher at the MIT Socio-Technical Systems Research Center. He previously worked for the W3C on security standards before quitting over the issue of DRM (Digital Rights Management). He is the author of *Social Semantics* and editor of *Philosophical Engineering: Toward a Philosophy of the Web*.

GIFT ECONOMY

Simone Wörer

Keywords: matriarchy, gift economy, need-orientation, mothering, capitalist patriarchy

A gift economy is based on the principle that material and immaterial goods and services are given or received without an immediate or future obligation to return the gift. This would be the common definition of the term. Historically, most of the academic writers who have been working on The Gift and the gift economy have been white men and patriarchal (Göttner-Abendroth 2007). They have ignored women as gift givers, denied motherhood, and denied nature as being within the scope of 'giving' as a fundamental principle. They split the Gift in order to create the idea of a controllable and predictable binding contract. Many of these authors, such as Mauss, Serres, or Bataille even interpret the Gift as a violent act and do not distinguish between the logics of giving and patriarchal exchange. Furthermore, most of those approaches are anthropological and focus on 'primitive' societies without realizing that a culture of the Gift and gift economy are still active inside today's globalizing capitalist patriarchy.

In recent years, we have witnessed a renaissance of theories of the Gift and the gift economy. For example, Charles Eisenstein's book *Sacred Economies* (2011) and the pioneering work of Genevieve Vaughan, both try to show that there would be no human life without the Gift. Eisenstein begins his analysis of the history of money with the words 'In the beginning was the Gift'; and Vaughan starts at the very beginning with children of a human mother and points out the crucial importance of need-oriented mothering and caring. From a semiotic point of view, she develops the concept of *homo donans*, portraying humans as gift-giving and gift-receiving beings, emphasizing the maternal roots of the gift economy. We nevertheless need an integral comprehension. A radical alternative worldview requires a profound analysis of patriarchy and critique of its economic and technological systems. This would combine findings of the Modern Matriarchal Sciences network and ecofeminist insights into linkages between feminism and ecology. This integrated understanding, should help re-discover the maternal roots of the gift economy and its life-affirming culture.

The mother's womb is the original topos, a place where we experience interconnectedness and direct satisfaction of needs. It might be considered a model of the gift economy. From a psychological point of view, at this very beginning of our lives the mother is our first Other, but we are not separate from her. On the contrary, in this stage of life, the world is experienced in entirety. Once born, we experience the same world, but for the first time, feel distance, in separation from the original topos. The urge to overcome this distance may well be the beginning of tenderness, of dedication and of Other-orientation.

Matriarchal societies, as mother-centred and gift-friendly communities and gift economies, collectively remember this origin. Furthermore, matriarchies – and indigenous societies in general – recognize Mother Earth as a giving and receiving entity. The study of Modern Matriarchal Sciences has shown that matriarchies are deeply committed to principles of balance and interconnectedness. They respect diversity between humans and nature, genders and generations. There is no private property in matriarchal gift economies and the well-being of all members is a priority. Production is based on principles of subsistence and cooperation, and instead of accumulating and hoarding goods, everything in matriarchies circulates as a gift. The basic elements of a gift economy are the following:

- balance by circulation

- interconnectedness and diversity
- abundance and ego-limitation
- mothering and need-orientation.

In his classic book *The Gift: Creativity and the Artist in the Modern World* (1983), Lewis Hyde pointed out that when the Gift stops moving and becomes 'capital', hunger appears. Hyde is not the only thinker to describe the movement of the Gift as a sort of circular nourishing flow. By contrast, the patriarchal logic of exchange is based on the motto *do ut des* – giving in order to get back. Researchers in the Critical Theory of Patriarchy[1] show that in its capitalist form, patriarchy appropriates and transforms the Gift as the original logic of life and bonding practice. Here, gift-oriented connections within nature and the human community are brutally transformed by machine-like systems, consisting of exchangeable and infertile parts. They function as man-made products: commodities, machinery, weapons, and money. A capitalist patriarchy produces scarcity by killing life materially and immaterially. The abuse of gift giving in the interest of accumulation, profit, and structures of dominance, leads to a generalized suspicion of need-oriented and Other-oriented gift practices. A capitalist patriarchy tries to control gift giving completely by unpaid work and enforced consumption. However, a gift economy would be based on the need-oriented nourishing flow of the Gift in its material and immaterial expression, and as von Werlhof (2011) calls it, a principle of the 'Interconnectedness of all Being'.

Note

[1] Forschungsinstitut für Patriarchatskritik und alternative Zivilisationen, www.fipaz. at.

Further Resources

Eisenstein, Charles (2011), *Sacred Economies*. Toronto: Evolver Editions.
Genevieve Vaughan and International Feminists for a Gift Economy, www.gift-economy.com.
Göttner-Abendroth, Heide (2007), 'Matriarchal Society and the Gift Paradigm: Motherliness as an Ethical Principle', in Genevieve Vaughan (ed.), *Women and the Gift Economy: A Radically Different Worldview Is Possible*. Toronto: Inanna.
Hyde, Lewis (1983), *The Gift: Creativity and the Artist in the Modern World*. New York: Random House.
von Werlhof, Claudia (2011), *The Failure of Modern Civilization and the Struggle for a 'Deep' Alternative: On 'Critical Theory of Patriarchy' as a New Paradigm*. Frankfurt am Main: Peter Lang.

Simone Wörer, PhD, MEd, MPS, completed her dissertation *The Crisis of the Gift* in political science at the University of Innsbruck under Claudia von Werlhof. She is an independent researcher; author of *Politik und Kultur der Gabe: Annäherung aus patriarchatskritischer Sicht* (2012); and a member of the Planetary Movement for Mother Earth and Forschungsinstitut für Patriarchatskritik und alternative Zivilisationen (FIPAZ), Austria.

———•••———

GROSS NATIONAL HAPPINESS BHUTAN

Julien-François Gerber

Keywords: alternative development, alternative indicator, Bhutan

Gross National Happiness (GNH) is a contested notion in Bhutan. It encompasses different interpretations and implementations, which are more or less radical. For some Bhutanese people, GNH is a loosely defined 'green growth' and/or a brand name to fit every occasion – a position I call 'commercial GNH'. For others, GNH is a new holistic indicator which should replace GDP in guiding development policies – a position I call 'narrow GNH' (Ura *et al.* 2012). And for others still, GNH is not just a new indicator but also a philosophy of social flourishing integrating exterior and interior needs and seeking sufficiency with respect to economic growth once everyone's basic needs have been met – a position I call 'deep GNH' (Royal Government of Bhutan 2013). Currently, it seems that it is 'commercial GNH' which tends to dominate the decision-making arena. But overall, Bhutan's policies are the outcome of these different forces at play (Hayden 2015).

Among the GNH-oriented policies of Bhutan, one could cite free education and healthcare for all; severe restrictions on foreign investments; no WTO membership; no outdoor advertising; heavy taxes on car imports; limits on mass tourism and ban on alpinism; strict cultural norms for architecture and official clothing; half of the country under protected areas; a constitutional mandate for 60 per cent forest cover; land distribution to landless farmers; willingness to shift to 100 per cent organic agriculture. Though the pro-modernization sectors of the Bhutanese society put pressure on these policies, there seems to remain so far a wide public support for them.

The expression 'GNH' was coined by the fourth king in the 1970s.

When a journalist asked him what was Bhutan's GNP, he replied: 'We are not concerned about Gross National Product; we care about Gross National Happiness'. His answer might have been reminiscent of a 1729 Bhutanese legal code stating: 'If the government cannot create happiness for its people, there is no purpose for government to exist' (Ura *et al.* 2012).

The modern concept of GNH, however, was formulated later, during the turbulent period of the 1990s. On the one hand Bhutan's internal situation was extremely tense. The Ngalopruling elites (western Bhutan) were dealing very poorly with the country's other linguistic and cultural groups – not only with the Lhotshampa (southern Bhutan) but also with the Sharchops (eastern Bhutan) and the Tibetans – to the point that a violent crisis exploded, leading to the establishment of Lhotshampa refugee camps in Nepal. On the other hand, the country's external situation was tense too. Its debts to India were swelling while India turned neoliberal. As the government saw the multiple negative impacts of neoliberalism, there was an urgent need for an alternative discourse. And so the idea of the GNH was developed, partly as a form of resistance to neoliberalism and partly, perhaps, as a tentative way to alleviate several internal wounds.

Gross National Happiness takes many forms. We will briefly examine two forms of 'narrow GNH' and two of 'deep GNH'. Firstly, the GNH Index, as a new integrated indicator, was developed in the mid-2000s under the leadership of Karma Ura of The Centre for Bhutan Studies. It covers nine domains – living standards, education, health, environment, community vitality, time-use, psychological well-being, good governance, and cultural resilience and promotion. These are measured by thirty-three indicators. It has been used so far in one pre-survey (2006) and three full surveys across the country (2008, 2010, 2015). Its purpose is to measure well-being holistically in order to assess its evolution and guide decision-making accordingly. Interestingly, the GNH Index includes a 'sufficiency threshold' for each indicator. These thresholds are benchmarks of how much is enough for a 'good life' and are based on international or national standards, normative judgements or the outcome of participatory meetings. For example, citizens working more than eight hours daily are considered time deprived and falling short of sufficiency.

Second, the GNH Policy Screening Tool has been designed to help the GNH Commission. It is a powerful organ orchestrating the planning process of Bhutan's economy in assessing policies and projects for their impacts on twenty-two variables reflecting the nine domains mentioned earlier. This screening tool played a significant role in Bhutan's rejection of WTO membership as well as on its limited exploitation of mineral reserves.

As examples of 'deep GNH in practice', one could mention the Samdrup Jongkhar Initiative (SJI), launched in 2010 in a neglected rural area of the country by a progressive religious teacher, Dzongsar Jamyang Khyentse Rinpoche. The SJI is currently led by an influential woman from the region, Neten Zangmo, and was intended to be a bottom-up initiative expandable to the entire country. It aims at establishing food sovereignty and self-sufficiency based on organic agriculture, protecting the environment with a zero waste campaign, strengthening communities, doing youth education, and fostering citizen involvement in decision-making.

Another example of 'deep GNH' could be the Universal Human Value training – remarkably consistent with aspects of Buddhist ethics and depth psychology – offered in all the higher education organizations of the country. The multiday training offers a systematic method encouraging participants to better understand themselves and their relationship with others and the environment. It intends to show how right understanding and feeling within oneself, fearlessness in society and co-existence with nature are foundational to harmony and thus 'happiness'. Unfortunately, deep-rooted confusion abound, so it is argued that money and the accumulation of material goods are central, which leads to various forms of domination, fear and exploitation.

From a post-development perspective, I would say that the primary original contribution of GNH – and perhaps also its most difficult challenge – is to incorporate existential 'interior' questionings as they relate to psychology and spirituality. At a time when most of the pressing 'exterior' problems are relatively well understood by many people, 'interior' interrogations – such as those regarding 'true needs', 'existential fulfilment', 'false consciousness', 'social conditioning' – are much more complex and largely under-addressed, and yet essential for effective research and action. There is a crucial need for more inner understanding in radical activism.

Further Resources
Hayden, Anders (2015), 'Bhutan: Blazing a Trail to a Postgrowth Future? Or Stepping on the Treadmill of Production?'. *Journal of Environment Development*. 24 (2): 161–86.
Royal Government of Bhutan (2013), *Happiness: Towards a New Development Paradigm*. Thimphu: Royal Government of Bhutan.
The Centre for Bhutan Studies and GNH Studies, http://www.bhutanstudies.org.bt/.
Ura, Karma, Sabina Alkire, Tshoki Zangmo and Karma Wangdi (2012), *An Extensive Analysis of GNH Index*. Thimphu: The Centre for Bhutan Studies.

Julien-François Gerber is Assistant Professor in Environment and Development at the International Institute of Social Studies in The Hague, Netherlands. Prior to that, he taught at Sherubtse College in Kanglung, Bhutan, for a year. His research is in economic anthropology, political ecology, and heterodox economics.

⦿•⦿

HINDUISM AND SOCIAL TRANSFORMATION

Vasudha Narayanan

Keywords: environment, India, religion

Hinduism is the predominant religion in India and also a global religious tradition. There is considerable diversity among the many traditions and groups that come under the rubric of 'Hinduism', or called 'Hindu', and they may look to a plethora of texts and authorities for guidance. In this essay, the term will be used in retrospect for the many beliefs, practices, and communities that affiliate themselves with the Sanskrit texts including but not limited to the Vedas, the epics, the Puranas, *dharma shastras*, and the many vernacular traditions in the Indian sub-continent, as well as in many other parts of the world.

Three conceptual frameworks inform this essay. The first is the notion of *karma* (action), but referring more broadly to the consequences of good and bad deeds which affect the quality of the current and future lifetimes, an idea that is shared with Jainism, Buddhism, Sikhism, and other South Asian traditions. Most Hindus would understand this to be a complex concept and not one that accounts directly for suffering and inequities in life.

This leads us to a second set of ideas connected with the idea of *dharma* (righteousness, duty, law). There are hundreds of passages in the Sanskrit and vernacular texts on the importance of justice, social welfare, and the importance of happiness for all other animals. In addition, there are several exhortations to maintain the purity and integrity of forests, lakes, rivers, and nature as a whole. The passages are not simply 'information' texts; they are injunctions for human beings to act in a manner that will multiply happiness and prosperity in this world. Specific instructions are given, for instance, to rulers to act with righteousness and for people to do their *dharma* which may be incumbent on several factors such as caste and station in life. Overriding the specific duties incumbent on one by way of one's station in life are statements on *samanya* (general, universal) or *sanatana* (continual, eternal) *dharma*, or virtues that ought to be common to all human beings. These include compassion, non-violence, generosity, patience, gratitude, and so on. Recent social movements based on the Hindu religious traditions cite these virtues to justify social action.

The third set of ideas and practices are connected with the idea of *bhakti* or devotion. Devotion to the deity, conceptualized in many forms, and/ or commitment to religious teachers and to other devotees has in some cases helped rectify iniquities connected with the hierarchical caste system as well service to improve social problems. For instance, an emphasis that a person's *bhakti* is more important than the circumstances of birth has led to periodically dominant ideas that devotion and piety ought to create a different hierarchical system which is based on faith and virtue and not on birth or occupation. Poets such as Nammalvar said that he would serve every devotee of the deity Vishnu, irrespective of caste. Several religious communities or leaders such as Vivekananda (1863–1902) also supported the idea that service to human beings is service to God, '*maanavasevemadhavanseva*' or '*bhagavatasevabhagavatseva*'.

There may be literally thousands of initiatives for social transformation which spring from Hindu roots. A look at the large Tirumala-Tirupati temple, the Ramakrishna Mission and the many organizations it has spawned or inspired, the Swaminarayan community, the works of the Ma Amritanandamayi, the worldwide network of activities inspired by Sathya Sai Baba, and many others give us an idea of the range of initiatives and enterprises that exist to help address issues of social injustice, ecological unsustainability, and economic exploitation. An example is their *jankalyan* (human welfare or happiness) schemes focused on providing medical facilities ranging from simple neighbourhood clinics to sophisticated 'super-specialty' hospitals, as well as educational opportunities for people who have no access to them because of birth/caste issues, geographic location, or financial constraints.

In addition, the Tirumala-Tirupati Devasthanam – one of the wealthiest religious institutions in the world – also has a focus on environmental problems. Encouraging activities such as planting trees, organizing educational seminars on conservation and biodiversity, stopping deforestation and saving water-bodies from further pollution are all part of their Haritha Project (Project Green). Several religious leaders have also worked actively with the general public, raising consciousness on very important projects such as cleaning major rivers like the Ganga and the Yamuna, planting trees in the mountains to prevent erosion.

Hindus face several challenges in their efforts to achieve social equity, maintain human rights, and remedy the degradation of the environment. In some instances, religion itself is an impediment in these endeavours. Those who think of the river Ganga as an agent of purification do not believe that the river itself needs purification. Hindus perpetuate and are also the target of activities rooted in intolerance. Extreme adherence to religious causes and aggressive proselytizing efforts by many religions in India go against human rights. Intolerance of others eating meat products, intolerance of other religions, religious vendettas, lingering caste-violence, especially against Dalits, in many religious traditions in the subcontinent, and the suppression of women, all go against the basic virtues of non-violence and generosity that are integral to *sanatana-dharma*.

The Vedic injunction *sarvebhavantusukhinah* ('may all people be happy') has been the lofty but elusive goal for many millennia but one which is certainly a guiding light for individuals and institutions in the Hindu traditions.

Further Resources

Alley, Kelly D. (2002), *On the Banks of the Gaṅgā: When Wastewater Meets a Sacred River*. Ann Arbor: University of Michigan Press.

Davis, Don (2010), *The Spirit of Hindu Law*. Cambridge and New York: Cambridge University Press.

Kane, PandurangaVamana (1968–1974), *History of Dharmasâstra (Ancient and Mediaeval Religious and Civil Law)*, five volumes. Poona, India: Bhandarkar Oriental Research Institute.

Narayanan, Vasudha (1997), '"One Tree Is Equal to Ten Sons": Some Hindu Responses to the Problems of Ecology, Population and Consumption', *Journal of the American Academy of Religion*. 65 (2) (June): 291–332.

Nelson, Lance E. (1998), *Purifying the Earthly Body of God: Religion and Ecology in Hindu India*. Albany, NY: State University of New York Press.

Sanford, A. Whitney (2012), *Growing Stories from India: Religion and the Fate of Agriculture*. Lexington, KY: University Press of Kentucky.

Tirumala Tirupati Devasthanams, http://www.tirumala.org/SAHarithaProject.aspx.

Vasudha Narayanan is Distinguished Professor, Department of Religion, and Director, Center for the Study of Hindu Traditions (CHiTra) at the University of Florida, US. She was educated at the Universities of Madras and Bombay in India, and at Harvard University, US. She is a former president of the American Academy of Religion and the Society for Hindu–Christian Studies.

——•••——

HUMAN RIGHTS

Miloon Kothari

Keywords: human rights, dignity, solidarity, justice, participation

Recognition of the inherent dignity and of the equal and inalienable rights of all members of the human family is the foundation of freedom, justice and peace in the world.

> – Preamble to the Universal Declaration of Human Rights (UDHR)

All human beings are born free and equal in dignity and rights.

> – Article 1, UDHR

Human Rights have been conceptualized, implemented, violated, and fought over for centuries. Today and in a post-development world, the universal principles and instruments of human rights offer a powerful moral, ethical, and legal basis with which to navigate justice through an increasingly unjust and devastated world.

In 539 BCE, the armies of Cyrus the Great, the first king of ancient Persia, conquered the city of Babylon. But it was his next actions (explained in the Cyrus Cylinder discovered in 1879) that marked a major advance for humankind. He issued a decree that freed slaves, declared that all people had the right to choose their religion, and established racial equality. The principles of human rights also emerged from the great texts of a number of the world's main religions. The documents that emerged from the American and French revolutions also espoused a human rights spirit. Through the centuries great minds articulated the content of human rights as a cornerstone for humankind to embrace democracy and freedom – Jean Jacques Rousseau, Mahatma Gandhi, Martin Luther King, Eleanor Roosevelt, Nelson Mandela, and the Dalai Lama. The lessons from centuries of thought and action, but also embodying a deep expression of humanity as a response to the horrors of world war, were manifested in the adoption by the UN General Assembly in 1948 of the Universal Declaration of Human Rights (UDHR).

Following the adoption of the UDHR and the period of decolonization, national constitutions emerged from, among others, the freedom struggle in India, and the anti-apartheid movement in South Africa. These constitutions are were primarily based on the notion of fundamental human rights. The

UDHR also sparked off the evolution of UN human rights instruments on cultural, civil, economic, social, and political rights, and on the specific rights of women, children, indigenous peoples, migrants, and the disabled.

The entire edifice of human rights is built on the preservation and upholding of people's dignity. The edifice of human rights is also, significantly, premised on meeting the needs of the most vulnerable first. Such a point of departure for the realization of human rights is particularly urgent at a time when the world's most historically, economically discriminated communities are increasingly impacted by neoliberal economic policies and the adverse impacts of phenomena such as climate change.

Human rights instruments protect economic, social and cultural rights, including not only the right to food, livelihood, health, housing, social security, and civil and political rights, but also the right to life, freedom of religion and belief, the right to peaceful assembly and association, and the right to participation in political and public affairs.

The human rights approach is a radical alternative that directly challenges injustice, exploitation, and discrimination of millions across the world in the following ways:

- The upholding of human rights presents a direct and powerful challenge to the global hegemonic forces that seek to reduce people's rights to commodities and seek to financialize what should be basic entitlements – water, land, housing, and so forth.
- It rests on foundational principles such as non-discrimination, gender equality, public participation, and the right to redress from human rights violations. These principles complement and reinforce environmental principles. The right to free prior-informed consent, for example, can be derived from international human rights law.
- Human rights are 'universal'; they apply to everyone across the world regardless of their race, religion, and economic status. The rights contained in the UDHR are accepted by all 193 member states of the UN. They are underpinned by a robust and constantly evolving international system, including the UN Human Rights Council, its Universal Periodic Review, Special Rapporteurs and Treaty bodies that are continuously monitoring the behaviour of state and non-state actors to ensure accountability to human rights commitments. Such comprehensive mechanisms are both radical and practical. This may be the primary reason that the human rights approach has found resonance among social movements and campaigns across the world, in addition to the long-standing system of Special Rapporteurs on the rights of indigenous peoples, extreme poverty,

housing, health, and food. Recently appointed Special Rapporteurs for water and sanitation, solidarity and human rights, and the environment.

- New instruments, to address current global realities, are currently being formulated through a transparent, inclusive process. These are the Declaration of the Rights of Peasants and other persons working in rural areas, and a legally binding instrument to regulate the activities of transnational corporations and other business enterprises. These are largely a result of remarkable cross-sectoral and interdisciplinary global initiatives led by international civil society coalitions and social movements working with progressive governments.
- At the national and sub-national levels, the protection system involves national courts, human rights commissions, tribunals and ombudspersons, civil society organizations, campaigns, and movements. The evolution of human rights at the national level, in Constitutions and laws, such as the right to land, right to a healthy environment, and the rights of nature are also important developments.

Notions of solidarity, fraternity, cooperation, and trust, which go beyond the predominantly individualistic approach which Western paradigms have been criticized for, are also increasingly acknowledged as cornerstones of an alternative, radical worldview based on human rights.

These values are best articulated in the human rights paradigm that places a premium on truth, is a powerful expression of ethical, moral and legal values and imperatives, and upholds the dignity of the individual, and the collective identity of community. It international, national, and regional contexts, human rights principles and instruments should be rigorously applied by post-development activists in their resistance to hegemonic forces.

Further Resources
Defending Peasants Right, https://defendingpeasantrights.org/.
International Service of Human Rights, http://www.ishr.ch/.
Treaty Alliance, http://www.treatymovement.com/.
UN Office of the High Commissioner of Human Rights,
    http://www.ohchr.org/EN/pages/home.aspx.
United Nations Declaration on the Rights of Peasants and Other People Working in Rural Areas, https://undocs.org/A/C.3/73/L.30
Via Campesina, https://viacampesina.org/en/international-peasants-voice/.

Miloon Kothari is a leading scholar-activist on human rights. He is the President of UPR (Universal Periodic Report) Info and was the former Special Rapporteur on Adequate Housing with the UN Human Rights Council. During his tenure, he led the process that resulted in the UN Basic Principles and Guidelines on Development-Based Evictions and Displacement – this being the current global operational human rights standard on the practice of forced evictions.

———•••••———

HURAI

Yuxin Hou

Keywords: indigenous well-being, naturalism, shamanism, cultural revival

Hurai ('all the best things'), expresses the logic of transforming from nature to animals and then to human beings, in accord with Chinese Tuvan's people's cosmology. It sustains the belief that human beings are capable of continuously receiving the *Hurai* and blessing from nature (Hou 2014). This term is strongly connected with indigenous well-being and traditions of naturalism and shamanism revolving around the sustainable interaction among Tuva people in China, animals, nature, as well as gods.

The Tuva people are ancient hunters and nomadic peoples whose lineage can be traced back to the Tang dynasty. Today, the Tuva people are scattered across Xinjiang of China, outer Mongolia, and Russia. Chinese Tuva people mainly live in the three villages of Kanas, Hemu, and Baihaba, and they have a total population of around 2,500 (Hou 2013). During the process of nomadic and hunting life, *Hurai* as an ancient concept prevailed in the daily life of Tuva people in China with a broad meaning of blessing humans, animals as well as other elements of nature by the gods, within a cultural context of Shamanism and naturalism. For example, the fungus *Ophiocordyceps sinensis*, found in high altitude areas, is of high economic and medical value but has rarely been dug by Tuva people in China because they believe it to be the protector of the grassland and one of the sources bringing *Hurai* for Tuva people in China. However, the huge economic value has attracted more and more outsiders to secretly dig it up without restraint, causing ecological damage. In order to deal with this problem, the Tuva community leaders in China have spontaneously organized a ranger team. In addition, the concept of *Hurai* applies in many other aspects of life.

Although modernization, sedentarization, and tourism have tremendously impacted on traditional livelihoods and cultures, Hurai is now playing an important role in cultural revival and in strategic ways of opposing modern development. From the indigenous concept of Tuva people in China, there is no concept that corresponds to the linear idea of development. *Hurai* as a holistic concept accommodates features such as good life, health, sustainability, love, sanctity, and so on. It is an anti-anthropocentric concept that puts nature, animals, and gods ahead of humans; it is reflected in a notion of well-being which implies that when nature, animal, and gods feel happy, humans are able to experience happiness naturally. In order to gain the *Hurai*, humans need to devote activities to safeguarding well-being as a whole, not only in rituals but on a daily basis.

Tuva community leaders, through daily life practices and ceremonies or festivals, educate young people to be fully involved in nature worship and shamanism in a manner that contrasts with traditional school education. Through a body of practice, they learn about and perceive the close relationship between humans and nature. In this way, 'natural relations' are established between humans and nature, animals, and spiritual beings. The ecological perspectives of the Tuva people in China are reflected in taboos; children are educated to abide by natural rules without breaching taboos, they are taught not to harm plants and animals and to live in harmony with their own well-being and happiness. A few examples would be the belief that pock marks will appear on a person's face if bird eggs are broken; people will suffer from tic disorder if a living tree is cut down. With respect to harvesting natural resources, Tuvan people in China strictly observe taboos such as no extensive or deep digging into the earth. In addition, Tuva people in China utilize a variety of annual ceremonies and rituals to educate youngsters on how to deal with the relationship between humans and nature.

Under the impact of modern development, the concept of *Hurai* has experienced a strategic transformation from the traditional indigenous concept to a concept intended not only to guide daily activities but for reviving traditional culture and resisting modernization. This transformation is expedient in uniting Tuva people in China and for revitalizing the natural bonds between humans and nature, animals and gods. In the 'bone smashing ceremony', the magic and oft-repeated word *Hurai*, accompanied by dancing and singing, creates collective jubilation and contributes to reviving the old hierarchy and structure between humans and nature. The bone smashing ceremony stresses the importance of blessing and receiving and experiencing well-being by means of consuming marrow (Hou 2016). The bone smashing ceremony, as other rituals or ceremonies in Chinese

Tuva society, plays a role of renewing the importance of the ancient concept of *Hurai*. In reality, the concept of *Hurai* helps the Tuva people in China to face up to the threats posed by external challenges, modernity, as well as the social and cultural predicaments in Chinese Tuva society. The efforts and strategies adopted by Tuva community leaders in China provide hope for the continuity of ancient concepts of *Hurai*, empowered by the cultural revival movement in Chinese Tuva society.

Further Resources
Hou, Yuxin (2013), 'Saving Our Identity: An Uphill Battle for the Tuva of China',
    Cultural Survival Quarterly. 37 (4): 24–25.
——— (2014), 'The Analysis of Current Situation of Tuva people in China', *The
    New Research of Tuva*. 1: 96–103.
——— (2016), 'Ritual and Cultural Revival at Tuvan Sacred Natural Sites
    Supports Indigenous Governance and Conservation of Nature in China', in Bas
    Verschuuren and Naoya Furuta (eds), *Asian Sacred Natural Sites: Philosophy and
    Practice in Protected Areas and Conservation*. New York: Routledge.
Friends of Tuva (FoT), http://www.fotuva.org/index.html.
Tuva Asia, The New Research of Tuva, http://en.tuva.asia/.
Yuxinhou, http://www.yuxinhou.com/.

Yuxin Hou is a Visiting Scholar at the Graduate Center of the City University of New York and also Research Fellow at the Institute for Philanthropy Tsinghua University. He received his PhD in Anthropology in 2012 from Peking University – research funded by the Ford Foundation.

IBADISM

Mabrouka M'Barek

Keywords: Islam, Ibadism, asceticism, community empowerment, polygamy

Many alternatives draw their inspiration from the spiritual realm or cosmovisions such as *sumakkawsay* or *ubuntu*. Can Islam also be a source of inspiration for building alternatives to neoliberalism, neo-colonialism, and predatory relations to nature? There have been ample conversations about what Islam stands for in terms of the economy and human relation to nature. The Koran and Hadiths have been subject to a wide range of interpretation but proven insufficient in providing a clear answer for today's challenges.

As a matter of fact, often the search for a more complex and flexible interpretation is an attempt to fit a capitalist way of life. Thus Islamic finance, presented as an alternative, is merely a mechanism that invents a new vocabulary to fit the precepts of Islam while sustaining neoliberalism. Political actors, such as the Islamist parties in Turkey or Tunisia, often argue that the Prophet was a merchant and therefore a capitalist, but this unsubstantiated connection is a pretext to justify Islamist obedience to the International Monetary Fund. Worse, countries of the Gulf exemplify waste and materialism to such a point that it is hard to believe that Islam can meet today's challenges.

To find an alternative model to capitalist society, we need to search beyond petro-monarchies, especially beyond the Sunni versus Shia divide, and take a look at Ibadism, a little known school within Islam. This discussion will situate that ideology historically and describe its main characteristics: a community-driven mode of governance, asceticism as a way of life, and cultural pluralism.

Ibadites are followers of Abdullah ibn-Ibad al-Tamini and form the oldest school of thought within Islam predating the Sunni versus Shia divide. The Ibadite faction was created after the death of the Prophet as its followers believed that the supreme religious leader did not have to descend from the Prophet's blood line, nor come from a certain ethnicity. Ibadism is the dominant religion in Oman, and has spread to Libya, Algeria, Zanzibar, and the Island of Djerba in Tunisia. Tunisian anthropologist Walid Ben Omrane has noted that the architecture of Ibadite cities indicates the relationship of the community to the notion of power. Indeed, contrary to traditional Islamic architecture, where the mosque represents the centre of power, here it is at the periphery. In addition, Ibadite mosques are smaller and more modest.

Ibadites follow egalitarian Berber family structures – where men and women's economic roles are similar. As Walid Ben Omrane (personal communication, Oct 10, 2017) notes:

> Berbers have adopted the Kairouan marriage contract which has blocked polygamous procedures for men. Moreover, Ibadite women have the right to repudiate their husband if he does not give any sign of life during his journey as a trader and this after the period of two years. Polygamy prospered in warrior, tribal, and nomadic societies. However, Ibadites have often been communities

of social peace to ensure prosperity of their trade; communities based on the traditional family and sedentary societies.

An Ibadite Imam remains the representative of the community but he is chosen from among the most 'deserving' individuals, which means that the elected individual must show no personal ambition, be modest and committed to serve his people. Ibadite political structure remains patriarchal as the Imam is chosen from among men. Nevertheless, he rules with community consensus, so if he becomes autocratic, he is removed even before his term ends.

In addition to the simplicity and modesty of their architecture, the Ibadite lifestyle rests on the principles of sobriety, modesty, and rejection of everything that is ostentatious. Independence is thought to be achieved by limiting needs and rationing what nature offers. This philosophy of life emulated the first caliphs in Islam, but became a polar opposite to the luxury that caliphs adopted since the Umayyad dynasty. This simple lifestyle still endures today. Inequality between the representative of the community and the people is considered an indication that the mode of governance is slipping towards the model of a king and his subjects.

As followers of an Islam against tyranny and domination, Ibadites celebrate ethnic plurality and interfaith dialogue. Because they believe anyone deserving could be chosen as Imam, the Berber population of North Africa has been more receptive and supportive of this egalitarian school of Islam. On the Island of Djerba in Tunisia, for example, Ibadites value their Amazigh or Berber origins as well as their African identity. By virtue of its religious tolerance and its defence of both ethnic plurality and community autonomy, they are shielded from global capitalism, Ibadism is viewed nowadays as an opportunity for peace in a Middle East torn by ethnic divide and the curse of oil.

Ibadites have resisted central power by organizing their communities at the margins of power. Ibadism undoubtedly owes its survival to its strategy of discretion or *el kitman*. However, in Tunisia, Ibadites have come out to fully engage in the public space during the post-independence struggle between Youssefist and Bourguibist tendencies and, more recently, during the revolution of Dignity 2010–11. In 2012, the Guellala Ibadite community on Djerba island successfully organized a resistance movement against the state authorities who wanted to continue operating a trash dump site that was overloaded and posed a risk to scarce water sources.

As the Ibadites of Djerba show, the strategy of *el kitman* is about occupying local space, organizing a community independent of the state, and, when the state interferes, resisting it. They were criticized for being silent during the dictatorship of Zine al Abedine Ben Ali which lasted from 1989 to 2011, but Ibadites manage to resist by demonstrating a mesh of multiple community-based modes of resilient self-governance.

Further Resources

Aillet, Cyril (2012), 'L'ibâdisme, uneminorité au cœur de l'islam', *Revue du monde musulmanet de la Méditerranée*. 132: 13–26, Aix-en-Provence, France: Presses Universitaires de Provence, https://journals.openedition.org/remmm/7752.

Baptiste, Enki (2016), 'Aux marges du califat, pouvoirs et doctrines dissidentes: retour sur le développement de l'ibadisme', *Les Clefs du Moyen Orient*. http://www.lesclesdumoyenorient.com/Aux-marges-du-califat-pouvoirs-et-doctrines-dissidentes-retour-sur-le-2260.html.

Ben Omrane, Walid (2012, March), 'La Communauté Amazighophone de Jerba et les Révolutions au Maghreb', paper presented at the Center for Maghreb Studies in Tunis, Tunisia.

——— (2018), Maghrib in Past and Present: podcast episode 39, https://www.themaghribpodcast.com/2018/06/blog-post.html.

Mabrouka M'Barek is a former elected member of Tunisia's National Constituent Assembly 2011–14, and currently a participant of the Brussels based Rosa Luxemburg Stiftung's Global Working Group Beyond Development.

ICCAs — TERRITORIES OF LIFE

Grazia Borrini-Feyerabend and M. Taghi Farvar

Keywords: territory, life, governance, indigenous peoples,
local communities, commons

At all times and in all world cultures, a phenomenon appears so strong and so natural as to be nearly invisible. This is the association – unique, profoundly rich, at times visceral – that ties a specific people or community to its own territory: the land, water, and natural resources on which and of which it lives. Around the world, many different terms are used to describe that special bond: *wilayahadat, himas, agdals, territorios de vida, territorios del buenvivir, tagal, qoroq-e bumi, yerliqorukh, faritraifempivelomana,* ancestral domains, country, community conserved area, sacred natural site, locally managed marine area, and many others – representing unique meanings for unique peoples and communities. In this second millennium, this phenomenon has been singled out as an essential feature of humanity and offered a *lingua franca* name as ICCAs – territories of life, which can be used across languages and cultures.

In a nutshell, ICCAs are 'territories and areas conserved by indigenous peoples and local communities' – those unique natural spaces where a strong community – territory association is combined with effective local governance and conservation of nature (Borrini-Feyerabend *et al.* 2010; Kothari *et al.* 2012). ICCAs thus encompass, but should never submerge, a wealth of local terms, which is a value in itself. To be sure, for custodian indigenous peoples and traditional communities the association that connects them to their territory is richer than can be expressed in words. It is a bond of livelihood, energy, and health, and a source of identity, culture, autonomy, and freedom. It connects generations, preserving memories and practices from the past and linking those to the desired future. It is the ground on which communities learn, identify values and develop self-rule. For many, 'territory' also bridges visible and invisible realities, material and spiritual wealth. With territory and nature go life, dignity and self-determination as peoples.

The presence of an 'ICCA – territory of life' implies a 'local governance institution' – a council of elders, village assembly, spiritual authority, ingrained cultural norms. It develops and ensures respect for the rules of access to and use of the natural commons, with positive outcome for both nature and people. It thus describes the presence of three characteristics:

- a 'strong and profound bond' between an indigenous people or local community and a territory or area
- the concerned people or community 'makes and enforces decisions' about that territory or area
- the decisions and efforts of the people or community lead towards the 'conservation of nature' and 'associated life, livelihoods, and cultural values'.

Of course, the socio-ecological phenomena are complex. There may be 'defined ICCAs' (exhibiting all three defining characteristics), 'disrupted ICCAs' (fulfilling the three in the past, but failing today because of disturbances, which can be reversed or counteracted) and even 'desired ICCAs' (having only one or two defining characteristics, but also possessing the potential of developing the third) (Borrini-Feyerabend and Campese 2017).

An ICCA can only be self-identified and kept alive by the people or community that governs and manages it. The fisherfolk who engage in surveillance operations for their estuarine conserved area in Casamance, Senegal, and the indigenous pastoralist communities of Iran, who take momentous decisions about when to migrate to their summering and wintering grounds – 'they' know they have an ICCA. The indigenous peoples of the Amazon region, who strenuously resist disruption by dams, roads, and mining operations, and the rural communities of Spain, whose commons are at the heart of local identity and culture – 'they' know that their bond is strong and effective enough. The forest dwellers of Borneo who recognize hundreds of plant and animal indicators, and the Malagasy women who regulate octopus collection to secure abundance in the next fishing season – 'they' can recognize and discuss conservation outcomes.

Today, the term 'ICCA' has taken on a life of its own. It is now used by conservationists and government agencies as a type of governance for the conservation of nature (UNEP WCMC's Protected Planet Reports 2012–16). ICCAs are recognized as protected areas of a specific governance type or as 'conserved' areas (Borrini-Feyerabend *et al.* 2013; Borrini-Feyerabend and Hill 2014), and/or through arrangements appropriate in overlaps with protected areas under various governance types. In coverage and conservation contribution, ICCAs are in all likelihood equally or more important than official protected areas, and therefore crucial to achieve global conservation goals.

ICCAs deliver 'lasting patterns of conservation' that depend on local integrity and capacities rather than external expertise and funding. They sustain livelihoods, peace and security, and cultural identity and pride. They are a non-market based mechanism to mitigate climate change and to help in adapting to it. They help achieve most of the targets of the Strategic Plan for Biodiversity 2011–20 (Kothari and Neumann 2014). And they contribute to most of the goals in the UN Sustainable Development framework. For custodian indigenous peoples and local communities, however, ICCAs remain essential for sustaining life and livelihoods, enjoying collective rights and responsibility to land, water, and natural resources, and ensuring respect for the knowledge, practices, and institutions essential to culture. These are crucial reasons why hundreds of indigenous peoples and community organizations and civil society supporters and individuals have joined forces in the ICCA Consortium[1] – an international association that, around the world, defends ICCAs against several pervasive threats, and fosters their appropriate recognition and support as 'territories of life'.

Note

[1] ICCAs – Territories and Areas Conserved by Indigenous Peoples and Local Communities. See www.iccaconsortium.org.

Further Resources

Borrini-Feyerabend, Grazia with Barbara Lassen, Stan Stevens, Gary Martin, Juan Carlos Riascos de la Pena, Ernesto F. Raez Luna and M. Taghi Farvar (2010), *Bio-cultural Diversity Conserved by Indigenous Peoples and Local Communities: Examples and Analysis*. Tehran: ICCA Consortium and Cenesta.

Borrini-Feyerabend, Grazia, Nigel Dudley, Tilman Jaeger, Barbara Lassen, Neema Pathak Broome, Adrian Phillips and Trevor Sandwith (2013), *Governance of Protected Areas: From Understanding to Action*. Gland (Switzerland): IUCN/WCPA.

Borrini-Feyerabend, Grazia and Rosemary Hill (2015), 'Governance for the Conservation of Nature', in *Graeme L. Worboys, Michael Lockwood, Ashish Kothari, Sue Feary and Ian Pulsford (eds), Protected Area Governance and Management*. Canberra: ANU Press and Australian National University.

Borrini-Feyerabend, Grazia and Jessica Campese (2017), *Self-strengthening ICCAs*. ICCA Consortium.

Kothari, Ashish with Colleen Corrigan, Harry Jonas, Aurélie Neumann and Holly Shrumm (eds) (2012), *Recognising and Supporting Territories and Areas Conserved By Indigenous Peoples and Local Communities: Global Overview: Global Overview and National Case*. CBD Technical Series No. 64, Montreal: Secretariat of the Convention on Biological Diversity, ICCA Consortium, Kalpavriksh and Natural Justice.

Kothari, Ashish and Aurelie Neumann (2014), *ICCAs and Aichi Targets*. Policy Brief of the ICCA Consortium, Issue No.1, Tehran: ICCA Consortium, Kalpavriksh, CBD Secretariat, CBD Alliance and Cenesta.

Grazia Borrini-Feyerabend is a co-founder and elected Global Coordinator (2010–present) of the ICCA Consortium, iccaconsortium.org. After developing the Social Policy Programme for the IUCN in the early 1990s, she has been active in the IUCN CEESP and WCPA Commissions ushering and leading the discipline of 'governance for the conservation of nature'. She has worked in five languages in over sixty countries and published 25 volumes.

The late **M. Taghi Farvar** was a co-founder and in 2010 elected President of the ICCA Consortium, www.iccaconsortium.org. Son of a Shahsevan nomadic tribe in the Iranian Azerbaijan, Taghi defended the rights of indigenous people's ancestral domains and promoted understanding of indigenous nomadic tribes as the original conservationists: https://www.iccaconsortium.org/wp-content/uploads/2018/07/Mohammad-Taghi-FARVAR-24-July-2018-1.pdf.

ISLAMIC ETHICS

Nawal Ammar

Keywords: *Qur'an*, interdependence of creation, mercy, *Khalifas*, moderation, *hay'a*

Scholars agree that the source of Islam is the *Qur'an* or Holy Book, which is God's Word transmitted through the Angel Gabriel to Prophet Muhammad. It implies *Sunnah* or the Prophet's traditions; *Hadith* or oral sayings attributed to the Prophet; and *Shari'ah* paths of daily action, including *Fiqh* jurisprudence and *Madahib* Schools of Law. The *Qur'an* is the most authoritative teaching on how God created the order of nature. Yet while nature must be respected as God's creation, it is outside the realm of divinity as such. It is against the prime principle of Islam, *Tawhid* and the One-ness of God to attribute sacredness to nature – animals, humans, and other life forms.

All of God's creation is interdependent, but dependent on God. Humans have been chosen by God to be protectors of the Earth – not because they are superior, but rather as a challenge. They are *Khalifahs* or stewards of the Earth, but not its owners. Faruqi emphasizes this protection as a human destiny; a sacred trust or *ammanah* designed to develop people's moral capacities. The ultimate devotional challenge for Muslims is to pass on from one generation to the next the bounties of the Earth, intact if not improved. Humans are part of the ecological community – and all creation derives from the same element, water. 'There is not an animal (that lives) on Earth. Nor a being that flies on its wings, but (forms a part) of a community like you' (*Qur'an* 6:38). Ecological justice in the *Qur'an* means utilizing Earth resources in a balanced way, as well as actively negating evil and doing good deeds.

Haq (1997:9) argues that Islam 'promulgates what one can call a cosmology of justice' to deal with the human dilemma of protecting, yet using, the Earth. The *Qur'an* addresses issues relating to dignity of the disabled (80:1–9); the rights of orphans (93, entire section; 89:17–18); honesty in exchange and barter (83:1–13); it condemns greed and the hoarding of wealth (100:6–11); calls for feeding the poor (89:17–23); just interaction (11:85); abstention from usury (2:161); distributive justice through taxation (2:267), just leadership (88:21–22, 18:29, 4:58, 5:8, 16:90, 42:15, 38, 49:9, 13); and respect for social differences as God's will

(10:99, 99:18). It is through this cosmology of justice that humans can fulfil their destiny as custodians of the Earth.

Islam does not stress the spiritual at the expense of the material, or vice versa; rather, it brings them into harmony. It is not an ecstatic religion commanding detachment from worldly goods; it encourages engagement in human pleasures. In *Qur'anic* verse, 'Wealth and children are *zinat* (beauty, decoration) of this worldly life' (18:46). In its social outlook, Islam is communitarian rather than individualistic; and as a natalist perspective, it does not favour abstinence. Parts of the *Qur'an* even address family planning. In the moral dimension of obedience and fulfillment of God's covenant, Muslims are expected to adhere to the principle of equilibrium (42:17, 55:7–9, 57:25). This is the theological trial for reaching the Gardens of Heaven – moderation in everything (2:143, 17:110, 25:67).

If Judaism is described as based on a paradigm of fairness, Christianity on love, and Buddhism on non-violence, scholars such as Al Idrissi suggest that Islam's ethical paradigm draws on mercy as an attribute of God. However, the world's economic dependency on oil has demoralized societies of the Middle East – a glaring wealth gap exists between the elites and lower classes; both oil extraction and war are devastating water, soils, and bodies; women are debased without rights; terrorist responses threaten everyday life. While the *Qur'an* speaks of respect for diversity, including for non-Muslims, and of protection of the environment as a devotional duty, many Muslims and Muslim majority leaders now fail to observe this.

For historical reasons, Islamic identity has become fragile. There are over a billion and a half Muslims living in the contemporary world from Australia to Greenland, speaking more than 80 languages. Many of these communities find themselves on the wrong side of post-colonial globalization, modernist development, racism, and demonization. There is no univocal monolithic understanding of Islam, and it should be noted that where social and educational deprivation exists, Muslim practice is often simply ritualistic, as distinct from reflective. Some Muslims even argue that if humans fail to fulfil their eternal destiny, it reflects the predestined will of God.

Other *Qur'anic* verses and Prophetic traditions enjoin their followers to be deliberative actors of good deeds toward the human community and other creatures. Thus usury or *riba* is discouraged, while the ideal tax system or *zakat* is socially redistributive, based on the ability to pay. Among more progressive Islamic notions, the Arabic word *hay'a* denotes an approach of environmental respect or dignified reserve in utilizing nature and in other human behaviours. The term has eco-spiritual potential for activists involved in inter-faith and social movement dialogue. In the *Qur'an* and Hadith, *hay'a* implies respect for women – traditionally rendered invisible under many local interpretations of Islam. This aspect of gender justice resonates, in turn, with ecological justice, wherever the population question is a concern.

Further Resources

Al Faruqi, Ismail R. and Lois Lamya Al Faruqi (1986), *The Cultural Atlas of Islam*. New York: Macmillan.

Al Idrissi, A. (2004), 'The Universality of Mercy in Islam', *Portada*. 273: 1–15.

Ali, Abdullah Yusuf (1989), *Holy Qur'an Translation*. Brentwood, MD: Amana Corporation.

Ammar, Nawal H. (2000), 'An Islamic Response to the Manifest Ecological Crisis: Issues of Justice', in Harold Coward and Daniel C. Maguire (eds), *Visions of a New Earth: Religious Perspectives on Population Consumption, and Ecology*. Albany: State University of New York Press.

Haq, S. Nomanul (2001), 'Islam and Ecology: Toward Retrieval and Reconstruction', in Mary Evelyn Tucker and John A. Grim, *DAEDALUS: Journal of the American Academy of Arts and Sciences*, US: MIT Press, https://www.amacad.org/publication/islam-and-ecology-toward-retrieval-and-reconstruction.

Khaled, Fazlun and Joanne O'Brien (1993), *Islam and Ecology*. New York: Cassell.

Nawal H. Ammar is a Professor of Law and Justice and Dean of the College of Humanities and Social Sciences at Rowan University, N.J., USA. She has published extensively on justice and human rights and her latest work focuses on Muslims in courts and prisons.

JAIN ECOLOGY

Satish Kumar

Keywords: non-violence, restraint, simplicity, reverential ecology

The highest and most profound principle of Jainism is *ahimsa*, the principle of total and comprehensive non-violence of thought, speech, and action. This means non-violence to oneself, to others, and to nature. Of course Jains realize that complete non-violence is not possible but we are required to be mindful of our mental, verbal, and physical activities in order to minimize any harm or damage we may inflict on ourselves or on other living beings. This constant awareness is key to maximizing compassion and minimizing harm. The importance of non-violence is recognized by many religions but is oriented mostly towards other humans, whereas Jains preach and practise non-violence to all living beings including humans.

Jains recognize that earth, air, fire, water, plants, forests, and animals, in other words, the entire natural world is alive. Nature is soulful and intelligent. Therefore, all life is sacred and must be treated with reverence. So, Jain ecology is reverential ecology. Many people believe that somehow human life is superior to non-human life, therefore, we can sacrifice non-human life for our use. That is why the production and consumption of meat is so prevalent and the destruction of rainforests and the overfishing of the oceans are so widely practised. But Jains are required to have equal reverence to human life and non-human life. Therefore, not only the production and consumption of meat and fish is sacrilege for Jains, but they are also required to limit their consumption even of plants. For example, my mother would not eat potatoes or carrots or any other root vegetables because she believed that disturbing the soil and uprooting the plants was a subtle form of violence. We should only take what the plants give us as ripened fruit. She would limit the number of vegetables and fruits she consumed. She would say that practice of non-violence necessitates the practice of restraint.

And through the practice of restraint we make peace with ourselves, peace with people, and peace with nature. Keeping animals in cruel conditions in factory farms, poisoning the soil with chemicals, the destruction of rainforests and the overfishing of the oceans are acts of war against nature. The principle of *ahimsa* necessitates peace with planet Earth.

Life is interdependent and interconnected. As in a family, parents and children, husbands and wives, brothers and sisters take care of each other, in the same way we should treat people of all nations, all religions, races, and colours as our brothers and sisters, and practise compassion towards all. Before we are Americans or Russians, British or French, Indians or Pakistanis, Hindus or Muslims, Christians or Jews, Buddhists or Jains, blacks or whites, we are all humans. We are members of one human family. This sense of the unity of all life goes beyond human life. The birds flying in the sky, animals wandering in the forests and earthworms working under the soil are all our relations and therefore we may not harm them. Our sacred duty is to practise compassion and enhance all life.

Another Jain principle of equal importance is *aparigraha*. It is a very beautiful word but not easy to translate. It means freedom from the bondage of material possessions. It is an ecological principle, of reduction in consumption and minimal accumulation of material possessions. If I can manage with three or four shirts why do I have to have ten or twenty shirts? Why do I need to accumulate a cupboard full of shoes? This goes for every material possession. Jains are required to use material objects to meet their need and not their greed, and free themselves of the burden, worry and anxiety of owning stuff.

The principle of *aparigraha* is just the opposite to the modern idea of the economy where maximization of production and consumption is the driving ideal. Even at the time of religious festivals, shopping and consumption takes priority. People become so obsessed with buying and selling that they are left with no time or very little time for themselves and their spiritual nourishment.

In a consumerist society, the majority of people have no time for poetry, art, or music. No time for family or friends. No time to go for a walk in solitude and appreciate nature, no time to celebrate. This kind of life is the antithesis of *aparigraha*. If we were to restore and rejuvenate the principles of *ahimsa* and *aparigraha*, we would have no ecological crisis, no social injustice, and no exploitation of the weak.

Ahimsa and *aparigraha* emphasize the quality of life over quantity of material possessions. With good care of the Earth all humans can have a good life, good food, good housing, good education, and good medicine. For Jains, the question is not how much you have but is your life good, happy, and fulfilled? Less is more as long as that less is of nourishing and nurturing quality.

As often happens, religions get stuck in traditions, becoming literal and dogmatic. They lose their initial inspiration, and that has happened with Jainism. Possibly most Jains today do not practise *ahimsa* and *aparigraha*. However, there are several radical Jains who are going back to their roots and discovering that practices based on these principles are not only good for themselves personally, but also good for society and planet earth.

For example, a monk called Hitaruchi and his followers in Gujarat (western India) are practising ecological principles of non-violence and restraint in daily life. They avoid the use of plastic altogether, they minimize the use of industrial products in favour of handmade local products. There is a movement initiated by the Terapanthi sect of Jainism in the late 1940s called A Movement of Small Vows (*anuvrat*). They actively promote a simple lifestyle free of greed and corruption and living lightly to minimize any harm to the natural world. This movement has gained a substantial following in India. There are many individual Jains also rediscovering their roots and discarding the idea of development which is wasteful and polluting. But much more of such rediscovery and practice based on it is needed if Jain ecology is to be part of the movements responding to the ecological and social crises we face.

Further Resources
Anuvrat movement, http://www.anuvibha.org/about/.
Kumar, Satish (2002), *You Are, Therefore I Am: A Declaration of Dependence.*
    Cambridge, UK: Green Books.
Terapanth.com, *http://terapanth.com/Anuvart.htm.*

Satish Kumar is the Editor-in-Chief of *Resurgence & Ecologist* magazine, www.resurgence.org, and the founder of Schumacher College, England.

——•••——

JUDAIC TIKKUN OLAM

Rabbi Michael Lerner

Keywords: Judaism, capitalism, Sabbatical Year, Global Marshall Plan, environment

If the ancient wisdom of the Jewish tradition – Tikkun Olam or world repair – was merged with the insights of radical economists and environmentalists today, it could provide a badly needed new foundation for sustainable development.

In Hebrew, Tikkun Olam means transformation and healing the world. Jewish prayer links it with a vision of a world rooted in nurturing 'feminine' energy and caring for the Earth. The core problem we face today is that the globalizing capitalist system needs endless expansion in order to survive. Thus it exploits the limited resources of the planet without regard for the fact that doing so pollutes the planet with poisons and endless garbage, simultaneously robbing future generations of resources recklessly consumed. People whose consciousness has been formed within this paradigm come to believe, and feel that there is 'never enough' and that they actually need more and more things, experiences and opportunities. It is critical to overcome the distorting role of money in social life, politics, and the education system, and to overcome a media industry that is subservient to the wealthy. Democratic processes will never achieve a new consciousness transcending the internal sense of 'not-enoughness' unless our politics is intrinsically tied to a spiritual transformation of society.

Spiritual traditions can foster the inner recognition that there is enough; we are enough; as well as the courage to stop seeking more. Instead, we can focus on how best to share that which we already have in a spirit of generosity. One place to start is to popularize the notion of a New Bottom Line. Corporations, government policies, and the education system might be judged 'efficient, rational or productive' to the extent that they maximize human capacities to be empathetic and generous. We need to be able to respond to the Earth not primarily as a 'resource' but as a living being that evokes awe, wonder, radical amazement, and appreciation.[1]

This raises the question of 'homeland security' and the trillions of dollars in public moneys spent on a strategy of 'power over' others. The domination is exercised at many levels – military, economic, as cultural penetration or diplomacy and by plain bullying. Rather we need a Strategy of Generosity, manifested in a new Global Marshall Plan that avoids the key errors of past development programmes. Such a plan would require that the advanced industrial societies dedicate 1 to 2 per cent of their gross domestic product (GDP) to life-affirming causes each year for the next twenty years. This is all it would take to eliminate – not slowly ameliorate – global poverty, homelessness, hunger, inadequate education, inadequate health care. Meanwhile, two hundred years of environmentally irresponsible growth and development brought about by capitalism and socialism might be repaired.

This kind of Global Marshall Plan differs from all previous plans for aid or development. To get to it, we would need an Environmental and Social Responsibility Amendment to the US Constitution mandating public funding of state and national elections. Corporations selling goods or services in the US would have to prove a satisfactory history of responsibility to a jury of ordinary citizens every five years; and in this they would be aided by environmentalist representatives. Testimony would be required not only from the corporation, but from people around the world affected by its operations, employment policies, ecological, and societal impacts. The new Constitution would also make environmental education compulsory at every educational level from K through college, university, graduate, and professional schools.

All of these steps would be necessary but not sufficient to create a society 'caring for each other and caring for the Earth'. The Torah institutes a practice, which we will need to revive in modern form: the Sabbatical Year. Every seventh year the entire society stops producing goods and focuses instead on celebrating the universe. In our modern context, we would shut down most production of goods, sales, and use of money – introducing a Sabbatical currency distributed equally to every person in society.

In the proposed Sabbatical Year, at least 85 per cent of the population could stop work while retaining vital services such as hospitals, food distribution, and a few others. Meanwhile, for the participating 85 per cent, here is a glimpse of what it could look like. Schools would be closed and students from grades six through high school would work on farms to care for animals, working with the earth to experience closeness to nature through doing. Though some people would mourn the absence of computers, cell phones, automobiles, or new styles of clothing and accessories, most would find themselves free to participate in community decision-making and shaping the next six years when people don't have enough time to engage in planning. Other benefits might be re-training for a different work path or sharing skills; time for play, dance, exercise, meditation, prayer, hiking, swimming, painting, writing poetry, novels, movies, or television shows. In short, a Sabbatical could give us a chance to celebrate the grandeur and mystery of the universe.

The central teaching of the Sabbatical Year would sink in: 'there is enough and you are enough' so you don't always need to do something more, or have more, to have a satisfactory life. With this consciousness, which should spread like wild fire, we have the psycho-spiritual foundation to build another kind of world based on a whole new conception of development. Tikkun Olam is the Jewish way of saying the world needs fixing, and it is our job to fix it. By the tradition of Mishna ethics, it is not necessary to finish this work – but you are not free to ignore it.

Note

[1] For a preliminary picture of what that could look like, see www.tikkun.org/covenant.

Further Resources

Environmental and Social Responsibility Amendment to the U.S. Constitution, www.tikkun.org/esra.

Lerner, Rabbi Michael (2005), 'A Path to a World of Love and Justice', www.tikkun.org/covenant.

Network of Spiritual Progressives, www.spiritualprogressives.org.

The Global Marshall Plan, www.tikkun.org/gmp.

Tikkun magazine, www.tikkun.org/.

Rabbi Michael Lerner is California based and edits *Tikkun* magazine. He is chair of the inter-faith Network of Spiritual Progressives; and author of many books including *Jewish Renewal* (1994); *The Politics of Meaning* (1996); *The Left Hand of God: Taking Back Our Country from the Religious Right* (2006); *Spirit Matters* (2000); *Embracing Israel/Palestine* (2012); *Revolutionary Love* (forthcoming); and with Cornell West – *Jews and Blacks: Let the Healing Begin* (1995).

KAMETSA ASAIKE

Emily Caruso and Juan Pablo Sarmiento Barletti

Keywords: conviviality, extractivism, Latin America, relationality, sociality

This entry describes *kametsa asaike* ('living well together in this place'), an indigenous philosophy for well-being pursued by Ashaninka people from the Peruvian Amazon. We argue that understanding personhood – the culturally inflected perception of the constitution of persons – is fertile ground upon which to grow and sustain practicable, radical alternatives to the dominant development paradigm.

Kametsa asaike has two main characteristics that challenge mainstream understandings of well-being:

1. subjective well-being is only possible through collective well-being, and the collective includes humans, other-than-human beings and the Earth; and
2. it is a deliberate practice – to live well everyone has to work at it.

Kametsa asaike demonstrates how measuring 'subjective wellbeing' in terms of health and/or consumption, all the vogue in conventional development circles, cannot capture the ethics of human sociality and ways of knowing and engaging with the world contained in such a practice. In knowing the world as a network of mutually constituted human and other-than-human actors, *kametsa asaike* implicitly questions the modern notion of the disembedded individual and the nature-culture dualism that underscores the development complex and sanctions large-scale extractivism, regardless of its consequences for life.

Depending on the context, Ashaninka means 'we the kin' or 'we the people'. Most humans, and other-than-human beings such as plants, animals, and spirits, are Ashaninka, that is, social actors. All of these perceive and act in the world in similar ways: they can be kind or mean; they get drunk, make mistakes and need shamans when they are ill; they laugh, cry, love, and fear. This understanding of other-than-humans is common among many indigenous peoples. Yet, this is 'not' an anthropocentric perspective: humans are one of the many different kinds of beings that share personhood.

In these contexts, personhood is not stable or given. From the moment of birth, Ashaninka babies enter into a lifelong process of being continuously 'made' into *ashaninkasanori* ('real Ashaninka people'). The art of making people involves, primarily, sharing substances and close living spaces, eating prescribed foods, and engaging in appropriate behaviour such as working hard, being generous with the product of one's work and sharing in socially constructive emotions. Living as an *ashaninkasanori* also requires following an ethos of conviviality in the relationship with the Earth, including respecting other-than-human beings; caring for the Earth through hard work; abiding by prescriptions regarding where to make gardens, and where, what and when to hunt, fish, and collect plants. In turn, animals, plants, spirits, and the earth provide what people need to live such as *ashaninkasanori*. Thus is established a cycle of interdependency and interconnection that continuously reinforces the 'ashaninkaness' of people and places and allows *kametsa asaike*.

Kametsa asaike comes into stark contrast with the boom of large-scale extractivist projects throughout Peru, which have become the backbone of reconstruction efforts in the wake of the Peruvian internal war (1980–2000). Large sections of Ashaninka traditional territory, the theatre for a particularly violent and enduring episode of the war, have been granted in concession to multinational companies by the Peruvian state for oil/gas extraction and the construction of hydroelectric dams. The continuum of violence, created by the war and followed immediately by extractivism, is experienced by Ashaninka people as a ripping apart of the fragile balance between people and the earth that allows them to live well. The last thirty years has seen the undoing of *ashaninkasanori*-ness and, our Ashaninka collaborators tell us, the Earth is angry. Having suffered the violence of war and of extractivism, she is turning her back on humans. Crops do not grow, trees do not bear fruit, the rivers no longer fill with fish nor the forests with animals, and the spirits that help shamans heal and protect forest animals are gone.

Now, Ashaninka people work tirelessly to expunge the memory of the war and halt the rise in extractive activity on their lands. They seek to rebuild their relationships of solidarity and interconnectedness with the Earth, other-than-humans, and each other, and thus re-establish the conditions for the practice of *kametsa asaike*. With the support of the Central Ashaninka de Rio Ene (CARE), an indigenous organization led by Ruth Buendía,[1] Ashaninka communities in the Ene Valley published a set of principles, based on everyday *kametsa asaike* practices, which they have called their Political Agenda (Central Ashaninka del Rio Ene 2011). The people represented by CARE expect that any individual or institution wishing to engage with them or their territories will abide by these principles. This manifesto had immediate application as CARE launched a series of projects in 2011, with support from international NGOs such as Rainforest Foundation UK. These projects were built upon the requirements for *kametsa asaike* expressed in the agenda. They have also used the agenda as an advocacy tool in their struggles against extractive projects.

While *kametsa asaike* may not be directly applicable in other social contexts, it presents radical solutions for the rebuilding of humanity, our relationship with the Earth, and an approach for transcending the excesses of the Anthropocene. First, it encourages us to examine what it is that makes us people – in relation to other humans and beings – in order to understand what will allow us to live well. Connectedly, it proposes that the pursuit of well-being is necessarily collective: we must recognize our interdependence and relationships with other beings and the earth in order to live well. It also suggests that achieving well-being might only be possible if people are given the tools to be truly human. For those lost in the false solutions of the development complex, in which human differences are obscured by discourses of 'improvement' and 'rationality', discovering what makes us human – and how it roots our constructions of well-being – is likely to be a complex task. However, it may be the only way we can build a more meaningful, respectful, and beautiful world. Finally, it shows us that the only way to sustain well-being, especially when it is under aggression, is to continue to practise it every day, in every big and small way.

Note
[1] http://www.goldmanprize.org/recipient/ruth-buendia/.

Further Resources
Central Ashaninka del Rio Ene (2011), *Kametsa Asaike: El vivir bien de los Ashaninka del Rio Ene. Agenda Política de la CARE*, http://careashaninka.org/wp-content/uploads/2013/01/AgendaKametsaAsaike.pdf.

Central Ashaninka del Rio Ene, http://careashaninka.org/.
Global Environments Network, www.globalenvironments.org,
Kametsa Asaike, el buen vivir de la Amazonía peruana, https://vimeo.com/88115558.
Sarmiento Barletti, Juan Pablo (2016), 'The Angry Earth: Wellbeing, Place, and
    Extractive Development in the Amazon', *Anthropology in Action*. 23 (3): 43–53.
When Two Worlds Collide, www.whentwoworldscollidemovie.com/.

Emily Caruso is the Director of Global Diversity Foundation, www.global-diversity.
org. A biologist and anthropologist by training, she carried out her doctoral fieldwork
among Ene Ashaninka communities. She is passionate about healing and sustaining
the relationships between humans, other-than-humans, and places, considering this to
be the starting point for individual and collective well-being.

Juan Pablo Sarmiento Barletti is a social anthropologist at the Center for International
Forestry Researchin Lima Peru. He has carried out extensive ethnographic research with
Ashaninka people on how they experience and know extractivism, the obvious impact
it has on their everyday lives, and the less obvious impact it has on their relations with
their other-than-human neighbours.

KAWSAK SACHA

Patricia Gualinga

Keywords: rainforest, Yachags, Pachamama, Llakta, Allpamama, Yakumama,
Kawsak Yaku, native people

The living rainforest, *Kawsak Sacha*, is a space where the life of a large
number of diverse beings flows, from the smallest to the largest. They
come from the animal, plant, mineral, and cosmic worlds; their function
is to balance and renovate the emotional, psychological, physical, and
spiritual energy which is a fundamental part of all living beings. It is a
sacred domain that exists in primary forests, in whose waterfalls, lagoons,
swamps, mountains, rivers, and millenary trees, supreme beings of nature
reside and regenerate ecosystems vital to all humans. *Kawsak Sacha* is also a
space where *yachags* (shamans) receive knowledge, where they connect with
the being and knowledge of their ancestors and living places, maintaining
the natural balance of the universe, cultural perpetuity, harmony, and the

continuity of life linked by the invisible threads of the entire universe, not from conventional dualist and monocultural perspectives.

For the native people who inhabit Amazonia, the rainforest is life. The entire world of *Kawsak Sacha* has energy and symbolizes the human spirit. Although the perspective of the living forest is a concept of the Amazon people, the Sarayaku people make its sacredness public, sharing it with other cultures and societies with the intention of contributing a deep knowledge of nature.

Other native peoples of the world have similar concepts, referred to as spirits and not as living beings who – given their purity – inhabit sacred places, such as the primary forests of Amazonia, whose existence is threatened by pollution and extraction. The disappearance of these beings causes environmental conflicts and the disappearance of ecosystems. The difference between forest spirits and forest beings is that while the former do not die, the latter, who are regenerators and guardians of nature, can die and disappear, putting the lives of native peoples at risk.

Each space and each element of the forest has its actors. In each one of these, there are parallel *runa ayllu llaktakuna* (peoples), with humans called *runa*, which also include the places of dwelling and shelter of sacred animals. Each mountain and the larger trees inter-communicate through invisible networks of threads where the *Supay*, or higher beings of the forest, mobilize, and communicate throughout the rainforest. The *kawsak yaku*, or living water, springs everywhere, from the waterfalls that act as entryways that connect lagoons and large rivers where *yakuruna* and *yakumama* (anacondas) circulate, to the great Amazon River. The *yakurana* conserve the bounty of the *icthyofauna*. When the *yakumama* leave their dwellings, the river and lagoons become sterile and deserted.

The *Allapamama* (Mother Earth) gives us everything, protects us, feeds us, and keeps us warm. The Earth and rainforest are what give us the energy and breath of life. From them we receive the wisdom, vision, responsibility, solidarity, commitment, and the emotions that keep humans together, with their families, their loved ones, and what they expect in the future as a result of their effort and ways of living.

Through the transmissions of *muskuy* (visions and dreams), the human spirit, which dwells in the *Kawsak Sacha*, receives the energy that symbolizes human life as much for its strength as for its grandeur and inner thoughts. Here, soul and life are one with the *pachamama*, making up part of our upbringing from the beginning of conception, so that we become part of the *runa kawsay*.

Regarding the *runakuna kawsay* (the life of a people), they have the following principles: to maintain and sustain community life, such as the family unit; to have a well-organized governing institution with participatory, collective decision-making which includes women, young people, children and the elderly; to practise, transmit, and perpetuate traditional cultural and spiritual knowledge with complete freedom; to have adequate human settlements as well as cultural infrastructures appropriate to the Kichwa; and to assure the population's food sovereignty.

The above implies the development of their own economy that is not based on the accumulation of goods, but on solidarity and reciprocity which strengthens sustainable production with appropriate Kichwa practices and technologies, always respecting nature's balance. This economy finds expression in strategies of collective labour and, in fishing communities, in the bartering of products and services, although currently they also involve planning to implement new ways of thinking about the economy.

Sumak Allpa (Healthy Earth) invites us to enjoy a healthy planet with a bio-diverse, fragile, unpolluted ecosystem while caring for and maintaining a level of consciousness regarding the forest's abundance of flora and fauna, conserving and respecting the sacred, living sites. At the same time, one would preserve the integrity of the territory through rules of management and use based on Kichwa people's conservation laws, through territorial zoning established by the people.

The fundamental principle of the *sacha runa yachay* (Kichwa knowledge of the forest peoples) is to perpetuate our ancestral knowledge, cultural and spiritual practices that allow us to continue to carry out our own traditional healing practices and the use of medicinal plants, together with our Yachags or wise men.

We value the resourcefulness of our traditional architecture, in harmony with the reality of our forest surroundings, and practise the art of ceramics and the construction of hunting instruments and fishing. Music, sacred chants, drumming, and dancing form an inseparable part of the *sacha runa yachay*. This is achieved by the transmission of our own techniques of preparation and use of soil for farming as well as techniques for hunting and fishing and fruit gathering. It is necessary to develop education with a human face and an open vision that values our own knowledge as a point of departure, and that respectfully interrelates with other cultures. We have to maintain solidarity among the Ayllus families, establishing relationships based on collective labour; we need to conserve the balance and relationship between human beings and nature, perpetuate the concept of a forest that is alive and living, with all its participants: its true owners and masters, as well as its ultimate representatives, the *yachags* or wise persons, true scientists with the highest understanding of nature and its beings.

The life and the cosmovision of the Sarayaku people are an integral, fundamental part of the knowledge necessary to continue to maintain our identity asa Kichwa People.

Patricia Gualinga, born in Sayaraku on the 21 September 1969, is a Kechwa native. She holds a baccalaureate degree in business, and studies Human Rights litigation, environmental management and communication. She is a consultant on Amazon matters, and Regional Manager for the Ministry of Tourism in Amazonia. She was formerly the Director of Women and International Relations for the Pueblo Sarayaku in Ecuador.

KYOSEI

Motoi Fuse

Keywords: *kyosei*, symbiosis, humanity–nature relations, conviviality

The word *kyosei* is used in the Japanese vernacular to mean symbiosis, conviviality, or living together. It can be used to describe relations between the sexes, different cultures, handicapped and non-handicapped people, humans and animals, humans and nature, and so on. *Kyosei* always covers human-to-human relations and humanity–nature relations. From the latter half of the twentieth century, it has been used to deal with ecological and social problems integrally. But used as a social ideal, *kyosei* attracted advocates both inside academia and out, subsequently introducing a range of meanings, even ambiguities. Thus Japanese political parties have applied the concept in both Right- and Left-leaning ways.

Given these circumstances, an Association for *Kyosei* Studies, aiming for transdisciplinary inquiry into social systems, was established in 2006 in Japan. The prospectus sets out to clarify concepts of *kyosei* and grounds these substantively in the real world. This association found a common denominator beneath the various understandings of *kyosei*. It established the fact that in general, *kyosei* aims to enhance equality and sustainability by respecting the heterogeneity of languages, cultures and climates.

The contemporary Japanese philosopher Shuji Ozeki (2015) has devised a framework for categorizing *kyosei* under three heads: 'sanctuary', 'competition', and 'communality'. The *kyosei* of sanctuary is pre-modern in orientation and represented in the ideas of architect, Kisho Kurokawa. It authorizes sanctuaries to protect and support conservative traditional societies and communities. The *kyosei* of competition is modernist in orientation and characterized by the ideas of legal philosopher Tatsuo Inoue. The principle of competition promotes heterogeneity, individualization, and denies communality. These two forms of *kyosei* are antithetical.

The third variety – *kyosei* of communality – is postmodern and expressed in the ideas of philosopher Kohei Hanazaki, ethicist Takashi Kawamoto, and theatrical director Toshiharu Takeuchi. It contrasts communalization, cooperation, and solidarity with the principle of competition as found in market fundamentalism and its social practices. Communality here is not simply traditional but emphasizes the needs and views of socially weaker and vulnerable people, exposing elements of inequality and subordination concealed in some usages of the word *kyosei*. The Hokkaido-based philosopher and activist Hanazaki has used *kyosei* to advocate rights for the indigenous Ainu people of Japan.

Concerning the third and communal type of *kyosei*, it must be added that there is another Japanese ideal of 'communality' called *kyodo*. However, whereas in communality, values, norms, and aims are shared, *kyosei* highlights the positive aspects of living together and experiences of mutual revitalization across differences. This *kyosei* respecting heterogeneity is counterposed to both the traditional communality of homogenization and the modernist struggle for existence through the market system. The third version of *kyosei* sublates the first two forms. It accepts conflict and rivalry as historical moments.

Turning to *kyosei* and the humanity–nature relation: as the ecological crisis becomes ever more serious, conventional ways of looking at this relation have been reconsidered, and currently, preference is being given to building sustainability in accord with the laws of nature. As an alternative orientation, *kyosei* has been applied in agricultural contexts by Ozeki, focusing on labour which mediates humanity and nature and activates the metabolism between them. In environmental theory, the conflict between anthropocentrism and ecocentrism can be overcome by applying the integral logic of *kyosei*. Here, both humanity and nature are regarded as subjects.

Further Resources

Association for *Kyosei* Studies, http://www.kyosei-gakkai.jp/.

Hanazaki, Kohei (2001), *Identity to Kyosei no Tetsugaku*. Tokyo: Heibonsha.

Murakami, Yoichiro, Noriko Kawamura and Shin Chiba (eds) (2005), *Toward a Peaceable Future: Redefining Peace, Security, and Kyosei from a Multidisciplinary Perspective*. Pullman: Washington State University Press.

Ozeki, Shuji (2015), *Tagenteki Kyosei Shakaiga Miraiwo Hiraku*. Tokyo: Agriculture and Forestry Statistics Publishing Inc.

Ozeki, Shuji and Yoshio Yaguchi, Sumio Kameyama and Koshin Kimura (eds) (2016), *Kyosei Shakai I*. Tokyo: Norin Tokei Shuppan.

Motoi Fuse was born in 1981 and gained his PhD from Tokyo University of Agriculture and Technology (TUAT) in 2011. He lectures at Tokyo University as well as at Tokyo Kasei University and Musashino University, Japan. He has published on environmental philosophy in the Japan based *Journal of Environmental Thought and Education*.

LATIN AMERICAN AND CARIBBEAN FEMINISMS

Betty Ruth Lozano Lerma

Keywords: Latin American feminisms, eurocentrism, modernity, decoloniality

'Latin American and Caribbean feminisms' have become a great transformative force working against patriarchal relations in the Continent, as women interrogate what it means to 'construct themselves' as such. Since the end of the nineteenth century, the profound quest of women to transform their societies manifested itself through socialist and anarchist women's movements rebelling against capitalism and patriarchy, Church coercion, repression by the State and in the family. Francesca Gargallo writes in *Las ideas feministas* that such movements appropriated the slogan: 'No god, no boss, no husband' and created others, such as 'Democracy in the country, at home, and in bed', coined by Chileans Julieta Kirkwood and Margarita Pizano. According to Yuderkis Espinosa and collaborators in their major volume on decolonial feminisms, *Tejiendo de otro modo* (2014), the feminist agenda included discursive contributions and political claims for the right to vote, education, and equality before the law, configuring a modernizing, liberal eurocentrism feminism in the first decades of the twentieth century. Since then, however, the history of Latin American and Caribbean feminisms has seen the confrontation of diverse feminist expressions with liberal feminism; this confrontation is deepening today with the powerful emergence of multiple other feminisms.

To speak of feminism in Latin America and the Caribbean, it is imperative to recognize indigenous, black, *mestizo*, and peasant women's struggles against submission since the Conquest in 1492 and colonial times. Hundreds of enslaved black women from *cimarrón* or maroon communities prosecuted their masters. Women of all ethnicities and backgrounds participated in pro-independence campaigns against colonial rule, in the construction of nation-states, and in twentieth-century revolutionary struggles, without necessarily transforming their subordinate situation.

Although feminism is part of modern critical thought, various expressions of Latin American feminism question modernity by posing an epistemic detachment from European knowledge, in order to think about the complex hierarchical relations of domination that are interwoven in 'the colonial matrix of power'. Thus, black, indigenous, working class, and lesbian women question the 'universal woman' of modern liberal feminism – Western, hegemonic, capitalist, bourgeois, and white – because such a concept is devoid of reflection on the inequalities that exist between women over heteronormativity and colonialism. In other words, from these other perspectives, liberal feminism responds to the historical ontology and episteme of the US and Europe rather than of the Americas.

The reconfiguration of Latin American feminism comes from a variety of women's spaces: artistic collectives with feminist agendas, lesbians, activists of an 'Other' feminism, popular feminisms, Bolivian communitarian feminism and the Xica Guatemalan community, autonomous feminism, black and/or anti-racist feminism, decolonial feminism, feminist liberation theology, ecofeminism, and social movements in defence of territory and Mother Earth led by black, indigenous, and peasant women. Many go beyond the criticism of western modernity and development in setting out to overcome capitalist patriarchal modernity. They consider 'the defence of territory' as a place to be and continue being women and people, as essential to their struggles.

Although Latin American and Caribbean feminisms have drawn from European and American feminist sources, their unique context and content reconfigures feminist theory by problematizing the concepts of gender, patriarchy, development, coloniality, and by incorporating raciality to the analysis of power. They challenge the notion of gender from the perspective of

Amerindian and Afro-descendant epistemes, questioning the liberal feminist view that interprets ancestral cultures as oppressive in order to propose Westernization as a path to gender empowerment. Decolonial feminisms state that gender and heterosexuality must be understood historically. They affirm an indissoluble interweaving of multiple oppressions, variously named as intersectionality, matrix of domination, fusion, co-constitution, overcrowding or *sobrecruzamiento*, and multiplicity of oppressions. Black and indigenous women use other categories in addition to race, gender, class, and sexuality – these include 'history' or 'long memory', 'territory', and 'collective rights'.

Latin American feminisms question both Western patriarchy and the subordination of non-heterosexual women and persons within indigenous and Afro-descendant cultures. They affirm the existence of pre-Hispanic patriarchies, giving rise to concepts such as 'ancestral original patriarchy' and 'low-intensity patriarchy', which show how women within the colonial context experienced an entanglement of patriarchies – *entronque de patriarcados* – and, for the Afro-descendent case, a 'black-colonial patriarchy', as I show in *Tejiendo con retazos*.

These feminisms thus propose the depatriarchalization of ancestral legacies and worldviews, the *cosmovisiones* that subordinate the 'feminine' *Pachamama* to the heavenly bodies of the 'masculine' universe. They question the ethnic essentialisms that do not object to the heterosexual norm. They emphasize the need to solve inequalities from within each culture, by drawing on the multiplicity of stories, subjects, and experiences making up the pluriverse, those that transcend the abstract universalism of Western culture and the modern eurocentric vision.

Women's movements in Latin America and the Caribbean are radically transforming their societies. Feminist activists advocate the simultaneous decolonization and depatriarchalization of society by interweaving the communal, the environmental and the spiritual. They contribute to create other worlds from ontologies that claim life beyond development. Through their multiple practices, they are building the post-patriarchal societies of the present.

Further Resources

Asociación La Cuerda de Guatemala, http://www.lacuerdaguatemala.org/.
Communitarian Feminism in Bolivia,
    https://www.youtube.com/watch?v=C6l2BnFCsyk.
Espinosa, Yuderkis, Diana Gómez and Karina Ochoa (eds) (2014), *Tejiendo de otro modo: Feminismo, epistemología y apuestas decoloniales en AbyaYala*. Popayán: Universidad del Cauca.

Gargallo, Francesca (2006), *Las ideas feministas latinoamericanas.* México: Universidad Autónoma de la Ciudad de México.

Lozano, Betty Ruth (2016), *Tejiendo con retazos de memorias insurgencias epistémicas de mujeres negras/afrocolombianas. Aportes a un feminismo negro decolonial.* PhD dissertation in Latin American Cultural Studies, Quito, Ecuador: Universidad Andina Simón Bolívar.

Sylvia Marcos's blog, https://sylviamarcos.wordpress.com/.

Betty Ruth Lozano Lerma holds an undergraduate degree in sociology and Masters in philosophy, both from the Universidad del Valle in Cali, Colombia, and a PhD in Latin American Cultural Studies from the Universidad Andina Simón Bolívar in Quito. She is an activist in black and decolonial feminism, and currently Research Director at the Fundación Universitaria Bautista at Cali.

LIBERATION THEOLOGY

Elina Vuola

Keywords: Latin American black, indigenous, and feminist theologies; spirit-body dualism; preferential option for the poor; Pope Francis

Liberation theology can be defined either narrowly or broadly. In the former sense, it is limited to the Latin American liberation theology (*teología de la liberación, teologia da libertação*), born out of a specifically Latin American context in the late 1960s. In the broader sense, liberation theology also includes black theology, feminist theology, and variations of Asian and African liberation theologies. The broad definition stresses the inter-relatedness of different structures of oppression and domination. Liberation necessarily involves political, economic, social, racial, ethnic, and sexual aspects. I will concentrate on Latin American liberation theology, which today includes all the above-mentioned variants, such as Latin American black, feminist, and indigenous theologies. It is possible to see all of them as the latest forms of liberation theology, even though most of them take critical distance from earlier liberation theology. However, methodologically and epistemologically they can be understood as part of the legacy of liberation theology.

An influential theory behind early liberation theology was dependency theory, according to which the main reason for the poverty and 'underdevelopment' of the Third World (periphery) was its dependency on industrialized countries (centre). The metropolitan centre 'developed' precisely because of its exploitation of the dependent peripheral regions. Liberation theology was also a radicalization and re-contextualization of European political theology and the prophetic denunciation of injustice and oppression in Christianity.

In the 1970s, liberation theology, the *iglesia popular* (popular church) – part of the Catholic Church and of some Protestant churches – were at the forefront of an emerging civil society in Latin America. Liberation theology offered the language, the rationale, and the legitimization for Christians to question economic policies that left large parts of society in poverty, and for direct opposition to military regimes and their egregious human rights violations. Liberation theology changed the historical role of the Catholic Church in Latin America.

The major accomplishments of liberation theology are, first, the creation of a radical Christian alternative from the global South after centuries of colonialism and evangelization by the global North; second, the critique from the impoverished South presented a prophetic voice within global Christianity that appealed to its tradition of seeing poverty and injustice as sins; third, it provided a new methodology for contemporary theology to consider issues of human suffering, (in)justice, and bodily realities as central.

Theologians such as Gustavo Gutiérrez, Leonardo Boff, Enrique Dussel, and Pablo Richard, among others, were influential in the early days; their work constituted the body of classical liberation theology. Quite early, there were also Protestant and feminist liberation theologians. Thus, it was never exclusively a Catholic and masculine enterprise.

Classical liberation theology claimed the poor to be its starting point in its critical reflection on the role of the churches in areas colonized by Christian Europe – what was called the preferential option for the poor. This starting point has been further developed since then. The category of the poor has been expanded and specified, especially due to the argument that poverty and the poor cannot be conceived merely in economic or material terms. Women, indigenous people, and Afro-Latin Americans have criticized liberation theology for having too narrow a view of poverty and the poor. Issues of racism and sexism have not been central in liberation theology, nor have they been elaborated into a deeper and more nuanced understanding of how poverty affects different people differently.

Today's liberation theology has failed to incorporate the social and political concerns of contemporary social movements, as earlier liberation theology did in the 1960s and 1970s. For example, political alliances between environmental, indigenous, feminist and LGBT rights movements, and liberation theological sectors of the churches are practically non-existent. Liberation theologians have not dealt effectively with these movements' practical and theoretical critique demanding concrete political changes at the intersections of race, gender and sexuality, class, ethnicity, race, and ecology. Issues related to sexuality and reproduction have been lacking in liberation theology since its beginning.

In addition, the distance between Christian liberation eco-theologians and indigenous cultures may reflect a fear of what for centuries has been perceived as syncretism and non-Christian ('pagan') beliefs. The rich diversity of Latin America's religious landscape is alive not only in formal religious institutions such as churches, but also in Africa-based religions such as *candomblé, umbanda, voudou, santería,* indigenous spiritual traditions and in different forms of popular Catholicism. In them, the body – rituals healing dance; the sacrality of life – land and water; everyday life – relationships, prayers, healing, votive offerings; these things are central. So too, emphasis is on the senses, concreteness, visuality, and often a space for women's voices.

In its assessment of the body, gender, nature, and indigenous cultures, liberation theology thus relies on unquestioned binaries of Christian theology and eurocentric philosophy despite its insistence to the contrary. The spirit–body dichotomy, as so many feminist theologians have pointed out, goes hand in hand with the male–female dichotomy, 'the West and the rest dualism', and the culture–nature dualism. The demonization of women and indigenous people (as classical and colonial theology did) or their romanticization, as much of eco-theology does, without taking seriously the real problems they are facing as bodily human beings, are problematic.

A new theology of the body and sexuality from a Latin American liberation theological perspective is attempting to bridge some of the gaps of analysis concerning poverty, racism, sexism, and ecology. This kind of work is today transnational and interdisciplinary, and goes well beyond Latin American liberation theology. In this sense, liberation theology as a framework does not have the same relevance as it did some decades ago. That, however, does not mean that it is or has been meaningless, quite to the contrary.

Liberation theology, including its feminist, eco and indigenous versions, has been bridging the legacy of Christianity in Latin America with the region's traumatic history, speaking with a truly Latin American voice on global issues. However, to some extent, liberation theology has remained

European, white and male, because of its lack of solidarity with post-dictatorship social movements. Understanding development and post-development from the perspective of the poorest of Latin Americans would entail a more substantial critique of racism, sexism, and colonialism as they are related to religion.

The first Latin American pope, Francis from Argentina, never had a close relationship with liberation theology, but has made some substantial changes in the Vatican's assessment of it, differing from his two predecessors.

Further Resources

Althaus-Reid, Marcella (ed.) (2006), *Liberation Theology and Sexuality: New Radicalism from Latin America*. London: Ashgate.

Boff, Leonardo (1997), *Cry of the Earth, Cry of the Poor*. New York: Orbis Books.

Gebara, Ivone (1999), *Longing for Running Water: Ecofeminism and Liberation*. Minneapolis: Fortress Press.

Vuola, Elina (2002), *Limits of Liberation: Feminist Theology and the Ethics of Poverty and Reproduction*. Sheffield and New York: Sheffield Academic Press and Continuum.

——— (2011), 'Latin American Liberation Theologians' Turn to Eco(theo)logy: Critical Remarks', in Celia Deane-Drummond and Heinrich Bedford-Strohm (eds), *Religion and Ecology in the Public Sphere*. London: T & T Clark/Continuum.

Elina Vuola is Academy Professor at the Faculty of Theology, University of Helsinki, Finland. She has been a visiting scholar at the Departamento Ecuménico de Investigaciones in San José, Costa Rica; at Harvard Divinity School; and Northwestern University, Evanston, USA.

———

LIFE PROJECTS

Mario Blaser

Keywords: good life, place-based collectives, pluriverse, ways of worlding.

The concept of Life Projects takes inspiration from a contrast drawn by Yshiro indigenous intellectuals of Paraguay between the 'good (modern) life' offered to them by 'development projects' and their own notions of 'good life' emerging from their own experiences 'in their place' (Blaser 2010). The key difference is in orientation, while development is oriented towards extending as universally valid its own vision of a good life premised on the primacy of the human, the Yshiro Life Projects are not only keenly aware of the specificity of the versions of a 'good life' they advance, but also see it as crucial for survival not to trample over other versions! Thus, Life Projects are oriented towards sustaining the heterogeneity of visions of a 'good life.' In drawing this distinction in orientation, Yshiro intellectuals hint at an emergent political possibility that counters the development mission of making One World (Sachs 1992). That is, they point at the pluriverse.

Borrowing the term from the Yshiro, Life Projects is thus a concept that contribute to the realization of the pluriverse by rendering visible and plausible the practices of a good life of the various place-based collectives that exist on the planet. Place-based collectives might sometimes be associated with 'indigenous peoples' but the terms are not synonyms. Place-based collectives are not the same as 'cultural communities' that live in natural places or 'territories'; rather they are very specific assemblages that 'take place' in specific locations, a change of location would make them something else. A place-based collective can be described as a network of (human and non-human) persons – i.e., entities endowed with their own dignity, will, and purpose – who are entangled through social and even familial bonds. It is not rare to hear place-based collective spokespersons refer to rivers, mountains, forests, or animals as 'grandfather, brother, spirit owner', and so on. These are not metaphors or 'beliefs,' rather the terms reflect the fact that what modern institutions would treat as territories composed of 'resources' and 'people,' Life Projects will treat as complex relational assemblages of humans and non-human persons.

This approach towards the entities that compose a place-based collective builds on a series of assumptions. These are often expressed in origin stories and ceremonies shared by many traditions of thought and practice across the Americas (Cajete 2000). For example:

- existence entails the unfolding of a creative force or principle ('the story of life') ungraspable in its full magnitude
- all existing entities are shaped out of this unfolding
- it is through their specific configurations, ways of being, and reciprocal relations that entities contribute to the story of life
- carelessly interfering with the trajectories of entities and abusing the relations that sustain the unfolding. For instance, when one party of the relation, usually the human part, takes advantage of another without regards for balance, this invariably has negative consequences for the quality of the story of life.

These assumptions speak to the relationality and inter-dependence that generate place-based collectives and imply that it is very difficult to privilege the needs of some of their components – at the expense of others without risking that the whole come undone. For example, from the perspective of a place-based collective to be offered a 'compensation' for the future damming of a river might feel like someone coming to you and saying 'we will kill your grandfather, but don't worry we will compensate you. Let's just figure out how much money is equivalent to the food he could give you, and the entertainment time he could provide' and assuming that you will be fine afterwards. This is why the Life Projects that emerge from place-based collectives are often at odds with development's assumptions about the primacy of the human and the universal validity of its vision of a good life. In fact, given their inherent heterogeneity, Life Projects are antithetical to a universe; they contribute to bring about a pluriverse. To cite the Zapatista: 'a world in which many worlds fit.'

The pluriverse can thus be seen as a collective place-based political proposal; one that is badly needed when facing the urgent planetary crisis evoked by the label Anthropocene! The Anthropocene is customarily presented as a side-effect of development, the result of 'humanity' having acted without full knowledge of the consequences. This implies that more and better knowledge is the way to solve it. Hence, the efforts to integrate Natural and Social Sciences in programs like the Earth System Governance Project. But if we transpose place-based collectives' assumptions regarding the proper relations between entities to the relations between different 'collectives' (place-based or not), the Anthropocene appears in a different light. It appears as the result of a practice (development) that conceives itself as universally valid and placeless and has no regards for the existence of place-based collectives. In this light, the mainstream 'solutions' to the Anthropocene are as universalist and unconcerned with place-based collectives as development. This point is made evident by growing conflicts between large 'green energy' projects for wind or solar versus local communities.

Thus, rather than universalist solutions, what the present crisis might require is the unleashing of heterogeneous practices for the good life associated with place-based collectives. Something of the sort is being pursued by many indigenous and Afro descendant peoples in the Americas who, even if not always without contradictions, are intent on pursuing their Life Projects. Refusing to propose universal visions, Life Projects point at the fact that the paradoxical common task of realizing the pluriverse requires from place-based collectives that they find their own unique ways of 'going on in difference together' with the other collectives they are entangled with (Verran 2013).

Further Resources
Blaser, Mario (2010), *Storytelling Globalization from the Chaco and Beyond*. Durham: Duke University Press.
Cajete, Gregory (2000), *Native Science: Natural Laws of Interdependence*. New Mexico: Clear Light Publishing.
Sachs, Wolfgang (1992), 'One World', in Wolfgang Sachs (ed.), *The Development Dictionary*. London: Zed Books.
Verran, Helen (2013), 'Engagements between Disparate Knowledge Traditions: Toward Doing Difference Generatively and in Good Faith', in Lesley Green (ed.), *Contested ecologies: Dialogues in the South on Nature and Knowledge*. Cape Town: HSRC Press.
Life Projects Network, https://www.lifeprovida.net.

Mario Blaser is an Argentinian-Canadian anthropologist and the Canada Research Chair in Aboriginal Studies at Memorial University of Newfoundland. He is the author of *Storytelling Globalization from the Paraguayan Chaco and Beyond* (Duke University Press, 2010) and co-editor of *Indigenous Peoples and Autonomy: Insights for the Global Age* (University of British Columbia Press, 2010), and *In the Way of Development: Indigenous peoples, Life Projects and Globalization* (Zed Books, 2004).

MEDITERRANEANISM

Onofrio Romano

Keywords: Meridianism, Mediterranean, Western culture

The idea of Mediterraneanism identifies and translates the historically and geographically specific logic of living and coexistence in the Mediterranean into a consistent cultural, political, and even ethical framework. As a vision, it stands as a systemic alternative or, more precisely, it recovers and radicalizes the alleged 'original' roots of the West, setting them against the perverse drifts of current Western civilization itself.

What is the Mediterranean? – Braudel (1985) asks – 'A thousand things

together. Not a landscape, but countless landscapes. Not a sea, but a succession
of seas. Not a civilization, but a series of civilizations stacked on each other
[. . .]. Traveling in the Mediterranean [. . .] means meeting ancient realities, still
alive, side by side with ultra-modernity.

This representation of the Mediterranean reality immediately inspires
Mediterraneanism – also called Meridian thinking – as a political ideal.
It coincides with a world in which it is possible for multiple cultures,
even belonging to different stages of civilization, to live together copying,
overlapping, affecting, and altering each other. So, before any specific cultural
identity or social model, Mediterraneanism alludes first and foremost to
multiplicity as a value in itself. The accidental historical coexistence of
multiple ways of living in a single basin becomes the deliberate design of a
political horizon of conviviality. The mutual acceptance and appreciation
of differences follows a general logic by which any culture tries to draw
what it lacks from the experiences carried on by other cultures. In this
sense, Meridianism opposes both universalism as the discovering of a
single humanity beyond any cultural crust, and communitarian nostalgia
responding with cultural seclusion to the anomic drifts of universalism.

While the Mediterranean area has been a constant subject of investigation
and reflection for social scientists – historians, anthropologists, economists,
sociologists – Mediterraneanism reaches its most accomplished and
systematic formulation in the mid-nineties of the last century, thanks to
the Italian philosopher and sociologist, Franco Cassano (2012). Conceived
within the cultural climate of post-modernism, post-colonialism, anti-
utilitarianism, the book is particularly inspired by the critique of the world's
Westernization proposed by Serge Latouche. According to Cassano, the
physical–geographical configuration of the Mediterranean is embodied
in the idea of Meridianism and its etymology of 'mediating the lands'.
A particular complicity between land and sea is staged here. The sea is a
constant presence for the people who live around the Mediterranean,
together with the awareness that beyond it they will find other lands, other
people, cultures, and diverse ways of life. Land becomes a general metaphor
of identity and rootedness; the sea, on the other hand, becomes a metaphor
of emancipation, liberty, escaping from the self and opening to the other.
Each one, by itself, runs into risks: land, without the presence of the sea,
will experience identitarian seclusion, refusal of the other and despotism.
Exclusively choosing the sea, however, might mean exposing oneself to
the vacuum of the ocean where all meanings are erased and all differences

flattened and reduced to a universalistic abstraction, under the exclusive domain of technique.

The two tendencies can be found in the biographies of Heidegger and Nietzsche. Against these two specular drifts, Mediterraneanism is connected with *phronesis* or measured wisdom – as the cultural attitude that allows for the coexistence of roots and emancipation, the sense of belonging and liberty, tradition and modernity, sense and sensibility. A measure without pacification, not aspiring to a synthesis, according to the Greek rhetorical tradition of the *dissòilógoi*, that is, the 'divergent discourses' that never melt in the uniqueness of modern *logos* or rational discourse.

Meridianism interprets the spreading of religious radicalization in our times as a reaction to the Western fundamentalism of growth and modernization – a form of *hubris* – exhibition of power and arrogance. In order to hold at bay every form of radicalization, we have to learn from the historical experience of cohabitation in the Mediterranean. It houses three monotheistic religions and multiple cultures belonging to three different continent). The Homeric character of Ulysses becomes the anthropological reference model for Meridianism: during his amazing circumnavigation of the Mediterranean, with its multiple worlds, Ulysses never loses the nostalgia for home, where he finally comes back. *Nostos* (the return) is thus highlighted as the key virtue. The desire to meet the Other is reconciled with the love for one's homeland.

Mediterraneanism was born in an academic environment. It mainly spread through the cultural debates in Southern Europe, particularly in Italy and France. The collective work entitled *The Mediterranean Alternative* (Cassano and Zolo 2007) represents an attempt to give the new *koinè* (common language) an accomplished cognitive and political form, bringing together in this debate prominent intellectuals from the two shores of the Mediterranean basin.

Meridianism has not been translated into a specific political movement, but it has surely inspired many experiences of social and cultural cooperation between different expressions of civil society in the Mediterranean countries. A considerable political impact of Mediterraneanism can be seen in the Italian 'Mezzogiorno' of southern Italy, where it has completely reformulated the issue of the South's development lag, inspiring a sort of civic Renaissance translated into many experiences of local government in the last decades.

Meridianism is nowadays in crisis. The rekindled conflicts and turbulence in the Mediterranean, after the Arab Spring, have affected the possibility of recognizing the area as a source of inspiration for social alternatives. Moreover, the core principles of Mediterraneanism have proven ineffective for the design of real political alternatives to Western modernity. Nevertheless, hope is not lost: the inability of the Mediterranean countries to meet the current standards of economic efficiency – they cannot compete either on technological innovation or on the cost of labour for structural reasons, makes them a favourable place of experimentation of 'delinking' and self-sufficient economies.

Further Resources

Braudel, Fernand (1985), *La Méditerranée. L'espace et l'histoire*. Paris: Flammarion.
Cassano, Franco (2012), *Southern Thought and Other Essays on the Mediterranean*. New York: Fordham University Press.
Cassano, Franco and Danilo Zolo (eds) (2007), *L'alternativa mediterranea*. Milano: Feltrinelli.
Il Militante Ignoto, http://www.ilmilitanteignoto.it.
Jura Gentium, http://www.juragentium.eu.

Onofrio Romano is Associate Professor of Sociology at the Department of Political Sciences, University of Bari, Italy. His research fields include social theory, modernity and post-modernity, the Mediterranean, degrowth, and anti-utilitarianism. Among his recent works is *The Sociology of Knowledge in a Time of Crisis* (Routledge, 2014).

MINOBIMAATISIIWIN

Deborah McGregor

Keywords: well-being, health, good life, Anishinaabe and Cree cultures, Earth rights

Minobimaatisiiwin (*m'nobi-MAH-t'see-win*), known in various dialects as *miyupimaatisiiun*, *bimaadiziiwin*, *pimatisiwin*, *mnaadmodzawin*, and *mino-pimatisiwin*, is a concept rooted in Anishinaabe and Cree cultures that conveys the idea of 'living a good life' or living in a 'total state of wellbeing' (King 2013). Though this practice of 'living well' has been in existence for thousands of years, its viability has been compromised through the devastating forces of colonial oppression and globalization that have undermined indigenous life in North America in every conceivable way. In recent decades, *minobimaatisiiwin* has emerged as part of a revitalization of indigenous healing systems. This re-emergence is in direct resistance to continued pressures stemming from these forces. Anishinaabe activist Winona LaDuke (1997) introduced the concept to environmental discourse as a response to environmental colonialism, racism, and injustice.

Minobimaatisiiwin is embedded with a holistic world view, and thus involves living on respectful and reciprocal terms with all of Creation at individual and collective levels. Thus, achieving *minobimaatisiiwin* is not possible without balanced and harmonious relationships with other beings. Reciprocal relations are required not only among people, but with all other 'relatives' as well – animals; plants; rocks; water; spirits; celestial beings such as the moon, sun, and stars; ancestors; and those yet to come. At the same time, all other beings and entities have to achieve *minobimaatisiiwin* in order to be healthy as well. The overall goal of sustaining life for all of Creation is a 'mutual' endeavour.

It was the ancestors' vision that their descendants would live according to *minobimaatisiiwin* and establish loving and caring relationships with the Earth and other beings in so doing. Their decisions were based on ensuring the well-being of future generations of people, as well as all of Creation. The concept of 'the good life', or 'living well', is guided by the seven original teachings – wisdom, love, respect, bravery, honesty, humility, and truth – to ensure balanced relationships among people, and with all of Creation. It is not possible to 'live well' if the Earth continues to suffer. In this theory and practice of relationships, humanity is obliged to care for its relatives, as they are obliged to care for us in reciprocity. *Minobimaatisiiwin* requires one to act sustainably: to take responsibility for and be spiritually connected to all of Creation, all of the time.

This way of living was supported by indigenous knowledge systems, principles, and laws that ensured that people's activities would affirm life, rather than denigrate or destroy it. Indigenous laws, based on harmonious ways of being, recognize the 'rights of the Earth' through a covenant of duties, obligations, and responsibilities (McGregor 2015). Anishinaabe law requires that people must cooperate with all beings of Creation. It is meant to allow for good relations and ultimately for each living being to achieve *minobimaatisiiwin*. It pertains to relationships among human beings as well as to the awesome responsibilities of co-existence with members of the other orders (King 2013: 5). Furthermore, *minobimaatisiiwin* recognizes that other beings or entities have their own laws that must be followed to ensure harmonious relations with Creation. These are natural laws. Enacting and living natural law requires vast knowledge of the environment and how it functions in ensuring survival for all.

At the present time, the Earth is continually treated in a manner that is diametrically opposed to the philosophy and enactment of *minobimaatisiiwin*. In Anishinaabe ontology, all elements of Creation are imbued with spirit and agency, including, for example, non-human life forms, rocks, mountains, water, and the Earth itself. Dominant societies commodify these same entities, exploiting them as resources and re-conceptualizing them as capital.

In Anishinaabe tradition, understanding the need to avoid a culture of commodification, consumption, and destruction of the planet is guided by the Windigo teachings. The Windigo is a cannibalistic being that is cursed with an overwhelming hunger that can never be satisfied, no matter how much it consumes. The Windigo wanders the Earth, destroying whatever it finds in its path, in an agonizing and unending quest for satisfaction. The example of the Windigo reminds us that we can choose a path of *minobimaatisiiwin*, or that of the Windigo, which will bring about the eventual destruction of all life.

As countless generations of Anishinaabe have done, dominant societies could also learn from the Windigo story about the consequences of eschewing *minobimaatisiiwin*. The social, economic, and environmental crisis in which global society finds itself results from a profound lack of respect for the Earth and for the requirement of reciprocity in its relations with all of Creation. *Minobimaatisiiwin*, as a life-affirming set of obligations and responsibilities with the living Earth, directly challenges the dominant neoliberal paradigm that regards nature as property and a resource to be exploited.

Minobimaatisiiwin offers a real and time-tested alternative. Indigenous nations around the world have denounced an economic world order that perpetuates inequity, injustice, and exploitation. Mounting resistance to this world order has been expressed recently in North America through such actions as the Idle No More movement and the Dakota Access Pipeline protest. At the same time, ancient indigenous ideologies, as expressed through international environmental declarations such as the Universal Declaration of the Rights of Mother Earth, are redefining 'sustainability' as 'living well with the Earth' in a mutually beneficial way.

Minobimaatisiiwin, along with other similar indigenous conceptions, offers a paradigm that is centuries old yet radical in a world of unrelenting industrial capitalism. Indigenous peoples, however, have been enacting political, legal, and governance systems founded on such a paradigm for countless generations. It can be postulated, then, that achieving *minobimaatisiiwin* for all, is the ultimate goal of indigenous autonomy and sovereignty.

Further Resources

Bell, Nicole (2013), '"Anishinaabe Bimaadiziwin": Living Spiritually with Respect, Relationships, Reciprocity, and Responsibility', in Andrejs Kulnieks, Dan Roronhiakewen Longboat and Kelly Young (eds), *Contemporary Studies in Environmental and Indigenous Pedagogies*. Rotterdam: Sense Publishers.

Hart, Michael (1999), 'Seeking Mino-Pimatasiwin: An Aboriginal Approach to Social Work Practice', *Native Social Work Journal*. 2 (1): 91–112.

Kimmerer, Robin (2013), *Braiding Sweetgrass: Indigenous Wisdom, scientific knowledge and the teachings of plants*. Minneapolis: Milkweed Editions.

King, Cecil O. (2013), *Balancing Two Worlds: Jean-Baptiste Assiginack and the Odawa Nation, 1768–1866*. Saskatoon: Dr. Cecil King.

LaDuke, Winona (1997), 'Voices From White Earth: Gaa-waabaabiganikaag', in Hildegarde Hannum (ed.), *People, Land & Community*. New Haven: Yale University Press.

McGregor, Deborah (2015), 'Indigenous Women, Water Justice and Zaagidowin (Love): Women and Water', *Canadian Woman Studies/les cahiers de la femme*. 30 (2/3): 71–78.

Deborah McGregor is of Anishinaabe descent. She is an Associate Professor in the Osgoode Hall Law School and Faculty of Environmental Studies at York University. She currently holds a Canada Research Chair in indigenous environmental justice. Her research focuses on indigenous knowledge systems, water and environmental governance, environmental justice, forest policy management, and indigenous food sovereignty.

NATURE RIGHTS

Cormac Cullinan

Keywords: eco-centrism, rights of nature, Earth jurisprudence, Wild Law

Most contemporary civilizations are organized to maximize Gross Domestic Product (GDP) in ways that degrade the environment and contribute to climate change. They are likely to collapse during the twenty-first century unless they can be reoriented to promote human well-being by enhancing the integrity and vitality of the ecological communities within which they are embedded. Advocates of Nature Rights, also known as the rights of Mother Earth, argue that in order to achieve this transition, legal systems must recognize that all aspects of nature are legal subjects that have inherent rights, and must uphold those rights. The legal recognition of Nature Rights both contextualizes human rights as a species-specific articulation of Nature Rights since people are part of nature and creates duties on human beings and juristic persons to respect Nature Rights.

Legal recognition of Nature Rights is an aspect of a wider discourse about Earth jurisprudence and other ecological approaches to governing human societies. Earth jurisprudence is a philosophy of law and governance which aims to guide people to coexist harmoniously within the Earth community instead of legitimizing and facilitating its exploitation and degradation.

Nature Rights, like human rights, are conceived of as inherent, inalienable rights that arise from the mere existence of the rights holder. This means that every being or aspect of nature, including people, must at a minimum, have the right to exist, the right to occupy a physical place and the right to interact with other beings in a manner that allows it to fulfil its unique role in ecological and evolutionary processes.

The most significant contemporary articulations of Nature's Rights is the Constitution of Ecuador, which was adopted in September 2008. Next, the Universal Declaration of the Rights of Mother Earth (UDRME) was proclaimed by a Peoples' World Conference on Climate Change and the Rights of Mother Earth in Cochabamba, Bolivia, on 22 April 2010. The Constitution of Ecuador states: 'Nature or Pachamama, where life is reproduced and exists, has the right to exist, persist and maintain and regenerate its vital cycles, structure, functions and its evolutionary processes' (article 72). The Constitution makes it clear that the recognition of the rights of Nature is intended to create a framework within which citizens may enjoy their rights and exercise their responsibilities in order to achieve well-being through harmonious cohabitation with nature. Further, it would be a framework that requires both the State and private persons to respect and uphold the rights of Nature and mandates the State to guarantee a development model that is consistent with doing so. New Zealand legislation recognizes the Whanganui River and the Te Ureweraarea as legal entities with rights. Courts in India have recognized the Ganga and Yamuna Rivers, the Gangotri and Yamunotri, glaciers from which those rivers arise, and related forest and watercourses as legal entities with rights.

The Constitutional Court of Columbia has recognized the Atrato River basin as a legal entity with rights to 'protection, conservation, maintenance and restoration'.

Modernity, capitalism, and consumerism arise from the deeply anthropocentric view that human beings are separate from nature and can transcend its laws. This human exceptionalism vews the Earth as a collection of resources which exist for the purpose of human gratification. Since resources are understood to be scarce, out-competing others in order to secure a greater share is understood to be of paramount importance. This worldview is the basis for most legal systems today. The law defines nature (other than human beings) as 'property' and grants the owner extensive decision-making powers in relation to these 'assets' and the power to monopolize the benefits from them. This provides the basis for economic and political systems that concentrate wealth and power and legitimize decisions that prioritize the short-term economic interests of a tiny minority of humans over the collective interests of the Earth community, and life itself.

The recognion that Nature has rights, on the other hand, is based on an eco-centric worldview that sees humans as a particular life form or aspect of Earth which has a unique, but not pre-eminent, role within the Earth community. For example the Preamble and first article of the UDRME refer to Earth as a self-regulating, living community of interrelated beings that sustains all beings and consequently prioritizes maintaining the integrity and health of the whole Earth community. Nature Rights advocates point to the findings of sciences such as quantum physics, biology, and ecology in order to provide evidence that every aspect of the cosmos is interconnected and to refute the widely held beliefs that human beings are separate from, and superior to, Nature. This approach also draws on ancient wisdom traditions and the cosmologies of indigenous people, which view Earth as a sacred community of life and require humans to maintain respectful relationships with other beings.

Earth jurisprudence and Nature Rights pose a fundamental challenge to every aspect of the mainstream 'development' discourse, and to capitalism and patriarchy. It posits a different understanding of the role of humanity, the fundamental purpose of human societies, and of how to promote human well-being. For example, from an eco-centric perspective 'development' is understood as the process whereby an individual develops greater depth, complexity, empathy, and wisdom through interrelationship or 'inter-being' with the community of life. This is the antithesis of the contemporary meaning of 'development' which involves exploiting and degrading complex natural systems to increase the GDP.

Since 2008, Nature Rights and Earth jurisprudence have become an increasingly prominent aspect of the discourse of social movements, environmental and social justice activists, and indigenous peoples throughout the world. These concepts have become a central theme of the discussions within the United Nations about 'Living in Harmony with Nature', and have been incorporated into the manifestos of several Green and eco-socialist political parties. Nature Rights and Earth jurisprudence address the deepest roots of contemporary environmental and societal problems. They provide a manifesto that transcends race, class, nationality, and culture, are based on an understanding of how the Universe functions – an understanding that is more accurate than anthropocentric, mechanistic, and reductionist worldview. Nature Rights provide a basis for a global rights-based movement that can shift the norms of acceptable human behaviour as has been done with human rights. These strengths mean that although the Nature Rights movement is still in its infancy, its influence is likely to continue to accelerate and it has the potential to have a profound global impact.

Further Resources
Cullinan, Cormac (2011), *Wild Law: A Manifesto for Earth Justice*. Second Edition. Cambridge (UK): Green Books.
Global Alliance for the Rights of Nature, https://therightsofnature.org/.
Harmony with Nature (UN), http://www.harmonywithnatureun.org.

Cormac Cullinan is a South African environmental lawyer whose ground breaking book *Wild Law: A Manifesto for Earth Justice* (2002) has helped inspire the global movement. He is a founder and executive committee member of the Global Alliance for Rights of Nature, a director of the Wild Law Institute and a judge of the International Tribunal on the Rights of Nature. He was one of the initiators of the Universal Declaration of the Rights of Mother Earth (2010) and the Peoples' Convention (2014) that established the Tribunal.

———◆◆◆———

NAYAKRISHI ANDOLON

Farhad Mazhar

Keywords: nayakrishi, community seed, community knowledge

Nayakrishi Andolon or the New Agriculture Movement led by farmers involving more than 300,000 diverse household ecological units in Bangladesh, is strategically focusing on 'seed' in its innovative farming practice. The objective is to demonstrate the *shohoj* way to joyful living, ensuring biodiverse ecological regeneration of nature to receive food, fibre, fuel wood, medicine, clean water, and many different bio-material and spiritual needs of the community. The word *shohoj* is grounded in the powerful spiritual tradition of Bengal, generally meaning an intuitively simple but transparent way of being in the universe. Philosophically, it implies learning to relate with internal and external realities with all our human faculties in a unity, allowing no hierarchy between our sensuous, intellectual or imaginative faculties. So, *shohoj*, in practice, explores the bio-spiritual potential of human communities in the real material world to transcend an oppressive, painful, and dehumanized existence.

The movement uses 'seed' as a powerful metaphor of continuity and history, and identifies regenerative space as the site where the invisible manifests as the visible, and potential is realized as reality. Agriculture is defined as the management of both cultivated and uncultivated space and not a 'factory' producing consumer products or commodities. As a practice, *Nayakrishi* celebrates the moments where we are sensuously engaged with nature as well as with our labouring bodies to understand and transcend the limit of abstract intellectualism. *Nayakrishi* is grounded in the powerful spiritual traditions of Bengal where Islam has met creatively with indigenous religious traditions and practices to give rise to *bhakti* movements such as that preached by Chaitanya (1486–1534), and at its apex has produced great saints such as Fakir Lalon Shah (1772–1890) (Sharif 1999: 241–73).

Since 1997, farmers are following ten simple rules to maintain and regenerate living and fertile soil, as also diverse life forms and eco-systemic variability, and develop the capacity of the indigenous knowledge system to engage and appropriate the latest advances in biological sciences. To be a *Nayakrishi* farmer, one must follow all ten rules. Rules 1 to 5 which include clauses such as 'absolutely no use of pesticide or any chemicals' and 'learning the art of producing soil through natural biological processes' are considered entry level practices to be a member. Rules 6 to 10 relate to surface- and aquifer-level water conservation, cultivating diverse fish species in ponds, and raising animals and poultry with farm-produced organic feed, as integrated and advanced practice. Developing integrated and complex ecological systems maximizes systemic yield and contributes to innovating interesting ecological designs, proving immense economic potential of biodiversity-based ecological farming as a successful practical resistance against globalization. Economy is considered as the site where social exchange takes place between life-affirming activities of diverse communities. *Nayakrishi* is a growing and expanding movement. Its success and consolidation pre-supposes the following:

1. Availability of a farmer-seed system is the key to the farmer-led innovation that has historically contributed to agro-ecological evolution and generation of agricultural knowledge.
2. Access and availability of community knowledge functioning through oral communication, community memory and conservation of the popular wisdom through stories and narratives.
3. Existence of a fairly functional system of culture related particularly to food and nutrition, which links agricultural consumption to production within specific agro-ecological systems. *Nayakrishi* is keen to transform hierarchical relations of class, caste, and patriarchy; consequently women and marginal farmers are natural leaders of *Nayakrishi*.
4. An informal or formal system of social exchange of farm-generated inputs, labour and knowledge capable to operate outside the capitalist market, including community management of common resources such as water, forests, and biomass.
5. An operative notion of common property taking into account the cultivated and uncultivated sources for food and livelihood, and commonly held moral values that ensure the right of the community members to use natural wealth.

An institutional innovation is the development of farmers' collective action called Nayakrishi Seed Network (NSN) with specific responsibility of ensuring in-situ and ex-situ conservation of biodiversity, with the farming household as the focal point. The NSN has three levels. First, *Nayakrishi* Seed Huts (NSH) are established by the independent initiative of one or two households in the village, belonging to the *Nayakrishi* Andolon, willing to take responsibility to ensure that all common species and varieties are replanted, regenerated, and conserved by the farmers.

Second, the Specialised Women Seed Network (SWSN) consists of women who specialize in certain species or certain varieties. Their task is to collect local varieties from different parts of Bangladesh. They also monitor and document introduction of a variety in a village or locality, and keep up-to-date information about the variability of species for which they are assigned.

Third, Community Seed Wealth (CSW) is the institutional set-up that articulates the relation between farmers within a village and between villages, in other districts and also with national institutions, for sharing and exchange of seeds. The CSW also maintains a well-developed nursery. The construction of CSWs is based on two principles: (a) they must be built from locally available construction materials and (b) the maintenance should mirror the household seed conservation practices. Any member of the *Nayakrishi* Andolon can collect seed from CSW with the promise that they will deposit double the quantity they received after the harvest. In the CSWs, there is a collection of over 3,000 rice varieties, and 538 varieties of vegetables, oil, lentil, and spices.

Nayakrishi encourages the growth of various plants including herbs that are uncultivated but are good as food sources for humans and other animals. The more environments are made free from chemicals, the more the uncultivated foods are found in the surroundings. Such assessment is done through cultural practices of celebrating *Chaitra Sangkranti*, last day of the Bengali calendar year when it is a custom to eat a meal that has at least fourteen different kinds of leafy greens or *shak*, mostly from uncultivated sources. This is a natural auditing that ensures renewable food sources for future. Resource-poor farmers are able to collect nearly 40 percent of their food and nutritional needs from uncultivated sources.

Nayakrishi Andolon represents peasant's resistance against the corporate takeover of the global food chain, an assertion that it is the farming community that feeds us. *Nayakrishis* regenerating the future by defending farming as a way of life, affirming agrarian activities and charting *shohoj* ways to *ananda*, that is, the joy of being in the world.

Further Resource

Sharif, A. (1999), '*Islam o Gaudiya Vaishanava* Motobad', in Abonty Kumar Sanyal and Ashoke Bhattacharya (eds), *Chaitanyadev.* Kolkata: P.M. Bagchi & Company Ltd.

Farhad Mazhar with a background in pharmacy and economics, is a well-known poet, writer-columnist, and founding member of the Bangladeshi association – UBINIG (Policy Research for Development Alternatives), as well as an initiator of Nayakrishi Andolon. He has been involved in major literary movements from the 1970s till present, and is author of more than twenty books published in Bangla on poetry, literature, and political issues.

—•◦•◦•—

NEGENTROPIC PRODUCTION

Enrique Leff

Keywords: ecological productivity, environmental rationality, entropy, sustainability

Negentropic production is a concept that synthetizes the scope of an alternative theory and practice to inhabit the planet; it aims at rethinking sustainability from the ecological and cultural conditions of people's territories.[1] Negentropy or negative entropy can be conceived as the overall process that continually creates, maintains, and complexifies life on the planet, based on the transformation of radiant solar energy into biomass by photosynthesis, the source of all life. Negentropic production is thus a response to the fact that economic growth transforms all matter and energy consumed in the process of production into degraded energy, ultimately as unrecyclable matter and irreversible heat.

Negentropic production is meant to counteract the dominant economic paradigm, based on a mechanistic view of productivity, labour, and technology that has negated the ecological and cultural conditions for sustainability, resulting in the planet's environmental crisis. Objectified nature, fed to the mega-machine of the global economy, is transformed following the entropy law into commodities, pollution, and heat. This degradation process manifests itself in deforestation, desertification, biodiversity erosion, and climate change that is producing the entropic death of the planet.

Humans are the major force transforming the biological basis of the planet's life-support system. Economic production is the means by which humans transform matter and energy from nature. The mode of production is the way by which humanity establishes the material conditions of its existence, deeply affecting the complex thermodynamics of the biosphere. Environmentalism questions the ineluctable 'entropic degradation of nature induced by the economic process underlying the prevailing productivist rationality.

The reversal of unsustainable production and the transition to a sustainable mode is not solved by reforming mainstream economics in order to internalize 'environmental externalities' such as ecological breakdown, pollution, biodiversity, climate change, greenhouse effect gases, environmental goods, and services. Nor can economic processes be made sustainable by harnessing economic behaviour to the ecological conditions required for the reproduction of nature in order to attain a 'steady state' economics. In contrast to these mainstream proposals, Georgescu-Roegen's 'bioeconomics' (Georgescu-Roegen 1971) confronted economic rationality with the ineluctable fact that economic processes destroy their own ecological basis. But he did not derive an economic paradigm based on the ecological basis of sustainable production.

What is required is a paradigm of eco-techno-cultural productivity built on the principles of an alternative environmental rationality (Leff 2004). In this conception, the 'environment' emerges from beyond the dominant rationality with its alleged unitary, universal ontology that constrains diversity and excludes otherness. This principle is guiding eco-Marxism, ecological economics and political ecology to the construction of an alternative mode of production, based on the conservation and redesign of the social and ecological conditions of production as embodied in diverse cultures.

The ultimate question for sustainability is the sustainability of life. Yet, the question arises: is a sustainable economy possible? – an economy that works with and through the creative forces of nature; an economy built on the ecological potentials of the planet; a human economy that recognizes the conditions of human life and cultural diversity? This implies deconstruction of the established economy and the construction of a new paradigm of production: a paradigm of eco-technological-cultural productivity – or 'negentropic production' – guided by the principles of environmental rationality.

Sustainability would appear in the horizon of those other possible worlds, if only we were able to unleash the potentialities of life that have been constrained by the rationality *anti natura* that drives 'sustainable development'. Sustainability is the outcome of the interplay of negentropic/ entropic processes enacted by different cultural beings in their modes of dwelling in their life territories. The construction of a sustainable future calls on us to envision the ecological potentials and the epistemological and social strategies required to construct an alternative mode of production, one based on the 'negentropic potentials of life'. What this means is a mode of production based on the thermodynamic-ecological conditions of the biosphere and the symbolic – cultural conditions of human existence.

The proposed negentropic paradigm of production is built on the articulation of three orders of productivity: ecological, technological, and cultural. Ecological productivity is based on the ecological potential of the different ecosystems. Ecological research has shown that the most productive ecosystems, those of the humid tropics, do produce biomass at natural yearly rates up to about 8 per cent. This ecological potential can be enhanced by scientific research, ecological technologies, and innovation of cultural practices – including high efficiency photosynthesis, management of secondary succession, and selective regeneration of valuable species in ecological processes; multiple associated cropping, agroecology, and agroforestry – to define and guide the cultural – economic value of the techno-ecological output of the productive process (Leff 1995).

This alternative paradigm of production is articulated from the perspective of non-modern cultural imaginaries and their ecological practices. The privileged spaces to deploy this strategy of 'negentropic production' are the rural areas of the world inhabited by indigenous peasant peoples who in their struggles to build their autonomous territories, enact this theoretical perspective through the reinvention of their identities and the innovation of their traditional practices.

This conception of sustainable production resonates with the struggles by social movements for the re-appropriation of their bio-cultural patrimony, such as with the ancestrality invoked by the Afro-Colombians of the Pacific rainforest region, the *sumak kawsay* of Andean peoples, the *caracoles* of the Zapatistas in México, and the imaginaries of sustainability of so many other traditional peoples, from the Mapuche and Guaranís in the South of Latin America to the Seri or Comca'ac in the arid North of Mexico. The privileged agents of a negentropic society are the traditional peasants and indigenous peoples of the world and the socio-environmental movements they are fostering. An emblematic example is the struggle of the *seringueiros* in the Brazilian Amazon region who have established their 'extractive reserves' as a strategy of sustainable development.

Note
[1] The concept is based on the notion of eco-technological productivity, which I first stated in 1975 and reworked in my 1986 book *Ecología y Capital*. See (Leff 1995).

Further Resources
Georgescu-Roegen, Nicholas (1971), *The Entropy Law and the Economic Process*. Cambridge, MA: Harvard University Press.

Leff, Enrique (1994), *Ecología y capital: racionalidad ambiental, democracia participativa y desarrollo sustentable*. México: Siglo.

——— (1995), *Green production. Towards an Environmental Rationality*. New York: Guilford.

——— (2004), *Racionalidad Ambiental. La reapropiación social de la naturaleza*. México: Siglo XXI Editores.

Enrique Leff is a Mexican environmental sociologist, political ecologist, and senior researcher at Instituto de Investigaciones Sociales, Universidad Nacional Autónoma de México. He was Coordinator of the Environmental Training Network for Latin America and the Caribbean at UNEP (1986–2008). His most recent book is *La apuesta por la vida: imaginación sociológica e imaginarios sociales en los territorios ambientales del Sur* (Siglo XXI Editores, México, 2014).

—•◦•—

NEW MATRIARCHIES

Claudia von Werlhof

Keywords: matriarchy, capitalist patriarchy, alchemy, matriculture

Contemporary matriarchal studies define matriarchies as life-oriented cultures, societies, or entire civilizations that have organized around the needs of mothers and children in a horizontal and egalitarian way without hierarchies, violence, the state, and a class structure. The term 'matriarchy' does not mean 'mothers' rule', but *mater arché*, in the sense of 'in the beginning of life there is a mother'.

Following extensive research, Göttner-Abendroth has demonstrated that matriarchies were normal in pre-patriarchal societies throughout the world. They have been peaceful and oriented towards the natural environment. Further, they have been subsistence economies, highly spiritual earth and life-related cultures, based on maternal competence and knowledge, and they have been non-discriminatory in relation to sex, gender, age, and activities. There have been multiple ways of organizing matriarchies in accordance with the surrounding conditions, but the basic principles have remained the same.

Patriarchies, on the other hand, started to develop some five or six thousand years ago. It seems that they were the outcome of climate catastrophes, forcing the populations of some regions, such as Siberia, to migrate to other places. They invented violence and war as a means to survive and to take control over matriarchies. States became the rule, especially where highly developed matriarchal civilizations had emerged. This happened in fertile regions around the globe such as the Indus, Mesopotamia, and the Nile River. A class structure, warrior societies, and exploitation of the original population emerged as a consequence.

Along with war, lords and rulers, gods and fathers were proclaimed. The *pater arché* came into being with its ideology and religion of 'in the beginning of life there is a father'. The origin of life with a father – instead of a mother – became the justification of his rule. The meaning of *arché* as origin/beginning was thus transformed; not only that, it was combined with the idea and exercise of rule, and thus with domination. It should be stressed that rule does not stem from nature; only under patriarchy does rule seem to be natural, like patriarchy itself! As patriarchal civilizations developed, matriarchal civilizations were conquered, oppressed, destroyed, replaced, and turned upside down everywhere.

We speak of matriarchy today as a 'second culture' within patriarchy, consisting of the remnants of matriarchal traditions still surviving. In many parts of the world, nevertheless, matriarchies have survived up to the present, particularly among indigenous societies. Many of them though have experienced local patriarchies and have lost their original traditions under the growing stress of globalization. Few have not experienced the brutal impacts of colonial patriarchy.

Modern colonization was a result of the further development of patriarchy into capitalism in Europe. The most well-known account of the first steps of this transformation is Frederick Engels' book, *The Origin of the Family, Private Property, and the State* (1992[1884]); however, matriarchy scholars have an alternative explanation. The European matriarchal civilization had been disrupted through invasions from the East, and later destroyed by the Roman Empire. With the advent of modern times, Europe applied this lesson to the rest of the world. The outcome was 'capitalist patriarchy' as we know it today, the world system developing into neoliberal globalization.

The relationship between capitalism and patriarchy – the subject of much debate – suggests that capitalism is not simply an economy backed by a patriarchal culture, but the latest stage of patriarchy itself. This insight became possible only when the question of 'development of the productive forces', and especially the machine as a technological system enabling the systematic destruction of nature, was included in the analysis. That realization demonstrated how since its inception patriarchy followed the utopian – 'alchemical' – idea of the 'creation' of a 'higher civilization' that would finally be independent of *mater arché*, having replaced it completely by *pater arché*. This male creation would end up in a pure patriarchy without any further need for either mothers or Mother Nature. The result however, has been the systematic destruction of life on the planet, which does not lead to a better world, as promised, but to a dead one.

If patriarchy is the problem, matriarchy is the answer. This means that we have to find a way out of patriarchy into a new matriarchy.

The solutions of the left appear to be the same as those propagated by 'development' itself. Capitalism and socialism prove to be two sides of the same coin, as both of them are oriented towards the modern alchemy of the destructive transformation of nature into capital. Still existing matriarchies and new ones that are emerging – even if they do not call themselves matriarchies, such as the Zapatista and the Kurdish Rojava movement – have to be recognized as alternatives to modernity as an alchemical war system. They move beyond destructive and violent relationships with nature, women, children, and society in general. Only in this way can post-development movements avoid remaining within the alchemical patriarchal tradition or evolving into post-capitalist neo-patriarchies.

There are movements and approaches within the West – such as the ecofeminist subsistence perspective, the gift economy, and permaculture – that foster a new relationship with Mother Earth in order to protect her from desertification and other threats such as geo-engineering, produced by modern patriarchy's military industrial complex. We could call these movements 'matricultures', but we will have to see how far they can go in inventing a new way of life, building on the remnant memories of matriarchal logic everywhere.

We all have to become conscious of the 'hatred of life' as if this were normal. Instead we need to rediscover love for life and Mother Earth and organize in their defence. However, so far, this deep understanding remains largely absent from most western political alternatives.

Further Resources

Engels, Frederick (1992[1884]), *The Origin of the Family, Private Property, and the State*. Chicago: Charles H. Kerr & Co.

Enlace Zapatista, www.Enlacezapatista/ezln.org.mx.

Flach, Anja, Ercan Ayboga and Michael C. Knapp (2015), *Revolution in Rojava*. Hamburg: VSA.

Forschungsinstitut für Patriarchatskritik und alternative Zivilisationen, www.fipaz.at.

Göttner-Abendroth, Heidi (ed.) (2009), *Societies of Peace: Matriarchies Past, Present and Future*. Toronto: Inanna.

Planet are Bewegungfür Mutter Erde / Planetary Movement for Mother Earth, www.pbme-online.org.

vonWerlhof, Claudia (2011), *The Failure of Modern Civilization and the Struggle for a 'Deep' Alternative*. Frankfurt: Peter Lang.

———— (2016), 'The "Hatred of Life" as Patriarchy's Core Element', http://www.globalresearch.ca/the-hatred-of-life-the-world-system-which-is-threatening-all-of-us/5541269.

Claudia von Werlhof, PhD, born 1943, Berlin, Germany, is Professor of Political Science and Women's Studies, at the University of Innsbruck, Austria. She is a co-founder of the ecofeminist research group known as the Bielefeld School. She established the network Critical Theory of Patriarchy and the Planetary Movement for Mother Earth, and the journal *Boomerang* for critique of patriarchy. She also works as a Dorn-therapist.

NEW WATER PARADIGM

Jan Pokorný

Keywords: solar energy, plant evaporation, water cycles, climate, landscape management

Humans existed on Earth as hunters and gatherers for hundreds of thousands of years and the carrying capacity of a forest is one to three persons per square kilometre. But the civilizations, characterized by agricultural overproduction to supply cities and armies, developed in the last ten thousand years, dried out their environments; archaeologists find their relics buried under sand. Population growth led to the conversion of forest into agricultural land. Crop plants such as grains, corn, and potatoes did not tolerate floods, so farmers would drain wetlands and fields. Rainwater was also collected and drained from towns. The old civilizations of Mesopotamia, Indus Valley, the South American Incas, and North Africa did not burn fossil fuels increasing the carbon dioxide concentration in the atmosphere; rather, they collapsed due to lack of precipitation and high salinity of soil. It was bad management of land and water that led to soil fertility loss, droughts, and sandstorms. Industrialization would introduce further anthropogenic disturbances.

The UN Conference on Climate Change 2015 in Paris (COP21) set a goal for limiting global warming at less than 2°C of global average temperature (GAT) compared to pre-industrial levels. According to the Intergovernmental Panel on Climate Change (IPCC), the quantifiable criterion of climate change is GAT and the reason for global warming is increasing greenhouse gas (GHG) concentrations, particularly CO_2 and CH_4. Water vapour is considered only as a passive 'feedback agent', rather than an active agent of climate change. The IPCC minimizes water and land cover as controlling factors of climate, yet the amount of water vapour in the air is one to two orders of magnitude greater than that of CO_2 and CH_4. Water vapour forms clouds, which prevent the passage of solar energy to Earth, reducing temperatures substantially. The transition between the three phases of water – liquid, solid, and gaseous – is linked to heat energy. But landscape management patterns – deforestation, wetland drainage, urban soil sealing – change the distribution of solar energy such that it cannot be used in the cooling process of atmospheric water evaporation.

On a sunny day up to 1000W of solar energy falls on each square metre of Earth. Dry land, city surfaces – roofs, road asphalt, pavements – will heat to about 60°C, whereas beneath tree shade temperatures do not go over 30°C. About 50 per cent of wetlands were drained in the USA (45.9 million ha) releasing a huge amount of heat into the atmosphere. A tree also actively cools itself and its environment by evaporation of water. A tree supplied with water is an air-conditioning system driven by solar energy. Solar energy is hidden or latent in water vapour and is released in cool places as water vapour precipitates back into liquid water. The tree equalizes temperature gradients in a double way. It cools through evaporation and heats through condensation. Technological air-conditioning is flawed in comparison with vegetation: first, because it depends on polluting electricity generation; and second, even while it cools inside a room it releases heat outside, so increasing the surrounding temperature.

Conventional analyses of global warming, such as those offered by the IPCC, typify what can be called the Old Water Paradigm. This treats global warming impacts on the water cycle rather than examining water as an active determinant of climate. The Old Water Paradigm assumes the following:

- The increase of global average temperature is the main climatic problem
- The mitigation via decrease of GHGs can perhaps be expected within a horizon of centuries

- Drainage and the urban landscape have a minimal impact on the water cycle
- Water vapour acts as a GHG causing higher temperatures
- Vegetation has low albedo or solar reflectance capacity and so increases greenhouse effects

The New Water Paradigm described in the book *Water for Recovery of Climate* (Kravčík *et al.* 2008) treats water as the medium equalizing temperature differences in time and space, between day and night, here and there. The assumptions are the following:

- Extremes of weather, erratic drought, and cyclonic storms are the main climatic problem
- Deforestation, large-scale agriculture, and urbanization change the local water cycle, which, in turn, impacts global atmospheric weather patterns
- Transpiring vegetation alleviates air temperatures, cloudiness moderates the intensity of solar radiation coming to the Earth's surface
- Water vapour condenses by night and prevents infra-red radiation (IR) moving from the Earth's surface towards the sky
- With a new approach to water management, a possible recovery of the climate can be expected within decades.

New Water Paradigm principles have been demonstrated in Australia by Peter Andrews' method of Natural Sequence Farming. This emulates the role of natural water courses to reverse salinity, slow erosion, and increase soil and water quality, recharge subterranean aquifers, and enables native vegetation to restore the riparian zone. In India, the Tarun Bharat Sangh project pioneered by Rajendra Singh is based on the revival of traditional water reservoirs. The work is aimed at designing water harvesting structures or *johads*. These are simple, mud barriers built across hill slopes to arrest monsoon runoff. The height of the embankment varies depending on site, water flow and topography. A *johad* holds water for livestock and allows water percolation down through the soil, recharging the aquifer as far as a kilometre away. This water harvesting has irrigated an estimated 140,000 hectares and increased the water table from about 100–20 metre in depth to 3–13 metre. Crop yields have highly improved. Forest cover has gone from 7 per cent to 40 per cent. More than 5,000 *johads* were built all together and over 2,500 old structures rejuvenated by village communities in 1,058 villages since 1985. Similar projects in Slovakia have created employment opportunities and enhanced the sense of community.

Further Resources

Andrews, Peter (2006), *Back from the Brink: How Australia's Landscape Can Be Saved.* Sydney: ABC Books.

Kravčík, Michal, Jan Pokorný, Juraj Kohutiar, Martin Kováč and Eugen Tóth (2008), *Water for the Recovery of Climate: A New Water Paradigm,* www.waterparadigm.org.

Makarieva, Anastassia, and Viktor Gorshkov (2007), 'Biotic Pump of Atmospheric Moisture as Driver of the Hydrological Cycle on Land', *Hydrol: Earth Syst. Sci.*11: 1013–33, 10.5194/hess-11-1013-2007.

Pokorný, Jan, Petra Hesslerová, Hanna Huryna, and David Harper (2016),'Indirect and Direct Thermodynamic Effects of Wetland Ecosystems on Climate' in Jan Vymazal (ed.), *Natural and Constructed Wetlands: Nutrients, heavy metals and energy cycling, and flow.* Zurich: Springer.

Ponting, Clive (1991), *A Green History of the World: The Environment and the Collapse of Great Civilizations.* London: Penguin.

Schneider, Eric and Dorion Sagan (2005), *Into the Cool, Energy Flow, Thermodynamics, and Life.* Chicago: University of Chicago Press.

Jan Pokorný is a plant physiologist who graduated from the Charles University, Prague. He has researched the photosynthesis of wetland plants with the Czechoslovak Academy of Sciences and the CSIRO, Australia. Since 1998, he is Director of the research organization ENKI, dealing with the direct role of landscape conditions and plant activity in the distribution and interaction of solar energy, water cycles, and climatic effects.

OPEN LOCALIZATION

Giorgos Velegrakis and Eirini Gaitanou

Keywords: political subjectivity, refugee movement, sense of place, solidarity, transformative praxis.

In her famous article of the early '90s 'A Global Sense of Place', feminist geographer Doreen Massey argued that the idea of place as having a single 'essential' identity based on a bounded history of a territory is flawed. What gives a place its specificity is the fact that it is constructed 'out of a particular constellation of social relations, meeting, and weaving together at a particular locus'. This radical re-identification of place lends itself to understanding some creative European responses to the arrival of refugees from global conflict zones. While a new academic vocabulary is emerging for 'place, articulation of local, inter-local, and global dimensions, and radical socio-spatial practices', the term 'open localization' can describe new experiments in community building, radical politics, and democratization processes.

The current capitalist crisis and elite recipes for recovery only serve to extend and normalize new forms of social repression across various geographical contexts. Second, since 2011 a common pattern of popular protest has arisen all over the world, focused on radical democracy and accountability of representation. Moreover, solidarity initiatives have emerged as networks of localized struggle in and across urban spaces. These movements exemplify a political culture based on meeting everyday material needs by local commoning. Cultivating a collectively created and non-pre-assigned culture of solidarity, each movement gives content to what David Featherstone (2015) calls the 'dynamic geographies of subaltern political activity and the generative character of political struggle'. Third, the movement of refugees into Europe and other western countries in 2015 is a reminder that mobility, and control over mobility, both reflect and reinforce power. Western countries seek to 'manage refugee flows' and to securitize 'forbidden' borders, with militarized detention camps. Yet, refugees are in great need of shelter, safety, solidarity, and common places just as locals are.

Recent years have witnessed several grassroots movements modelled here and now common places for refugees and locals through provision of dignified housing inside the city. Local residents in solidarity and refugees improvise projects of self-organization and solidarity centres based on principles of anti-racism and inclusion; the right to free movement; decent living conditions and equal relationships. They actualize a conception of everyday life and common struggle that empowers from below, ultimately leading to the creation of open localities.

Such initiatives have a long presence in Italy, Spain, Sweden, Germany, and other mainly European countries, but here two examples from Greece are outlined. The first one is Refugee Accommodation and Solidarity Space City Plaza, which has operated since April 2016 as a self-organized housing project for homeless refugees in the centre of Athens. City Plaza emerged as a practical response to the dominant anti-migration policies in Greece and the EU, while over recent months has been developed as a new, open locality based on principles of self-organization, autonomy, and solidarity. The second exemplar is Social Solidarity Medical Centre of Thessaloniki (SSMC) operating since December 2010 as a social healthcare collective, giving primary medical and pharmaceutical treatment to uninsured residents. It started from a group of activists and doctors willing to provide health support to immigrants but soon developed as a health support to everyone – local or immigrant – who was unable to afford the National Health Care System.

Such initiatives give meaning to what we can identify as 'open localization', a process that transforms existing localities into open places for social and political solidarity. These common places encourage specific socio-spatial practices that address everyday life demands, while cultivating social experimentation, democratization, self-organization, and multicultural forms of coexistence.

Open localities are paradigms of 'political communities under construction', combining social and political struggle and moulding participants as political subjects/citizens. These forms of struggle propose a radical articulation of social interests, and of the ways to assert them. Their project speaks a new mode of politicization by expressing a need for the re-appropriation of collective participation, the creation of public spaces, spaces of social experimentation and alternative counter-institutions. What is at stake here is the need for a 'move to the political' outside its traditional forms of exercise, with street politics as a strong component.

According to the French philosopher Jacques Rancière, the political character of a movement is related to finding spaces of action, of discourse and thought, exceeding the mere affirmation of a concrete group. In this respect, the public sphere is expanded, the political is conceptualized as an 'active determination', and identities as 'political processes in progress'. In this regard, we can identify the above-mentioned solidarity initiatives as open localities that perform politics, and thus democracy as ways of life – 'as an art of life'. This expresses not only a means of transition and of organization of society, but also of constant self-transformation, democratization, and learning: processes of constituting the people as a collective subject. This kind of activism, strongly mediated by spatial practices and daily life issues and forms of organization, leads to the development of specific unifying processes, without necessarily inducing the construction of new, unified social bodies, but rather materializing a 'unity in diversity' scheme. Connections are not formed simply based on solidarity, but on common interests, common demands, and mobilization.

Simplistic accounts romanticize such solidarity movements and initiatives, but they are rife with tensions, limitations, and complexities. The open localities they create should be analysed as 'phenomena in constant motion' that transform their practices, methodologies, and internal relations. However, they conceptualize in a relational and sometimes controversial way 'a radical sense of place', which should always be open and democratic.

Further Resources

Featherstone, David (2015), 'Thinking the Crisis Politically: Lineages of Resistance to Neo-Liberalism and the Politics of the Present Conjuncture', *Space and Polity*. 19 (1): 12–30.

Massey, Doreen (1991), 'A Global Sense of Place', *Marxism Today*. 38: 24–29.

Rancière, Jacques (2011), *Hatred of Democracy*. Athens: Pedio.

Refugee Accommodation and Solidarity Space City Plaza, http://solidarity2refugees.gr/.

Social Solidarity Medical Centre of Thessaloniki, http://www.kiathess.gr/en/.

Giorgos Velegrakis is a PhD researcher in political ecology in the Geography Department, Harokopio University Athens. His research focuses on extractivism and socio-ecological movements, urban geographies, as well as theoretical debates on ecosocialism, degrowth, and Marxist political ecology.

Eirini Gaitanou has a PhD in European and International Studies from King's College London. Her research interests are social movements, political participation, political subjectivity, and consciousness, from a Marxist perspective. As an activist, she participates in both practical and theoretical debates.

PACIFIC FEMINISMS

Yvonne Underhill-Sem

Keywords: Pacific Feminist Charter, LGBTQI, gendered power relations, Pacific islands

Self-identified feminists in the Pacific continue to publically ask the simple question 'what about women and gender inequality?' The struggle to uphold women's right and to secure services for women and girls, let alone for gender non-conforming people, remains because gender inequality in the Pacific is a daily experience. This is despite the existence of femocrats and women's ministries, or The Convention on the Elimination of all Forms of Discrimination Against Women (CEDAW) reporting processes. Tradition and culture in the Pacific fuels the processes of exclusion and control of women in systematic and sustained ways. Consequently, individual and collective tragedies endure, these include everyday sexual violence, sorcery killings, unwanted pregnancies, unsolicited bullying and intimidation, unjust distribution of resources, and so on.

The result is a diverse manifestation of Pacific feminism – from self-declared and socially identified feminists, to those whose practices of social reproduction are aligned with sustaining gender-just livelihoods. The Pacific feminist movement includes radical LGBTQI groups (Haus of Khameleon), intellectual collectives (DAWN Pacific), creatively articulate advocates (DIVA for Equality), courageous and skilled human rights defenders (Voice for Change), dedicated legislators and policy makers (Regional Rights Resources Team), feminist human rights organizations (Fiji Women's Rights Movement), youth-based organizations (Pacific Youth Council), as well as others.

Amidst the cultural diversity of peoples that constitute the Pacific, the politics of indigeneity adds further complexities. But in terms of gendered power relations, patriarchal privilege prevails to the detriment of women regardless of the hue of their skin, the superiority of their skill or knowledge, the quality of their education or their leadership potential.

In November 2016, the Charter of Feminist Principles for Pacific Feminists was launched at the Inaugural Pacific Feminist Forum in Fiji. The Charter is loosely modelled on the African Feminist Charter, but its Pacific flavour was palpable as feminists from Oceania recognized our common bonds of the sea (*wansolwara*), land (*vanua*), and ancestors (*tauanga*). This initiative tapped into important sentiments of gender egalitarianism from the past but also challenged pervasive contemporary leadership systems that privilege men.

The progress entailed by the Pacific Feminist Charter signals a new attempt to organize across diverse lines of gendered power relations to challenge historical patriarchal privilege as well as new asymmetrical gendered power relations. The Charter highlights a fundamental strength of the Pacific feminist movement – its roots in contemporary struggles for social justice and its diverse manifestations. It provides further impetus for change alongside, for instance, the Pacific Feminist Sexual and Reproductive Health and Rights Statement of 2013 and the Pacific Women's Network Against Violence Against Women. The logic that underpins such efforts is shaped by a politics of knowledge, personal affective practices, and a shared assumption that gender equality in the Pacific is imperative.

Pacific people have long faced the often violent, increasingly militarized and deeply dogmatic devotion to global capitalist processes of exploitative accumulation that deepens gender inequality. Feminists in the Pacific have developed practices of radical alliancing (Fiji Women's Forum), awkward but nonetheless respectful engagement (Kup Women for Peace in Papua New Guinea), considerable creativity (Women's Action for Change), and fervent fortitude in the face of civil war (Nazarene Rehabilitation Centre).

Pacific feminism has deep roots in social justice struggles. In the 1960s, the Young Women's Christian Association in Fiji campaigned for just tax systems and fair wages. In the 1970s, the East Sepik Council of Women provided nutrition programmes in palm oil plantations. Pacific feminists also recognized the importance of working regionally to challenge the vestiges of colonialism such as the continuation of nuclear testing, the intensification of extractive industries and the dearth of women's voices in national parliaments. Sadly, the gendered power relations in these regional challenges too easily succumb to patriarchal leadership and privilege, regardless of the fine intentions of a select few and the rhetoric of many more.

At different times and places, feminism emerged differently. In Fiji, the rights-based feminism of the 1980s International Decade for Women largely overtook the welfare-based Pan-Pacific South East Asia Women's Association, established to support immediate colonial projects. Later, rights-based feminism informed the establishment of a network of organizations such as the Fiji Women's Crisis Centre, Punanga Tauturu in the Cook Islands, Women and Children's Crisis Centre in Tonga, Vanuatu Women's Center, and Women United Together Marshall Islands. Some of these re-focused their activities on peace and democracy projects, such as in post-coup Fiji and in the establishment of the Autonomous Bougainville Government in Papua New Guinea.

In the Pacific, people with non-conforming gender identities and women who do not biologically bear children are socially and culturally recognized for the contributions they make to raising children. This recognition challenges asymmetrical power relations, all of which are gendered. In the Charter, Pacific feminism embraces this diversity of people with explicit reference to LGBTQI and young women. In addition, women of faith, who have played critical roles in peace building, are also supportive of the calls for decriminalizing both LGBTQI and abortion.

As a political organizing concept, Pacific feminism is still weak and scepticism of its value prevails. The reasons include the post-colonial insistence on rejecting non-Pacific or 'Western' concepts and naming practices in Pacific struggles; the growing influence of conservative, faith-inspired ideologies of traditional gender egalitarianism; as well as personal ignorance, misogyny, and malice.

The clear ambition of those signing the Charter is for transformative change that upholds the rights of women, girls, and non-gender conforming people so that we get the 'best full lives for ourselves and our Pacific communities'. This is still a radical ambition, but with the Charter, one with renewed momentum.

Further Resources
Charter of Feminist Principles for Pacific Feminists (2016),
    http://www.fwrm.org.fj/images/PFF/PFF-Charter-Final-2Dec2016.pdf.
Emberson-Bain, Atu (ed.) (1994), *Sustainable Development or Malignant Growth?*
    Perspectives of Pacific Island Women. Suva: Marama Publishers.
Mishra, Margaret (2012), 'A History of Fiji Women's Activism (1900–2010)
    Repository USP', *Journal of Women's History.* 24 (2): 115–43.
Slatter, Claire and Yvonne Underhill-Sem (2009), 'Reclaiming Pacific Island
    Regionalism: Does Neo-Liberalism Have to Reign', in Bina D'Costa and Katrina
    Lee-Koo (eds), *Gender and Global Politics in the Asia-Pacific.* New York: Palgrave
    McMillan.
Teaiwa, Teresia and Claire Slatter (2013), 'Samting Nating: Pacific Waves at the Margins
    of Feminist Security Studies', *International Studies Perspectives.* 14 (4): 447–50.
Underhill-Sem, Yvonne (2012), 'Contract Scholars, Friendly Philanthropists,
    and Feminist Cctivists: New Development Subjects in Oceania', *Third World*
    Quarterly. 33 (6): 1095–112.

Yvonne Underhill-Sem is Associate Professor of Development Studies in the School of Social Sciences, Faculty of Arts, University of Auckland.

PACIFISM

Marco Deriu

Keywords: war, pacifism, degrowth

In its essential core, the pacifist ideal implies the unjustifiability of war and the commitment to resolve conflicts by nonviolent means. The first political organizations for peace were established in the first decades of the nineteenth century in the United States and England. It was however Émile Arnaud, president of the *Ligue Internationale de la Paix et de la Liberté*, who proposed, in 1901, the terms 'Pacifism' and 'pacifists' to indicate a specific political movement. In the contemporary age, pacifist movements have pursued their ideal not as a simple rejection of war, but through the promotion of positive goals such as justice, human rights, and development.

'Development is the new name for Peace', affirmed Pope Paul VI in 1967 with a slogan that subsequently became famous. The Encyclical *Populorum Progressio* summarized in one expression the expectations and the aspirations of an entire age that connected faith in the liberation from poverty, slavery, and violence to the promise of development. Peace and development were not conceived as the same thing, but the latter – in the shadow of the Cold War and decolonization – seemed an essential condition for the achievement of the former. In the aftermath, the connection hypothesized by *Populorum Progressio* appears very problematic for at least three reasons.

First among the reasons is the connection between consumption of resources and resource wars. Our development model is founded on a continuous flow of resources that we extract from territories that are often hidden to the eyes of the consumer. Many among the longer and bloodier wars revolve around natural resources. We can call it 'militarized extractivism', or even 'extractive militarism'. We find both private companies that, in order to gain access or defend the control of a strategic area, do not hesitate to finance armies or factions or hire private security forces; as also an increasing recourse to governmental military missions for the control or management of strategic sites and commercial channels.

Second is a symbiosis between military and civil industries, which in recent decades have been increasingly interweaving and fusing. The more advanced technology industries – mechanical, aerospace, electronics, computer science, that have as their core nanotechnology and new materials – find in the army one of their main customers. In fact, there is an increasing range of products and technology directly or potentially 'dual', that is, they can be used for both civil and military purposes. If we look at the list of the 100 largest companies producing weapons and military services, we can note that they involve sectors such as aerospace technology, satellite and security, aeronautics, engines, turbines and propulsion systems, naval mechanics, electronics, communications and information technology or conglomerates such as General Electric, Mitsubishi Heavy Industries, Kawasaki Heavy Industries, Hewlett-Packard that operate in very different economic sectors.

Third is a connivance between the banking system and the arms trade. Big banks make possible the enormous and profitable financial transactions related to this market thanks to their diffusion and the international presence, their speed and safety in payment, their knowledge of customers, the possibility of credit and a certain degree of confidentiality in banking operations.

Therefore, by observing the growth of the plunder of resources, the growth of production of goods, the growth in investment banking in arms trade, we can detect that there is a strong connection between the current capitalist development and the unfolding of violence at a global level.

For half a century, since Harry Truman's famous speech on the State of the Union in 1949, it was the 'poverty' of the 'underdeveloped' regions to be conceived as a 'threat', something that had to be fought and removed through aid, structural adjustments, development policies. But today in the face of looting and pollution of ecosystems, growing waste and greenhouse gases, loss of biodiversity, climate change and wars for ever more scarce resources, a new awareness. It is plain that the imbalance and the threat to peace come from uncontrolled growth of 'wealth', that is, by capitalist propensity to a continuous increase in extraction, production, marketing, and consumption.

For a host of reasons, therefore, we could say: 'Degrowth is the new name for Peace'. Today, degrowth in 'developed countries is in fact a 'necessary condition for peace', even if it is not a 'sufficient guarantee'. Indeed, we must ask on which conditions we will concretely reach a reasonable reduction of forms of production and consumption. Today the double challenge of a 'critical pacifism' and a 'democratic and non-violent degrowth' is to understand more deeply the connections between the 'means of production' and the 'means of destruction', between the ways in which we produce wealth and economic prosperity and the ever new ways in which wars and violence spread themselves (Deriu 2005). For the pacifist movement, it is not enough to criticize military action without building a sufficiently strong and organized opposition to the economic and political system that demands military operations to defend the fundamental economic interests of 'developed' countries. For the degrowth movement it is fundamental to explore how the mere evocation of socio-ecological catastrophe risks – in the absence of a strong participatory and democratic movement – leaves space for securitization and militarization of the environmental question (Buxton and Hayes 2016).

It is therefore necessary to develop strategies of nonviolent struggle (Engler and Engler 2016) against political, economic, and legal models imposed by national and transnational bodies. We must think about what forms the non-violent defence of territories and local community could take. The goal is to be able to represent through paths of local and international mobilization a 'public crisis' that reveals these injustices and obliges public opinion and governments to undertake adequate actions to defend democracy, equity, environmental sustainability in relations among people, countries, genders, and generations.

Further Resources

Buxton, Nick and Ben Hayes (eds) (2016), *The Secure and the Dispossessed: How the Military and Corporations Are Shaping a Climate-Changed World*. London: Pluto Press.

Deriu, Marco (2005), *Dizionario critico delle nuove guerre*. Bologna: Emi.

Engler Mark, and Paul Engler (2016), *This Is an Uprising: How Nonviolent Revolt Is Shaping the Twenty-First Century*. New York: Nation Books.

Environmental Justice Atlas, https://ejatlas.org/.

Global Witness, https://www.globalwitness.org/en/.

The SIPRI list of Top 100 Arms-Producing and Military Services Companies, https://www.sipri.org/sites/default/files/The-SIPRI-Top-100-2015.pdf.

Marco Deriu is a researcher at the University of Parma, Italy, where he teaches the sociology of political and environmental communication. He is also an active member of the *Associazione per la decrescita*.

———————

PEACEWOMEN

LAU Kin Chi

Keywords: violence, subaltern marginalization, women's relational skills, global peace

An initiative called 1,000 Women for the Nobel Peace Prize 2005 was launched in 2003 as a way to make the thoughts and actions of subaltern women more widely known. From Switzerland, a call went out around the globe and an international committee of twenty women from all continents was formed. After selection and documentation, 1,000 women from over 150 countries were collectively nominated for the Nobel Peace Prize in 2005.

Women were chosen for the project without any intention to essentialize 'the feminine' or any biological polarization of women versus men. Rather, the intention was to highlight women's nurturing relations in everyday life; their experience and learned skills so critical for overcoming violence and for fostering lasting peace. The historical marginalization of women, like that of indigenous, peasant, and other groupings, needs to be viewed against the complex economic and cultural forces that keep them in social subjection. At the same time, women's unique initiatives and resistance need to be made visible. PeaceWomen fosters conversations from different quarters of the globe on alternative modes of life and subjectivities, and new imaginings about ways of becoming.

One of the driving forces behind PeaceWomen Across the Globe (PWAG) is the desire to build a platform for women to tell their stories to one another. These are stories of how women have turned limiting conditions into openings by actualizing connections outside the forces of capitalist commodification. PeaceWomen's 2005 nomination for the Nobel Peace Prize was an intervention to make women visible; for their creative efforts at the margins are generally deemed minor, local, and fragmentary. By the logic of globalization and modernization, that which resists must be dismissed. The globally dominant political centre excludes what it cannot control.

As co-president of the project, Dr Ruth-Gaby Vermot-Mangold, member of the Swiss National Council and Council of Europe, remarked:

> In January 2005, we presented their names to the Nobel Prize Committee in Oslo, in the belief that, 100 years after Bertha von Suttner, the first woman to be awarded the Nobel Prize in 1905, it is time that more women be honoured for their efforts, their courage and their determination in building peace.

One among the 1,000 women was the 'Anonyma' 'to stand for all the nameless women whose work has been overlooked, and for those endangered women who set things in motion and bring about change but who must remain nameless'.

The project evoked much interest and enthusiasm in many parts of the world. The Nobel Peace Prize Committee even endorsed a special process for this nomination which normally should remain confidential. Here, confidentiality could not be adhered to because the very point of the project was to make visible the contributions of 1,000 women who represented millions of other marginalized women. Even though the project lost to the International Atomic Energy Association (IAEA) for the Nobel Peace Prize 2005, it continued, and was renamed PWAG in 2006. A 2,200-page book was produced, and over 1,000 exhibitions featuring the 1,000 PeaceWomen's stories in various languages were staged across the globe in the next five years.

Two key themes have been pursued by PWAG: first, to promote the United Nations Resolution 1325 so as to enable women to be engaged in peace processes; second, to promote interconnectedness between peace, livelihood, and ecology concerns. Projects taken up include:

- peace dialogues in Egypt
- dozens of peace roundtables in different continents

- inter-regional learning of women from Argentina
- Brazil and Indonesia to combat violence
- women peace mediator courses
- engagement in the #WomenSeriously
- the One Billion Rising campaigns
- exchanges among women farmers from Africa, Latin America, and Asia.

In 2015, to celebrate its 10th anniversary, PWAG launched a 'visibility–connectivity–expertise' project: Wiki PeaceWomen. It aims to expand the knowledge and skills of millions of PeaceWomen working in all fields of human security, conflict resolution, ecological security, environmental justice, health, education, legislation, and others. Their expertise is to be disseminated at different levels beyond their current spheres of influence, ranging from community to global associations. This campaign invites stories from a Million PeaceWomen, to be written and translated, fostering a global awareness of women's contribution to combatting violence.

PeaceWomen defy the powerful, the greedy and the mean; they are dedicated to making this world a better one. Their stories must be told for future generations to listen to and to take pride in. PWAG is a global project of hope, of the alliance of hope.

Further Resources

1000 Women for the Nobel Peace Prize (2005), *1000 PeaceWomen Across the Globe*. Zurich: Scalo.

Chan Shun Hing, Lau Kin Chi, Dai Jinhua and Chung Hsiu Mei (eds) (2007), *Colors of Peace: 108 Stories of Chinese PeaceWomen*. Beijing: Central Compilation and Translation Press.

Lau Kin Chi. (2011), 'Actions at the Margins', *Signs: Journal of Women in Culture and Society*. 36 (3): 551–60.

PeaceWomen Across the Globe, http://www.1000peacewomen.org.

Wiki PeaceWomen: Visibility,Connectivity, Expertise, Mission, http://wikipeacewomen.org/wpworg/.

Lau Kin Chi is an Associate Professor in the Department of Cultural Studies at Lingnan University, Hong Kong, China. She is an International Board Member of PeaceWomen Across the Globe; a Coordinator of the Wiki PeaceWomen project; and a founding member of the Global University for Sustainability.

PEDAGOGY

Jonathan Dawson

Keywords: education, pedagogy, empowerment

Pedagogy: from the middle French (16ᵗʰ century), from Latin paedagogia, 'education, attendance on boys', from paidagogos 'teacher'.

Pedagogy may be considered the Cinderella of the educational world – largely overlooked despite the critical role it plays in the educational household. This is especially evident in the field of economics where the focus of student protests in recent years has been on the curriculum, with demands centred on the need to teach schools of thought other than neoliberal economics. The implicit assumption here is that switching one set of textbooks for another will be enough to redress the current dysfunction within the discipline. On closer inspection, this begins to look like an unsatisfactory, superficial analysis.

A number of core assumptions dominate current educational practice, so deeply entrenched that we are scarcely aware of their existence. This is what Stephen Sterling calls 'the subterranean geology of education' – dominate current educational practice. The assumptions include the belief that there is a fixed body of knowledge to be transmitted in subject-specific silos by the 'expert' teacher; that the intellect is the only legitimate faculty of learning; and that learning is an individual and competitive process, in which collaboration is denounced as cheating.

Thankfully, these core assumptions are increasingly being called into question and we are seeing a new wave of educational experimentation built on an altogether more enlightened ethic.

There is a growing recognition that knowledge, rather than being fixed and in any sense representing objective truth, is in fact socially constructed, with meaning-making emerging from an ongoing iterative process of experimentation, questioning and reflection within the learning community. In this ongoing learning journey, and building on insights from Paolo Freire, there is a recognition that the language we use to interpret the world is not a true representation of some objective reality but rather that it emerges from structural power relations which, when unquestioned and unchallenged, it tends to insidiously perpetuate (Freire 2007).

Seen in this light, the centre of authority needs to shift from the teacher to the learning community, with the role of the educator being 'to draw out from', recast as a resource, mentor, provocateur and, in some sense, peer rather than only a transmitter of information.

This has been an especially marked development in those contexts where the role of education in the process of ideological colonization has been particularly strong. Universidad de la Tierra, or Unitierra, in Oaxaca, Mexico, for example, was created in response to a belief that '[t]he school has been the main tool of the State to destroy the Indigenous people'. Unitierra has created a learning ethic that is more closely based on Indigenous educational practice and that emphasizes informal, peer-supported project-based education over the hierarchical model of the conventional teacher–student relationship.

Conventional scientific method leans heavily on rationality and validation based on empirical evidence. The student/researcher is presumed to stand outside the field of study as an impartial observer, able to reach objective conclusions on the basis of cognitive reasoning alone. Our emerging understanding of the learning process relocates the learner as embodied and deeply implicated in the world, exploring it with the full range of human faculties: rational and cognitive, experiential, intuitive, relational, and embodied.

The validation of the subjective that this entails brings the classroom back to life. Students are no longer required to park their emotions, their intuition and their bodies at the classroom door. Rather, they are invited into a space that welcomes their creativity and playfulness, their passions and their tears. The student's role shifts from that of object to be operated on to a subject within relationships.

For example, as part of their masters level course, students of economics at Schumacher College in Devon, England, engage in a Theatre of the Oppressed workshop and constellate various politico-economic contexts, using their bodies to map complex systems and to experiment somatically with potential courses of action (Dawson and Oliviera 2017).

A growing number of studies and surveys of innovative educational initiatives are finding that this more holistic approach to education – one that stimulates 'head, heart and hands' – is significantly more effective in catalysing behaviour change and enabling students to engage critically and creatively with the values, skills, and knowledge requirements of the sustainability-related challenges we face.

Finally, conventional pedagogic practice defines the student as a self-contained and essentially independent learner, in competition with his/her peers in the race for marks. An emerging alternative interpretation recognizes that the student is embedded in multiple relationships within the human and other-than-human world and that it is precisely these relationships that enable and catalyse the emergence of knowledge.

Given this, it is unsurprising that a common feature of many of today's pioneering educational initiatives is their rootedness in community. Centres such as the many ecovillages worldwide that host the Gaia Education curriculum, and institutions such as India's Swaraj University and the Barefoot College are designed explicitly to be embedded within 'living and learning' communities, deeply embedded in the Gandhian ashram and *nai talim* tradition (Sykes 1988). In these programmes, students and staff work side by side managing the education centre: growing food, cooking, washing dishes, cleaning and maintaining the buildings. This expands the 'living classroom' to include all dimensions of life of the college, enabling a breaking down of the artificial boundaries that conventionally exist between the theory and practice of sustainability. Students learn to grapple with issues relating to decision-making and conflict resolution, sourcing and cooking food, and relating to others in respectful and regenerative ways.

For too long, students' vibrantly creative subjectivities have been locked out of the classroom. The story that has been imposed – that of a lonely, competitive, hyper-rational being dissecting the world so as to more effectively manipulate it for his own benefit – has left us bereft of meaning while the Earth bleeds. We need to weave ourselves back into the fabric of life. A recasting of the form and purpose of learning is as good a place to start as any.

Further Resources

D'Alisa, Giacomo, Federico Demaria and Giorgios Kallis (2015), *Degrowth: A Vocabulary for a New Era*. New York: Routledge.

Dawson Jonathan and Hugo Oliviera (2017), 'Bringing the Classroom Back to Life', *EarthEd (State of the World): Rethinking Education on a Changing Planet*, Washington DC: Worldwatch Institute.

Freire, Paulo (2007), *Pedagogy of the Oppressed*. New York: Continuum.

Lakoff, George, and Mark Johnson (1980), *Metaphors We Live By*. Chicago, IL: University of Chicago Press.

Sykes, Marjorie (1988), 'The Story of Nai Talim', http://home.iitk.ac.in/~amman/soc748/sykes_story_of_nai_talim.html.

Jonathan Dawson is an educator at Schumacher College in Devon, England, where he coordinates and teaches in the innovative Masters programme on Economics for Transition, https://www.schumachercollege.org.uk/courses/postgraduate-courses/

economics-for-transition. He is one of the core team that created the Gaia Education curriculum, https://gaiaeducation.org/and a former president of the Global Ecovillage Network.

•••••

PERMACULTURE

Terry Leahy

Keywords: agriculture, permaculture, sustainable living

'Permaculture' is a term coined by Australians, Bill Mollison, and David Holmgren (1978). It is variously rendered as 'permanent agriculture' and 'permanent culture'.

The permaculture movement has evolved through three phases, each represented by a key book. *Permaculture One* (1978) emphasizes the replacement of annual crops with perennials. Mollison and Holmgren say that permaculture is 'an integrated, evolving system of perennial or self-perpetuating plant and animal species useful to man [*sic*]'. Agriculture is a 'food forest'. The fertility of agricultural soils depends on humus laid down by centuries of forest cover. Once this land is cleared to grow cereals, this topsoil is progressively used up. Permaculture attempts to escape this trap, using perennials to provide food and to build soil.

By the second phase, this definition by perennials is quietly abandoned. In *Permaculture: A Designers' Manual* (1988) Mollison defines permaculture in two sentences:

> Permaculture (permanent agriculture) is the conscious design and maintenance of agriculturally productive ecosystems which have the diversity, stability, and resilience of natural ecosystems.

This equates to a definition in terms of agricultural sustainability – stability and resilience. The next sentence broadens permaculture substantially:

> It is the harmonious integration of landscape and people providing their food, energy, shelter and other material and non-material needs in a sustainable way.

Here we are not just talking about agricultural systems but about every kind of technology that humans may use to relate to nature. It includes energy, the use of metals, pottery, even computers, so long as all these things can be produced sustainably! But in fact, the *Designer's Manual* is almost entirely about agricultural strategies, with a brief discussion of solar passive design for housing. Mollison also adds four 'permaculture ethics' – care for the earth; care for other people; limits to population and consumption; distribution of surplus. These are ethical positions common to the environmentalist movement as a whole.

The most recent phase of the permaculture movement comes out of Holmgren's influential book, *Permaculture: Principles and Pathways Beyond Sustainability* (2002). This continues the drift away from permaculture defined purely as an agricultural strategy. Instead it develops a set of 'design principles' relevant to all decision making – personal, economic, social, and political. For example, 'produce no waste' and 'obtain a yield'.

This broadening of the permaculture concept can obscure the grassroots focus of the movement as permaculture becomes a movement for the popular science of sustainable agriculture and settlement design. Permaculture strategies show us the direction in which we will all have to move in order to reduce energy use. The principles are also relevant to any kind of non-capitalist agriculture; to situations where industrial inputs are not paid for by high profits from cash crops; and to situations where labour costs are not a determining factor.

What follows is a summary of the lessons of permaculture – the accumulated wisdom of the movement:

- Permaculture favours organic agriculture – synthetic chemical fertilizers, pesticides and herbicides damage the soil, our health, and other species.
- Permaculture designs must include and emphasize perennial crops – to maintain and retain soils, to provide fodder, fuel, and food.
- A poly-culture is the best agricultural strategy – to maximize biodiversity and to deal with pests and diseases without using harmful chemicals. Here, an integration of livestock and cropping is required so that resources from both can be readily interchanged.
- We must do without machinery, transport and inputs dependent on fossil fuels. We are already running out of such resources and global warming is a big problem.
- Agriculture must surround and interpenetrate settlements – so food transport can be on foot or by animal traction.
- A local agriculture permits recycling of nutrients in human and animal manure and avoids the need to refrigerate meat or vegetable foods.

- Permaculture emphasizes plants and animals that are robust in a particular locale, not ones that depend on irrigation and synthetic inputs.
- Agricultural jobs must be diverse and labour intensive, requiring knowledge of a range of species and their interactions.
- Permaculture emphasizes built structures to retain and use water in the landscape rather than pumping water over long distances, and using fossil fuel energy.

These lessons make good sense for a post-development strategy. Integration into a global high energy economy cannot rescue the poor. Permaculture strategies are democratic, with jobs accessible to all. Here, subsistence farmers who cannot afford commercial inputs have an organic strategy. Perennial crops feed animals, fix nitrogen, and provide mulch. Variety prevents a pest species from wiping out the whole harvest. A local agriculture does not depend on long supply chains and polluting oil-based transport. Proximity makes it easy to link crops and animals, recycling nutrients. Local earthworks can save rain where it falls, and manage water supply for domestic use and crops. Permaculture makes for an engaging experience of agricultural work.

Examples of permaculture design have been appearing around the world. The Cubans saved their country from starvation after they lost access to oil from the Soviet Union. Permaculture volunteers established a local agriculture that did not depend on oil or industrial inputs. The Loess Plateau of China, turned into a desert by generations of farming, has been restored using permaculture techniques. In Niger, World Vision has pioneered 'farmer managed natural regeneration' to re-establish a mixed agricultural regime of woodland and cropping. In the Philippines, peasant farmers driven into debt by cash cropping have moved from high-input agriculture into food security. Their movement MASIPAG means 'diligent and energetic' in Tagalog language. In Zimbabwe, the Chikukwa clan has restored the food security of six villages with a permaculture project named in Shona language 'Strong Bees'. It was initiated by local people and still continues after twenty years (Leahy 2013). More experiments such as these will proliferate as the growth economy falters.

Further Resources

Birnbaum, Juliana and Louis Fox (eds) (2015), *Sustainable Revolution: Permaculture in Ecovillages, Urban Farms and Communities Worldwide.* New York: Random House.
Holmgren, David (2002), *Permaculture: Principles and Pathways Beyond Sustainability.* Hepburn: Holmgren Design Services.
Leahy, Terry (2013), 'The Chikukwa Permaculture Project (Zimbabwe) – The Full

Story', https://permaculturenews.org/2013/08/15/the-chikukwa-permaculture-project-zimbabwe-the-full-story/
Mollison, Bill (1988), *Permaculture: A Designer's Manual.* Tyalgum: Tagari Publications.
Mollison, Bill and David Holmgren (1978), *Permaculture One.* Uxbridge: Corgi.
Permaculture Research Institute, www.permaculturenews.org.
The Chikukwa Project, www.thechikukwaproject.com.

Terry Leahy is an activist sociologist, recently retired from the University of Newcastle, Australia. His site, www.gifteconomy.org.au demonstrates the relevance of permaculture to food security projects in Africa.

POPULAR SOLIDARITY ECONOMY

Natalia Quiroga Díaz

Keywords: popular economy, non-capitalist economies, solidarity, Latin American economy

Popular economy arose in the 1980s in Latin America as an academic concern in response to the widespread use of the concept of informality proposed in 1972 by the International Labour Organization (ILO), and by academic and governmental institutions. 'Informality' was coined to explain the persistence of unsalaried work susceptible to modernization. Popular economy, however, had existed for a long time, particularly, although not exclusively, in multiple persisting forms of economic, social, and cultural indigenous practices.

In subsequent decades, neo-liberal policies deepened social inequalities and the worsening of living conditions. In this context, and with a strong Marxist influence, a polysemic concept of popular economy emerged. It was based on the premise that the basic organization of the popular economy was not small enterprises, but the domestic unit, from which people develop labour strategies, whether remunerated or not, intended for the satisfaction of needs. Thus understood, the concept underlined the work of social reproduction carried out predominately by women in various guises.

It is well known that hegemonic social theory fragments society into spheres: economy, society, culture, and politics. The economy is defined as a self-regulating market system which, left to its own devices, will optimally resolve the allocation of resources. For critical thinkers such as Karl Polanyi, however, the economy is embedded in society the diversity of social, cultural, and political ties are not external, but integral to it. The economy is thus qualified as 'social economy', given that it is seen from the perspective of the entire society, according to the ultimate ethical principle of the reproduction of life.

The economy is further conceived as a plural system of institutions, norms, values, and practices which organize and coordinate the process of production, distribution, circulation, and consumption, whose purpose is to generate the material basis for meeting the legitimate needs and desires of all, and to live with dignity and responsible freedom, democratically and in harmony with nature. It assumes an ineluctable ethical principle opposed to the mercantile project with its endorsement of unrestricted accumulation at others' expense.

Regarding 'solidarity social economy' in Latin America, it is as much a special way of carrying out the everyday economy, as a project of collective action intended to counteract the destructive tendencies of capitalism, with the potential for constructing an alternative economic system. Solidarity comes into being in the disposition of each individual or community to recognize others and safeguard their necessities without having to renounce their own interests. It implies responsible cooperation instead of destructive competition, the sharing of resources and responsibilities, participation in the redistribution of wealth and the collective fostering of preferable ways of sociability.

For the social economy, the market is only one of a multiplicity of economic principles; others include: the social division of productive labour in the generation of living conditions in the exchange with nature; autarky or self-sufficiency, reciprocity, primary distribution/appropriation, the redistribution of surplus, non-market exchange, ways of consumption, and the coordination of the entire economic process. Far from being universal, these axes are unique to each society and historical moment.

Latin American social and solidarity economy has an important specificity: the existence of a large sector of popular economy that in some areas might involve two thirds of the national demand: for example, farmers satisfying local food needs. The vitality of these economic forms shows that society is sustained by various rationalities and relations that do not have profit making as their only goal. The various forms of popular, social, and solidarity economy maintain a necessary imbrication with diverse rural and urban movements struggling for land, popular habitats and community economies, as well as with the feminist movement (Quiroga 2009).

By criticizing the entrepreneurial and capitalist vision of the world, the concept of popular economy thus offers a contextualized understanding of economy. Razetto's foundational work (Razetto *et al.* 1983) underscores the contributions of the impoverished sectors themselves in constructing responses to their subsistence problems through a popular and solidarity economy geared to overcoming poverty. Coraggio (1987) examines the diversity of forms taken by work and the fragmentation imposed by capital, emphasizing the necessity of overcoming atomization in order to strive for a reproductive rationality of life as the central purpose of an economy capable of producing transition alternatives to capital. Quijano (1998) suggests the concept of 'marginal poles' which characterize structural heterogeneity, be it of economic activities, organizational forms, or use, and level of resources and technology and productivity. He highlights the importance of conditions for reproductive autonomy – land and services – more than that of labour and salary conditions.

The practice of constructing and sustaining a different economy takes place not only within social, cultural, and political conflicts, but also in the midst of conceptual contradictions that become relevant when they find their way into popular forms of solidarity. Thus the term 'social economy' has been used by governments to name focalized policies of 'assistentialism' that promote precarious self-employment among the poor. Moreover, oftentimes activities that are considered part of solidarity economies end up being temporary measures, carried out only until the crisis is overcome and a return to the 'modern' economy becomes feasible; these activities, therefore, do not have a transformational potential. That said, there have been in Latin America effective public policy experiences of supporting processes of rural and urban self-management and organization of production and reproduction; these have widened the space for the institutionalization of the economy beyond the market, thus limiting the unchecked expansion of the commercial logic promoted by neo-liberalism.

With its long history of struggle and organization, popular, and solidarity economies continue to confront the technical understanding of the economy, by prioritizing the material and symbolic conditions of reproduction of those who, in exercising their labour, produce the conditions of their territorialized existence with one another.

Further Resources
Coraggio, José Luis (1989), 'Política económica, comunicación y economía popular', *Ecuador Debate*. CAAP, 17, Quito. Bogotá: Revista Foro.

————— (2015), 'Para pensar las nuevas economías. Conceptos y experiencias en América Latina', in Boaventura de Souza Santos and Teresa Cunha (eds), *International Coloquium Epistemologies of the South*. Volume 3, Coimbra: Centro de Estudos Sociais (CES).

Hinkelammert, Franz, and Henry Mora (2009), *Economía, sociedad y vida humana: Preludio a una segunda crítica de la economía política*. Buenos Aires: UNGS/ Altamira.

Quijano, Aníbal (1998), *La economía popular y sus caminos en América Latina*. Lima: Mosca Azul Editores.

Quiroga, Natalia (2009), 'Economías feminista, social y solidaria. Respuestas heterodoxas a la crisis de reproducción en América Latina'. *Iconos: Revista de Ciencias Sociales*. 33, Flacso-Ecuador.

Razetto, Luis, Apolonia Klenner, Arno Ramirez and Roberto Urmeneta (1983), *Las Organizaciones Económicas Populares*. Santiago: Ediciones PET.

Natalia Quiroga is academic coordinator for the Masters in Social Economy at the Universidad Nacional General Sarmiento, Argentina, and co-coordinator and co-cofounder of CLACSO's working group for emancipatory feminist economics. She has degrees in economics (BA, Universidad Nacional de Colombia); regional planning and development (Universidad de los Andes, Colombia); and social and solidarity economy (MA, Universidad Nacional General Sarmiento).

POST-ECONOMIA

Alberto Acosta

Keywords: degrowth, post-extractivism, progress, buenvivir, transdisciplinarity, rights of nature

The dominant capitalist patriarchal civilization suffers a multi-faceted and systemic crisis. Never before have so many pressing questions appeared simultaneously. This crisis of civilization occurs across many fields: politics, ethics, energy, food, in culture and everyday life. We are living not only with severe environmental problems, but with a loss of historic meaning. This calls for great transformational solutions.

The capitalist patriarchal economy as we know it, cannot offer acceptable solutions. Its ostensible quest for well-being merely covers over a search for power, driving life on Earth to a precipice of no return. Despite this, powerful social forces are resistant to change – material and ideological change. An assortment of economic think tanks is central to these barriers. So, it is urgent to think outside of the economy – pillar of modernity that it is.

The discipline of economics has a troubling history. Its theory has evolved through a complex and contradictory process but with an incessant search for progress and its stepchild, development. Neither the goal of permanent optimization, nor balance of social and environmental relations, are effective. Economists try to understand and change the world through an instrumental faith, whether by maximizing the logic of the market or by a rationalization of state structures.

Worse, economics has an unhealthy tendency to seek legitimacy by distancing itself from other so-called social sciences, preferring to locate itself among the exact and natural sciences. The positivism and functionalism of the latter are envied by many economists who yearn to emulate their methods, including the use of a mathematical language never developed for understanding the complexity of human society.

The weak identity of economics as a discipline has given birth to a Prize in Economic Sciences in memory of Alfred Nobel. This serves to justify, even purify, an area of human knowledge buffeted by a complex battle of interests. It motivates some practitioners to look for balance, exactitude and measurability, while others reach for new ideas like chaos and complexity; still others end up vanquished in useless prose.

Repressive social structures, capitalist accumulation, and economic rationalist dynamics must be replaced if life on Earth is to survive. As Karl Polanyi conceded in *The Great Transformation* (1944), this may result in the death of economic science itself. But we need a post-economy, gathering together only those ideas that will guarantee a harmonious life between human beings, and between humans and nature.

The above-mentioned post-economy should abandon anthropocentrism. We must accept that all beings have the same ontological value no matter what their usefulness, or what work is required to maintain their existence. We need to recognize and celebrate non-instrumental values in the non-human life-world, overcoming the crass materialism of capitalist patriarchal economic thinking.

How do we build this post-economy free of 'use values' and 'exchange values'? A response is impossible in few lines, but one clear point is the need to create from the start, paradigms, languages, and methodologies focused on understanding the interconnection of social reality and natural reality.

We need a holistic social ecology to overcome capitalist patriarchalism, particularly now, in its era of speculative degeneration.

This post-economy is not an anti-economy. Actually, a post-economy would acknowledge that society needs production, distribution, circulation, and consumption to reproduce its material and socio-political life. Nevertheless, these processes must be regulated by a social-ecological reality, and not by capital, which drowns the planet in its own waste (Schuldt 2013).

This change can only be possible if we overcome the hurricane of progress, as Walter Benjamin (1920) understood it. It is also urgent to abolish the fetish of economic growth: a finite world does not accept infinite growth. Therefore, it is imperative to call upon degrowth, right across the northern hemisphere, to physically equilibrate the global 'economic metabolism' and to do away with any centre. What is wanted is community relationships, not individualistic; plural and diverse relationships, neither one-dimensional nor mono-cultural; as well as a profound decolonization (*Cuestiones y Horizontes* 2014). Simultaneously, a commitment to post-extractivism is called for in the southern hemisphere. A degrowth–post-extractivist convergence (Acosta and Brand 2017) means that the poor will not continue to sustain the opulence of the rich.

A post-economy demands:

- de-commodification of nature and common goods
- recognition of rights and harmonious relations among all living beings
- community-based criteria of evaluation
- decentralized and deconcentrated production
- a profound change in patterns of consumption
- a radical redistribution of wealth and power

and many other actions that will be considered collectively. It is crucial to recuperate alternative epistemes like *buenvvir, eco-swaraj, ubuntu,* or communitarianism – to understand and re-organize the world without falling into mandates such as *development.*

Finally, it is crucial that a post-economy should be trans-disciplinary – not uni- or multi-disciplinary, but drawing on the most complete knowledge possible through dialogues with diverse human wisdoms, contemplating the world as a question, an aspiration. We must critically learn from *social sciences* as well as natural sciences to integrate them in a systematic understanding of the world as a multi-faceted, constitutionally diverse totality – and this, without any zeal of superiority (Acosta 2015). The task implies constructing and reconstructing the pluriverse in discussion without dogma or imposition. We either continue to be dominated by misguided economic visions or construct another economy for another civilization in permanent epistemic subversion.

Further Resources

Acosta, Alberto (2015), 'Las ciencias sociales en el laberinto de la economía'. *POLIS Latinoamericana Magazine*. Número 41, https://polis.revues.org/10917?lang=en.

Acosta, Alberto, and Ulrich Brand (2017), *Salidas al laberinto capitalista – Decrecimiento y Postextractivismo.*Barcelona: ICARIA.

Benjamin, Walter (1974 [1920]), *On the Concept of History. Gesammelten Schriften I:2.* Frankfurt am Main: Suhrkamp Verlag.

Polanyi, Karl (2001 [1944]), *The Great Transformation: The Political and Economic Origins of Our Time.* Boston: Beacom Press.

Schuldt, Jürgen (2013), *Civilización del desperdicio: psicoeconomía del consumidor* [The Civilization of Waste-Consumer Psycho-Economic] Lima: Universidad del Pacífico.

Quijano, Annibal (2014), *Cuestiones y Horizontes – Antología Esencial – De la dependencia histórica-estructural a la colonialidad/decolonialidad del poder.* Buenos Aires: CLACSO.

Alberto Acosta, Ecuadorian economist, is former marketing manager of CEPE (Ecuadorian State Petrol Corporation); an officer of OLADE (Latin American Energy Organization); an international consultant; former Minister of Energy and Mining for Ecuador; and former President of the Constituent Assembly of Montecristi. He is presently a professor and author of numerous books and articles; a comrade-in-arms with grassroots struggles; and a Member of the Permanent Group on Alternatives to Development established by the Rosa Luxemburg Foundation.

PRAKRITIK SWARAJ

Aseem Shrivastava

Keywords: *swaraj, swadeshi,* democracy, liberty

Do not consider this Swaraj to be like a dream.
— M.K. Gandhi

Speaking of *swaraj* in the twenty-first century, one feels assured that one is aiming to recover and revitalize a vision which belongs to a strong indigenous stream of Indian philosophical thought, culture, and political practice.

Let us contemplate the word *swaraj*. Its etymological origins in Sanskrit are simple and obvious: *swa* (self) + *rajya* (rule) = *swaraj* (self-rule). The adjective *prakritik* can be understood as 'natural', or as expressing human nature so as to remain in rhythm with the natural world around us.

A notion such as *swaraj* did not swim in a historical and cultural vacuum. There is evidence of face-to-face political assemblies – including at the village level – in ancient India. Sources – both oral and documented – reveal traditions of governance through discussion and consultation and sometimes of decisions taken by dialogue and consensus.

The important thing to remember is that notions such as *swaraj* – in Sanskrit or Pali, from which some of the vocabulary of modern Indian democracy is drawn. These words predate the colonial era by centuries, often by millennia, and are, by no means, translations of concepts imported into India from the Western world. It means that these were notions in use in one period of Indian history or another and became substantially dormant, especially with the coming of colonial rule in the modern period.

So Gandhi did not dream up the idea of 'village republics', or *gram swaraj*, out of thin air. In 1909, he published his most important work *Hind Swaraj*. Gandhi's use of the term built upon earlier usage during the freedom struggle. Tilak[1] deployed the term during the early phase of the Indian freedom struggle in the 1890s. *Swaraj* seemed to become virtually equivalent to the modern Western notion of liberty and independence. In 1906, when Dadabhai Naoroji, as President of the Indian National Congress, declared *swaraj* to be the goal of the national movement, he had this very limited meaning in mind.

Gandhi's vision went well beyond this. Aware of its ancient lineage, Gandhi, in 1931, wrote in *Young India* of *swaraj* as 'a sacred word, a Vedic word' (Gandhi 1931). He hoped that India and the world could recover the old idea of *swaraj*, and realize it one day.

For Gandhi, authentic self-rule is possible if and only if the self is capable of being its own sovereign. Gandhi was religious. He believed that without transcendence, it was impossible for the self to become sovereign over its life. For him the notion was as spiritual as it was political. But, importantly, the causation works only one way. Ultimately, for Gandhi, *swaraj* was a divine imperative, with fruitful consequences for human affairs. Spiritual mastery and self-possession can also yield the marvels of political sovereignty as a by-product, but not the other way around.

Politically, self-rule, as Gandhi understood it, was anything but modern parliamentary or representative democracy. In *Hind Swaraj*, he mocked modern parliaments as 'emblems of slavery'. It is unfortunate that *swaraj* is frequently translated as 'democracy'. In fact, in its representative form democracy has been adopted in most countries, but their cognitive premises could not be more different.

First, *swaraj* is inconsistent with mass politics, an everyday fact of democracies today. Where finite, face-to-face neighbourhood assemblies are not viable, *swaraj* cannot function. Crowds can serve as the grease for political parties in democracies, not for *swaraj*. Numbers and their comparisons are as crucial to modern democracies as they are irrelevant to *swaraj*.

Second, modern democracy is focused on the individual's direct, unmediated relationship to a state that guarantees her rights of citizenship by law. The setting 'assumed' for this relationship is one of an atomized society in which human alienation is normalized. What *swaraj* needs for its nourishment, by contrast, is a community in which the individual can come into her own through filial, cultural, social, political, economic, and ecological relationships with those including sentient beings other than the humans around her.

Third, in a modern democracy, an individual is, almost indifferently, and in the name of 'freedom', left to his tastes and desires – all of modern economics rests on this assumption, the community playing no part in making him/her scrutinize them. There is no obligation for the individual to consider his/her desires in a critical light, unless and until their realization interferes with the fulfilment of another's desires. In fact, such is virtually the very definition of 'freedom' in modern liberal democracies, often understood in terms of the notion of 'negative liberty'.

Gandhi's idea of *swaraj* has to do with an individual's or a community's autonomy to 'create' their choices, rather than passively accepting the menu from which they must 'choose'. Applied to our market-driven, media-prompted world, it would first require us to take ecological and cultural responsibility for our desires and explore their origins in passions stoked by advertising. Such a manipulation of desire, in which virtually everything is at stake, is antithetical to freedom for any advocate of *swaraj*. Desire, which is at the philosophical heart of the notion of freedom in modern consumer democracies, has to be critically scrutinized under *swaraj*, especially given the context of an ecologically imperiled world. One implication of this is that Gandhi's idea of *swaraj* is inevitably bound up with *swadeshi*, which brings in the necessity of economic localization.

Finally, it should be mentioned that the idea of *swaraj* continues to inspire social, political and ecological movements in India. The resistance against displacement by development, undertaken by several movements that are part of the National Alliance of Peoples' Movements, the recently formed Swaraj India party which aims to empower people at the grassroots, the movements for food sovereignty and *adivasi* or indigenous self-rule, and others, are initiatives attempting to creatively adapt the notion of *swaraj* in today's context.[2]

Notes
[1] Bal Gangadhar Tilak, prominent freedom fighter and social reformer of late 19th century.
[2] For Swaraj India, see https://www.swarajabhiyan.org/; for National Alliance of People's Movements (NAPM), see https://napmindia.wordpress.com/; see also Food Sovereignty Alliance https://foodsovereigntyalliance.wordpress.com/

Further Resources
Gandhi, Mohandas Karamchand (1931), *Young India*, March 19, Ahmedabad.
———— (2010), *Hind Swaraj: A Critical Edition*. Annotated, edited and translated by Suresh Sharma and Tridip Suhrud. Delhi: Orient Blackswan.
Muhlberger, Steven (2011), 'Republics and Quasi-Democratic Institutions in Ancient India', in Benjamin Isakhan and Stephen Stockwell, *The Secret History of Democracy*. London: Palgrave Macmillan.

Aseem Shrivastava is a Delhi-based writer and ecological economist. He holds a PhD in economics from the University of Massachusetts, Amherst. He is the author with Ashish Kothari of the book *Churning the Earth: The Making of Global India* (Penguin Viking, 2012). He is presently working on a project involving a study of the ecological thought of Rabindranath Tagore.

QUEER LOVE

Arvind Narrain

Keywords: queer, transformation, identity, love, LGBTI politics

Love is not an emotion that can be domesticated by the market, family, community, medical science, religion, or nation. Love instead has that indefinable quality of madness, which the Greeks called 'mania', and which can impel a person to defy tradition and break orthodoxies.

An act of love can begin to subvert the status quo. This idea of a subversive love is at the heart of the queer movement. The word 'queer' implies a questioning of the norms of gender and sexuality. For some it is a re-imagination of ideas of love and relationships; and for others a restructuring of the law, politics, and society, a challenge to the way we inhabit the world. It represents those who fall out and/or choose to stand out of the contours of the heteronormative social order. When 'pride' celebrations around the world adopt as their slogans 'the right to love', 'love is a human right', and 'don't criminalize love', they signal the radical potential of queer love. Queer politics is wider than the politics of LGBTI – Lesbian, Gay, Bisexual, Transgender, or Intersex. It moves beyond identity or single-issue politics. It is marked, as Leela Gandhi says, by a 'capacity for radical kinship' and the queer becomes the site on which the ideological scope of LGBTI politics is expanded to include 'unlikely affinities with foreigners, outcastes and outsiders'.

This form of radical love finds expression in the lives of thousands of people who choose to love the person of their choice, going against religious edicts, societal conventions and family expectations. Queer love is so threatening to the existing social order that some families would rather kill the lovers rather than allow love to flourish. In India, the roll call of martyrs includes the young Dalit boy, Ilavarasa who was killed for daring to marry a higher caste woman, Divya; Rizwanur Rahman who was killed for daring to fall in love with a Hindu girl, Priyanka Todi. Swapna and Sucheta from Nandigram preferred to commit suicide rather than be forced to live without each other.

By challenging these social codes, these lovers are making the social order more permeable and are laying the foundations for a more egalitarian world where differences of race, religion, and caste would cease to matter.

Queer politics aims to shatter the rigid and oppressive norms of gender and sexuality, which are encoded in social institutions as diverse as marriage, family, law, and the medical establishment. The queer political project is also a defence of a larger vision of social transformation which has its roots in the intimate desire of two or more people who want to be with each other so fiercely that they are willing to defy social strictures.

Queer love also implies the cultivation of a sensibility, which allows one to go beyond the love of a single other person to empathize with the suffering of strangers. To love a person regardless of gender or sexuality and to empathize with seemingly distant causes are sometimes two intertwined aspects of love. One of the exemplars of this form of love is Irish revolutionary Roger Casement, whose passions included not only sex with men, but also a deep concern over justice for the Congolese who suffered brutalities at the hands of King Leopold's men. He cared equally for the Indian tribes of the Amazon who were brutalized by colonial powers. Casement, apart from fighting for Ireland's freedom, documented these forms of suffering in two pioneering human rights reports, all the while, continuing to have sex with men.

In the contemporary world, another such remarkable figure is Private Bradley Manning, who moved from being a loyal member of the US army to one of its most courageous dissenters. On this journey from soldier to whistleblower, Manning also transitioned from a masculine gender (Bradley) to a feminine gender (Chelsea). As Chelsea Manning, she risked imprisonment by releasing confidential documentation of brutal military actions, because she was impelled by an uncommon empathy for Iraqis as human beings.

No less important than the public act of whistle blowing – an act of the deepest love, was Manning's decision to speak out about the intimate truth that she wanted to be recognized as a woman and not a man. There is not just a public and outer dimension to Chelsea's deep moral convictions, but also a private and inner dimension. There is a recognition that something is not quite right with the world as it exists, just as there is recognition that there is something not quite right about 'who I am'.

In the capitalist world, retreat into the private and personal is a default position for many people bewildered by forces that perpetrate injustice on a global scale. But a shift from the intimate dimension of love and sexuality to a public transformational love, as gestured by Manning, is vital. In the contemporary era, the very existence of life itself, human and non-human, is threatened. A politics that takes criminalization of the erotic seriously has to engage with that social system as it dominates and enslaves nature. A queer politics will build alliances proactively with the common struggle against any form of development that simultaneously marginalize queers, women, blacks, Dalits, and other oppressed beings.

Further Resources

Tóibín, Colm, 'A Whale of a Time',
    https://www.lrb.co.uk/v19/n19/colm-toibin/a-whale-of-a-time.
Gaard, Greta (1997), 'Toward a Queer Ecofeminism', *Hypatia*. 12 (1):137–156.
Gandhi, Leela (2006), *Affective Communities*. Delhi: Permanent Black.
Gupta, Alok and Arvind Narrain (eds) (2011), *Law Like Love*. New Delhi: Yoda Press.
Narrain, Arvind (2015), *Nothing to Fix: Medicalisation of Sexual Orientation and Gender Iidentity*. Delhi: Sage and Yoda Press.
Narrain, Arvind (2017), 'Imagining Utopia: The Importance of Love, Dissent and

Radical Empathy', in Ashish Kothari and K.J. Joy (eds), *Alternative Futures: Unshackling India.* Delhi: Authors Upfront.

Arvind Narrain is a founding member of the Alternative Law Forum in Bangalore, India and currently the Geneva Director of Arc International, working on LGBTI rights in international law and policy. His several books include *Queer: Despised Sexualities and Social Change* (2004); co-edited, *Because I Have a Voice: Queer Politics in India* (2005); *Law Like Love: Queer Perspectives on Law* (2011).

———•••———

RADICAL ECOLOGICAL DEMOCRACY

Ashish Kothari

Keywords: decentralization, localization, community, *eco-swaraj*

In the midst of the socio-economic inequities and ecological collapse we see around the world, there are a growing number of initiatives practising or conceptualizing ways of achieving human well-being that are just and sustainable. Some of these are assertions of continuing lifestyles and livelihoods that have lived in relative harmony with the Earth for millennia or centuries. Others are new initiatives emerging from resistance movements or encounters with the destructive nature of currently dominant economic and political systems. While incredibly diverse in their settings and processes, many of these initiatives and approaches exhibit some common features that enable the emergence of broad frameworks or paradigms.

One such framework that has emerged from grassroots experience in India, but is beginning to see more global resonance, is Radical Ecological Democracy (RED), locally also called eco-*swaraj*.[1] This is an approach that respects the limits of the Earth and the rights of other species, while pursuing the core values of social justice and equity. With its strong democratic and egalitarian impulse, it seeks to empower every person to be a part of decision-making, and its holistic vision of human well-being encompasses physical, material, socio-cultural, intellectual, and spiritual dimensions.[2] Rather than the state and the corporation, *swaraj* puts collectives and communities at the centre of governance and economy. It is grounded in real-life initiatives across the Indian subcontinent, encompassing sustainable farming, fisheries and pastoralism, food and water sovereignty, decentralized energy production, direct local governance, community health, alternative learning and education, community-controlled media and communications, localization of economies, gender and caste justice, rights of differently abled and multiple sexualities, and many others.[3]

Radical Ecological Democracy encompasses the following five interlocking spheres:

Ecological wisdom and resilience. This including the conservation and regenerative capacity of the rest of nature – ecosystems, species, functions, and cycles, and its complexity, building on the belief that humans are part of nature, and that the rest of nature has an intrinsic right to thrive.

Social well-being and justice. This including lives that are fulfilling and satisfactory physically, socially, culturally, and spiritually; where there is equity in socio-economic and political entitlements, benefits, rights, and responsibilities across gender, class, caste, age, ethnicities, 'able'ities, sexualities, and other current divisions; where there is a balance between collective interests and individual freedoms; and where peace and harmony are ensured.

Direct or radical political democracy. This is where decision-making power originates in the smallest unit of human settlement, rural or urban, in which every human has the right, capacity, and opportunity to take part; building outwards from these basic units to larger levels of governance that are downwardly accountable; where political decision-making takes place respecting ecological and cultural linkages and boundaries. This implies challenging current political boundaries including those of nation-states; and where the role of the state eventually becomes minimal, for functions such as connecting across larger landscapes, and whatever welfare measures may still be necessary.

Economic democracy. In this democracy local communities including producers and consumers, often combined in one as prosumers, have control over the means of production, distribution, exchange, and markets; where localization is a key principle providing for all basic needs through the local regional economy; larger trade and exchange, as necessary, is built on and safeguards this local self-reliance; nature, natural resources, and other important elements feeding into the economy are governed as the commons; private property is minimized or disappears; where non-monetized relations of caring and sharing regain their central importance; and indicators are predominantly qualitative, focusing on basic needs and well-being.

Cultural and knowledge plurality. In this democracy diversity is a key principle; knowledge including its generation, use and transmission, is in the public domain or commons; innovation is democratically generated and there are no ivory towers of 'expertise'; learning takes place as part of life and living rather than only in specialized institutions; and, individual or collective pathways of ethical and spiritual well-being and of happiness are available to all.

Seen as a set of petals in a flower, the core or bud where they all intersect forms the following set of values or principles, which too emerges as a crucial part of the alternative initiatives. These can also be seen as the ethical or spiritual foundation of societies, the worldview(s) that its members hold.

- Ecological integrity and the rights of nature
- Equity, justice, and inclusion
- Right to and responsibility of meaningful participation
- Diversity and pluralism
- Collective commons and solidarity with individual freedoms
- Resilience and adaptability
- Subsidiarity, self-reliance, and eco-regionalism
- Simplicity and sufficiency (or the notion of 'enoughness')
- Dignity and creativity of labour and work
- Non-violence, harmony, and peace.

The broad components and values of eco-*swaraj* have been under discussion across India through an ongoing process called Vikalp Sangam or Alternatives Confluence.[4] This process brings together a diverse set of actors from communities, civil society, and various professions who are involved in alternative initiatives across all sectors. A series of regional and thematic confluences that began in 2015, enable participants to share experiences, learn from each other, build alliances and collaboration, and jointly envision a better future. Documentation of alternative initiatives in the form of stories, videos, case studies, and other forms provides a further means of disseminating learning, and spreading inspiration for further transformation, through a dedicated website[5], a mobile exhibition and other means.

Beyond India, this approach is also linking up to radical alternatives in other parts of the world. In 2012, several civil society organizations and movements signed onto a Peoples' Sustainability Treaty on Radical Ecological Democracy;[6] subsequently a discussion list has kept alive the dialogue, and opportunities have been found for mutual learning with approaches such as degrowth, ecofeminism, cooperative societies, and social/solidarity economy in Europe, *buen vivir*, and its other equivalents in Latin America, and others.

Eco-*swaraj* or RED is an evolving worldview, not a blueprint set in stone. In its very process of democratic grassroots evolution, it forms an alternative to top-down ideologies and formulations even as it takes on board the relevant elements of such ideologies. This is the basis of its transformative potential.

Notes
[1] For the meaning of *swaraj*, please see essay Prakriti Swaraj in this volume.
[2] See Kothari 2014; Shrivastava and Kothari 2012.
[3] See www.alternativesindia.org for several hundred examples.
[4] Information on the process and its outputs is at http://kalpavriksh.org/index.php/ alternatives/alternatives-knowledge-center/353-vikalpsangam-coverage.
[5] Vikalp Sangam, www.vikalpsangam.org.
[6] Radical Ecological Democracy, http://radicalecologicaldemocracy.wordpress.com.

Further Resources
Kalpavriksh Environment Action Group, http://kalpavriksh.org/index.php/ alternatives/alternatives-knowledge-center/353-vikalpsangam-coverage.
Kothari, Ashish (2014), 'Radical Ecological Democracy: A Way for India and Beyond', *Development*, 57 (1): 36–45.
Shrivastava Aseem and Ashish Kothari (2012), *Churning the Earth: The Making of Global India*. New Delhi: Viking/Penguin India.

Ashish Kothari is a founder of the Indian environmental group, Kalpavriksh, He taught at the Indian Institute of Public Administration; coordinated India's National Biodiversity Strategy and Action Plan; served on the Boards of Greenpeace India and Greenpeace International; helped initiate the global ICCA Consortium; and chaired an IUCN network on protected areas and communities. Ashish has (co)authored or (co)edited over thirty books, including *Birds in Our Lives; Churning the Earth*; and *Alternative Futures: India Unshackled*. He helps coordinate the Vikalp Sangam, and Global Confluence of Alternatives processes, and is a Member of the Permanent Group on Alternatives to Development established by the Rosa Luxemburg Foundation.

REVOLUTION

Eduardo Gudynas

Keywords: revolution, development, ontology, capitalism, socialism

A revolutionary shift away from development is an idea whose time has come. It is indispensable to us as we face the current social and environmental crisis; it is urgent given the accelerated pace of destruction of the environment and people's livelihoods; and it is immediate in the sense that it is possible to practise it in the here and now. A new meaning of revolution must be capable of questioning radically the conceptual basis of development and moving beyond modernity.

The concept of revolution invokes a number of substantial political and cultural changes. Considering the French Revolution as the best known example, revolution is seen as indispensable to break away from an unfair order and to transform the institutions and forms of political representation, including the social and economic fabric of society. With different degrees and emphases, this concept was used to describe radical change in Mexico, Russia, China, and Cuba, among others.

The idea of revolution has also been instrumental in promoting conventional development practices. Such is the case of the industrial, technological, internet, and consumer revolutions. Such revolutions reinforced the core ideas of development even while achieving substantial changes in the structure of society.

More recent events confound the concept. In some regions there are still significant social movements defending traditional conceptions of revolution, for instance, as a means to break away from capitalism and move towards socialism. In Central and Eastern Europe, the exit from 'Real Socialism' was presented as a revolution, albeit in the opposite direction, towards market economies. Conversely, socialist revolutionary experiences, for example, in China or Vietnam, maintain such a discourse but their development strategies are functional to capitalism. And whereas Islamic revolutions reinforced the criticism of development by attacking its eurocentrism, they endorse economic growth.

Since the beginning of the twenty-first century, Latin America witnessed a left turn with several governments describing themselves as revolutionary – Venezuela, Bolivia, Ecuador, and Nicaragua. But these countries they adopted neo-developmentalist styles that fuelled economic growth through the intensive appropriation of natural resources.

Therefore, we are confronted with a variety of events that have been described as revolutionary, particularly referring to the political dimension, but also affecting cultural, economic, and religious aspects of society. In all of these cases, however, the basic components of development survived, such as economic growth, consumerism, the appropriation of nature, technological modernization, and democratic weakness. There is a paradoxical situation whereby classic revolutions such as in Russia or China, and recent revolutions such as twenty-first-century socialism in South America, whether secular or religious, all gravitated around the idea of development. Some of these revolutions showed positive results regarding political representation and social equality, but remained trapped in instrumental ends geared towards capturing the state (particularly the Leninist, Trotskyist, and Maoist versions). They all failed to promote alternatives to development.

This could be explained by the fact that all modern political traditions share the same background. Indeed, the idea of revolution matured along with other categories of modernity, such the state, rights, democracy, progress, and development.

The persistence of developmentalism has led many activists and academics to become disillusioned with revolutionary experiences, and to argue that the concept is no longer applicable to present-day realities, favouring instead a focus on local practices. Yet, this position creates an important hurdle, given that proposals for radical alternatives to development imply a set of revolutionary transformations.

Given that all of the current varieties of development are unsustainable, any radical alternative must question their shared conceptual bases in modernity. The radicalism involved in such effort requires a revolutionary practice and spirit. A revolution in the modern sense might foster, for example, a change in state regime, or replacing one variety of development with other. It thus becomes necessary to create a new interpretation of the idea of revolution capable of exceeding modernity and of imagining an alternative to its ontology.

This concept of revolution entails a rebellion vis à vis modernity, highlighting its limits while exploring alternatives to it; it summons an innovative imagination in order to outline and rehearse other rationalities and sensibilities, as well as an expanded politics involving multiple social sectors, practices, and experiences.

This understanding of revolution possesses substantive similarities with the Andean idea of *pachakuti*. *Pachakuti* refers to the dissolution of the prevailing cosmological order, while installing a state of disorder that allows for another cosmovision to emerge. Therefore, a revolution in terms of *pachakuti* does not aim at destroying modernity, but at provoking the disorganization and dissolution of its structures while generating other understandings and effects. It involves a significant re-creation.

The practices of this kind of revolution have many antecedents. The experience of disorder and re-creation is nourished both by rational ideas such as the overwhelming evidence of the social and environmental crisis, as well as by affective, artistic, spiritual, and magical experiences. This revolution does not endorse monocultures but a diversity of expressions, is collective, and requires personal transformation, particularly in restoring the value of life – Mahatma Gandhi or Ivan Illich, *zapatismo* or *buen vivir*, offer models of this. Revolution in this sense allows for a rupture with utilitarian values, re-claiming multiple ways of assigning value – aesthetic, religious, or ecological, while accepting the 'intrinsic value' of the non-human world.

As development is a performative construct, constantly produced and reproduced by all of us through daily practices, this revolution interrupts that performativity. For instance, it suspends the commodification of society and nature. These and other features of modernity thus become disorganized, leading to an unavoidable and sometimes uncomfortable consequence: a revolution that breaks away both from capitalism and socialism.

This revolution's prefigured political practices intertwine synergistically, while disseminating throughout society, becoming concretized in actions, affects, and other styles of doing politics, particularly through the interstitial rebelliousness that stems from dignity and autonomy. This is a revolution with the co-participation of non-human actors, including animals and other living beings. It reinterprets the meaning of society. Consider the possibility of an 'animal proletariat'.

This kind of revolution disorganizes the duality between society and nature, while allowing for the recreation of relational worldviews that re-embed society in nature and *vice versa*; it extends notions of 'the subject' to non-humans.

In sum, while modernity presents itself as a self-contained universal domain, hiding its limits and neutralizing the search for alternatives to it, this revolution disorganizes, exposes and fractures modernity's limits by opening them up to other ontologies. The revolutionary act consists in creating the conditions of possibility for new ontological openings.

Further Resources

Holloway, John (2003), *Change the World without Taking Power: The Meaning of Revolution Today*. London: Pluto Press.

Williams, Raymond (1983), 'Revolution', in Raymond Williams (ed.), *Keywords*. New York: Oxford University Press.

Eduardo Gudynas is senior researcher at the Latin American Center for Social Ecology (CLAES), Montevideo, Uruguay; associate researcher, Department of Anthropology, University of California, Davis; and advisor to several grassroots organizations in South America.

RURAL RECONSTRUCTION

Sit Tsui

Keywords: urban-rural integration, peasant Societies, contemporary China, *sannong*

As a response to the problems caused by industrialization and modernization in a developing country such as China, rural reconstruction has been designed as a political and cultural project to defend peasant communities and agriculture. These grassroots efforts are separate from, parallel to, or in tension with projects initiated by the state or by political parties. As an attempt to construct a platform for mass democracy and to experiment on participatory, urban-rural integration for sustainability, the Chinese model of rural reconstruction may become an alternative politics of 'de-modernity'.

From the 1920s to the 1940s, several well-known scholars of different visions were actively involved in rural reconstruction movements. James Yen who received a Western, Christian education, promoted a mass education movement and civil society in Ding County, north China, and later in southwest China. Liang Shuming, Confucian, and Buddhist, advocated rural governance through regeneration of traditional knowledge and culture in Zouping Township, Shandong Province. Lu Zuofu, owner of a shipping company, established social enterprises and public facilities to modernize Beibei town, southwest China. Tao Xingzhi combined livelihood education with communism. Huang Yanpei designed vocational training programs for rural people. After 1949, James Yen continued his rural reconstruction projects in Taiwan, the Philippines, and different countries in Asia, Latin America, and Africa.

The project of rural reconstruction is a response to the 1979 market reforms and push to export-led manufacture. The consequent demand for cheap labour, aggravated the urban-rural divide and other kinds of social polarization. The global financial crisis also impacted severely impacted the Chinese economy. Rural reconstruction as a necessary movement to defend the rural way of life was proposed in 1999 by Wen Tiejun, then a researcher of the Ministry of Agriculture and later Executive Dean of the Institute of Advanced Studies for Sustainability, Renmin University of China. He coined the term *sannong*, referring to the three rural dimensions of peasants, villages and agriculture. Since 2004, *sannong* issues have been officially accepted as being of 'the utmost important among all important tasks' in the Central Document No.1 of the Party and the State. While the government has prioritized rural development by investing over RMB 10 trillion (~1.2 trillion US$) in infrastructure and welfare for the last twelve years, rural reconstruction is committed to self-organization and mass democracy. Most of these local efforts are autonomous, operating on their own initiative, sometimes complementary to state policies.

Wen Tiejun has mobilized officials, villagers, scholars, and university students to work together for rural reconstruction. Particularly, rural women play an important role in organizing on the ground, and their engagement is extensively documented in the PeaceWomen Across the Globe Project conducted by Lau Kin Chi and Chan Shun Hing, professors from Lingnan University, Hong Kong. Of the diverse rural reconstruction endeavours, some notable events include the 'Rural Edition' of *China Reform*, a national journal which speaks for peasant interests. In 2001, the Liang Shuming Rural Reconstruction Centre was set up to provide training programmes for university students and peasant cooperatives. In 2002, the Beijing Migrant Workers' Home was set up to provide cultural and educational programmes for peasant workers. In 2003, the James Yen Rural Reconstruction Institute was established, which organized peasant training programs and advocated ecological agriculture. In 2005, the James Yen Popular Education Centre was established to promote localized popular knowledge and courses for peasant workers. In 2008, the Green Ground EcoTech Centre was established to promote rural–urban cooperation, community-supported agriculture, and ecological skills and techniques; it manages the Little Donkey Farm, a common project of the Haidian District Government and the Centre of Rural Reconstruction at Renmin University of China. In 2009, the first China Community Supported Agriculture Conference was held in Beijing. In 2013, the Association of Advancement for Loving Home Village Culture was set up to organize campaigns for recognizing grassroots efforts in defending rural heritage. In 2015, the Participatory Guarantee System of social organic agriculture was launched to build a national network of agro-ecological working groups. In addition, throughout China, there are rural reconstruction bases with diverse experiments. These include rural integrated development projects in Yongji of Shanxi province, Shunping of Hebei province, Lankao and Lingbao of Henan province; rural finance projects in Lishu of Jilin province; and popular education projects and community colleges in Xiamen and Longyan of Fujian province.

The new rural reconstruction movement has reached out to share experiences with popular movements in India, Nepal, the Philippines, Thailand, Indonesia, Japan, South Korea, Brazil, Peru, Mexico, Ecuador, Argentina, Venezuela, Egypt, Turkey, South Africa, and Senegal among others. These facilitations have paved the way for organizing three South-South Forums on Sustainability in Hong Kong and in Chongqing from 2011 to 2016.

Rural reconstruction promotes social participation, ecological agriculture, and sustainable livelihood. It is committed to the Three Peoples' Principles: people's livelihood, people's solidarity, and people's cultural diversity. It emphasizes peasants' organizational and institutional renewal – the implementation of local comprehensive experiments with the application of grassroots knowledge.

Throughout the twentieth century China had been through several political regime changes, yet regardless of who was in power, the main pursuit had been modernization, to the benefit of a small elite and to the detriment of the majority of the population. However, if rural China can be sustained for the cultivation of interdependent and cooperative relations within and among communities, not only will this protect the livelihoods of the majority of the population, but this will also function as a resistance to external crises derived from global capitalism. In that sense, the historical and contemporary manifestations of rural reconstruction, which are based on the small peasantry and village community, provide an alternative to destructive modernization.

Further Resources
Liang Shuming Rural Reconstruction Centre provides training programmes for
  university students and peasant cooperatives, http://www.3nong.org/.

Little Donkey Farm, 'Green Ground Eco-Tech Centre promotes rural-urban cooperation and agro-ecological knowhow', http://littledonkeyfarm.com/.

The Global University for Sustainability, a 'virtual' university with an online presence, facilitates an international networking for ecological and socio-economic sustainability with justice, http://our-global-u.org/oguorg/.

Tiejun, Wen, Zhou Changyong, Lau Kin Chi (eds) (2015), *Sustainability and Rural Reconstruction*. Beijing: China Agricultural University Press.

Tiejun, Wen, Lau Kin Chi, Cheng Cunwang, He Huili and Qui Jianshent (2012), 'Ecological Civilization, Indigenous Culture, and Rural Reconstruction in China', *Monthly Review: An Independent Socialist Magazine.* 63 (9): 29–35.

Wong, Erebus and Sit Tsui (2015), 'Rethinking "Rural China", Unthinking Modernization: Rural Regeneration and Post-Developmental Historical Agency', in Rémy Herrera and Kin Chi Lau (eds), *The Struggle for Food Sovereignty: Alternative Development and the Renewal of Peasant Societies Today*. London: Pluto Press.

Sit Tsui is an Associate Professor at the Institute of Rural Reconstruction of China, Southwest University, Chongqing, and a founding member of the Global University for Sustainability.

＊◦＊◦＊

SEA ONTOLOGIES

Karin Amimoto Ingersoll

Keywords: oceanic literacy, ontology, re-imaging, seascape

Oceanic peoples embrace specific knowledges and ways of being that offer a unique perspective for discussing development. Contemporary Native Hawaiians, for example, have a non-instrumental navigational knowledge about the ocean, wind, tides, currents, sand, seaweed, fish, birds, and celestial bodies as an interconnected system that allows for a distinct way of moving through the world. In this oceanic literacy, the body and the seascape interact in a complex discourse as a navigator's eyes, ears, muscles, and skin read the ocean's movements through a dynamic and indigenous connection to space and place. Oceanic literacy creates a politics and ethics that privilege interconnectedness as an alternative to the grand narrative of

Western thought-worlds which keep our 'selves' separate from land and sea, travelling to cross over an ocean rather than journeying inside it.

This oceanic literacy and ontology can be illustrated with the voyage of *Hōkūle'a*, the 62-foot double-hulled canoe, across thousands of miles of open ocean.[1] Native Hawaiian navigator, Bruce Blankenfeld, must expand his sense of 'self' in order to read the fluid seascape around him. 'Those stars have to be a part of you,' he states.[2] Blankenfeld becomes an aesthetic subject whose movement on the ocean articulates a Hawaiian ontology and epistemology through the stimulation of the senses: sight, smell, taste, sound, and touch. The kinesthetic body is included in his oceanic literacy through an active engagement, and 'place' takes on meaning as related to native identity, which has an origin in the mercurial sea.[3] Blankenfeld imagines his being as part of the coral polyp and sandpaper sharks below his canoe. The cold north wind blowing against his face and the rhythmic drumming of ground swelling in his ears link his mind and body to a deep identity moored in a sense of belonging.

Understanding oneself as related to the surrounding world fosters an acute spiritual awareness that can bring both joy and empowerment – power as defined through connection and purpose. For example, Blankenfeld can 'see' his island destination 'before' he leaves on a voyage. He feels the island inside his bones and blood, so that it becomes part of his body's expressions as he journeys towards it. Sight becomes corporeal: the feet, nose, and eyes all have a specific way of seeing, which become emotions burned into muscles that are interpreted by the mind. This constant interaction between body and mind generates a specific epistemological and ontological context.

Seeing thus becomes a political process: as the navigator's sight expands through oceanic literacy, so does the ability to think beyond a static mindset and a single reality. Blankenfeld finds his direction by extending his sight into his imagination, from which he deliberately creates a route of travel. If he does not imagine this route, it would not exist. If we don't imagine our connection to the earth, it would not exist either. In this way, oceanic literacy enables a journey within a new circulation of power as ethical rather than political, geographic,, or economic.

Merging the body with the seascape enables a reading of all memories and knowledge learnt within oceanic time and space but which have been effaced by rigid colonial constructions of identity, place, and power. Hawaiian place names, in both land and sea, have been re-named as a result of colonization. Kaluahole Beach on the south shore of O'ahu, for instance, is known today as Tonggs or Diamond Head Beach, named after the businessman Tongg, who bought a beachfront house on the now famous surf spot.[4] This oceanic space, however, was named Kaluahole by native Hawaiians because the oral history tells that there was a cavern filled by 'Ai'ai, the son of the Hawaiian god of fishermen', with *aholehole* fish just offshore from this beach. The knowledge of this specific fishing cavern and the cultural significance of this place is part of what can re-emerge through an engagement with oceanic literacy.

Much of the world proceeds without memory, as if the spaces we inhabit are blank geographies, and thus available for consumption and development. The human environmental and spiritual costs of this collective forgetfulness are evident in the erasure of cultures, through the overexploitation of earth's gifts, and through the recklessness that passes for free choice in our consumerist society. In our modern reality of capitalism, militarism, and ecologically challenging development, oceanic literacy becomes an ethical relationship steeped in a cellular and spiritual consciousness about our intimate connections to places. Humanity is found in the ocean. Native Hawaiian voyaging fosters a way of being and moving that potentially re-awakens this consciousness that can carry us beyond 'environmentalism' or 'conservation', into lived relationships of compassion and reciprocity. Recognizing and re-imaging our relationships can change the ways in which we move through this world together.

Notes

[1] Hōkūle'a was first conceived in 1973 by native Hawaiian watermen Herb Kawainui Kane and Tommy Holmes, and anthropologist Ben Finney, as a way of proving that Pacific Islanders were able to travel across great distances over eight hundred years ago, purposefully settling the islands of the Polynesian Triangle, and finding their way back home using only oceanic navigational methods. In 1976, Hōkūle'a's first deep-sea voyage from Hawai'i to Tahiti succeeded. For a history of the Hōkūle'a, see Finney's book in the Resources section.

[2] Interview by author, March 20, 2008, Honolulu, Hawai (tape recording).

[3] The predominant Hawaiian creation story is a chant called *He Kumulipo*, composed by a native Hawaiian priest around the eighteenth century. This chant tells that darkness spontaneously gave birth to a son and a daughter, who in turn gave birth to the coral in the sea, and many other creatures followed, first in the sea and then on land. Hawaiian origin, then, is in the ocean.

[4] 'Diamond Head' is the name given to the crater in 1825 by British sailors exploring the mountain when they thought they had found diamonds, which were in fact calcite crystals. Native Hawaiians named the now dormant volcano *Le'ahi* or *Lae'ahi*, the original meaning of the name is uncertain and thus the spelling of the name has never been agreed upon. For more on Kaluahole Beach and other Hawai'i place names, see John R.K. Clark's, *The Beaches of O'ahu* (1977).

Further Resources
Clark, John. (1977), *The Beaches of O'ahu*. Honolulu: University of Hawaii Press.
Finney, Ben (1994), *Voyage of Rediscovery*. Berkeley: University of California Press.
Ingersoll, Karin Amimoto (2016), *Waves of Knowing: A Seascape Epistemology*. Durham: Duke University Press.
Lewis, David (1994), *We, The Navigators: The Ancient Art of Landfinding in the Pacific*. Honolulu: University of Hawaii Press.
Sullivan, Robert (1999), *Star Waka*. Chicago: Independent Publishers Group.

Karin Amimoto Ingersoll is an independent writer and scholar from O'ahu, Hawai'i. She received her BA from Brown University, her MA and PhD from the University of Hawai'i at Mānoa, and has held a Hawai'i-Mellon Postdoctoral Fellowship. She recently published her first book, *Waves of Knowing: A Seascape Epistemology*.

SENTIPENSAR

Patricia Botero Gómez

Keywords: genealogies of the river, Afro diasporic peoples, socio-territorial theories in movement, relational ontologies

Sentipensar is a word that Afro-descendant people and fishermen enunciate in many river communities in Colombia. '*Sentipensar* means acting with the heart using the head', as a fisherman from the San Jorge River in the Colombian Caribbean expressed to the sociologist, Orlando Fals-Borda in the mid-1980s.[1]

Sentipensar constitutes an affective lexicon of these *pueblos* (peoples) who, by linking together experience and language, create a revolutionary promise, a grammar for the future. 'The heart, as much as or more than reason, has been to this day an effective defense of the spaces of grassroots peoples. Such is our secret strength, still latent, because another world is possible' (Fals-Borda 2008: 60).

Sentipensar is a radical vision and practice of the world, insofar as it questions the sharp separation that capitalist modernity establishes between mind and body, reason and emotion, humans and nature, secular and sacred, life and death. It is a powerful element in the dictionary of the peoples that we find in the genealogy of river and amphibian cultures. It can be perceived in those other histories and geographies that survive, as the communities of the Patía river say, in 'living libraries' inscribed in the heart and in intergenerational forms of inhabiting the world. These ways of existing reveal the grounded relational worldviews between human and non-human worlds that these peoples have been able to defend, in the midst of ferocious attacks by the modern ontology of separation (Escobar 2014). As a leader from the black community of La Toma in Southwest Colombia put it, referring to her community's struggle against the proposed diversion of their beloved river Ovejas to feed the large Salvajina dam, 'the river is not negotiable; we honour our traditions as we learned them from our grandmothers, ancestors and elders, and we hope it is what our *renacientes* learn'.[2]

Sentipensar conveys an active resistance to the triad of capitalism by dispossession, war and corruption that efface the long-standing – sometimes millenarian – cosmovisions that accompany peoples' struggles. Grassroots peoples have a clear understanding of this; as they say, *Para que el desarrollo entre, tiene que salir la gente* ('For development to enter, people have to leave'). The Process of Black Communities, PCN, a large network of black organizations, explains it in terms of the interrelation that needs to exist between being black (identity), the space for being (the territory), autonomy for the exercise of being, and their own vision of the future, linking these principles to the reparation of historical debts caused by persistent racist policies. These principles were at play at the recent Convergence of Black Women Caretakers of Life and the Ancestral Territories, which stated that 'our policy is founded on collective affection, love and kindness'.[3]

Globalization has accentuated ontological conflicts between worldviews or cosmovisions. In the San Jorge river, for example, fishing communities co-existed with the colonial expansion of livestock; today, legal and illegal mining destroy the mangroves and causes young people to abandon their knowledge and their trade. As they say, they will continue to plant mangroves, because without mangroves there are no fish and without fish there are neither fishermen nor fisherwomen.[4]

Sentipensar inhabits the ancestral knowledge and people's economies, as seen in the projects of young people from Afro descendant communities in southwestern Colombia such as the Yurumanguí River and La Alsacia. In these autonomous projects, youth and women confront the patriarchal capitalist model of education and economics that has amputated the communal forms of embodied knowledge and life-worlds. It is from these zones of affirmation of Being that people create their own socio-territorial theories in movement that allow us to visualize the collective and plural autonomies rooted in the territories and a host of transition alternatives that conventional disciplinary categories, functional to the death system of capitalist modernity, make invisible.

These forms of Afro-Latin resistance informed by *sentipensar* constitute a politics of hope that re-imagines the world from realities that were not entirely colonized by modern categories. They 'sentipiensan' and imagine worlds free from dependence on the capitalization of life-worlds, the state, and discourses of progress.

Between silences, forgetfulness and eloquent wordlessness, *la palabra* (the word) is sung by drums born from the pain of black enslaved peoples who well know of the existence of the beautiful worlds embedded in waters, birds, and trees. As they express it, 'If there is no inspiration there is no life, that's why music and joy come from the songs and languages of the river.'[5] To the rhythm of the drum and the earth, these groups create references for our time that would allow us to pass from policies of death towards policies of life. They call on us to *sentipensar* with the land and to listen to the 'sentipensamento' of the territories and their peoples, rather than to the decontextualized categories of development and growth.

Sentipensar takes place between *mingas* and *tongas* indigenous and Afro-descendant traditional collective work forms, respectively oriented towards post-development and *buen vivir*. *Sentipensar* with the territory implies thinking from the heart and mind, or co-reasoning, as people inspired by the Zapatista experience say. Thus, in the interstices and against racist and patriarchal discourses and practices and conventional academic knowledge, there survives an area of affirmation of Being that heals the primary bond with land and territories; herein lies one of the most fertile sources of food sovereignty and of people's cultural and political autonomy.

Notes

[1] Fals-Borda, Mompox y Loba, p. 25b. The term was originally reported by Fals-Borda as used by people of the rivers and marshes of the Caribbean coast region; it was later popularized by Eduardo Galeano and used more recently by Escobar (2014).

[2] *Renacientes* is a category used by black communities that suggest that everything is constituently being reborn. Here it refers to the coming generations.

[3] See PCN, *Declaración*.

[4] Statement by women of Cejebe (Sucre) fishing village, October 2016.

[5] Statement by Álvaro Mier (Fals Borda's co-researcher), Festival de la Tambora, San Martín de Loba, Novemberde 2016.

Further Resources

Escobar, Arturo (2014), *Sentipensar con la tierra*. Medellín: Unaula.

Fals-Borda discusses the notion of *sentipensar*, https://www.youtube.com/watch?v=LbJWqetRuMo.

Fals-Borda, Orlando (1986), *Mompox y la loba. De retorno a la madre tierra*. edited by Carlos Valencia, Bogotá: Universidad Nacional.

———— (2008), *El Socialismo Raizal.* Caracas: El Perro y la Rana.

Hacia el Buen Vivir Ubuntu, https://buenvivirafro.wordpress.com.

Machado, Botero, Mina, Escobar comp. (en prensa). *Buen Vivir Afro*. Manizales: Proceso de Comunidades Negras (PCN) y Universidad de Manizales.

Proceso de Comunidades Negras (PCN) (2016), *Declaración del Encuentro Nacional e Internacional de Mujeres Negras Cuidadoras de la Vida y Los Territorios Ancestrales*, https://renacientes.net/2016/11/25/primer-encuentro-nacional-e-internacional-de-mujeres-negras-cuidadoras-de-la-vida-y-los-territorios-ancestrales/.

Patricia Botero is Professor of Social and Human Sciences at Universidad de Manizales, Colombia. She is a member of the Group of Academics and Intellectuals in Defense of the Colombian Pacific (GAIDEPAC) and she collaborates with the Proceso de Comunidades Negras (PCN) campaign – *Otro Pazífico Posible*.

❖

SLOW MOVEMENT

Michelle Boulous Walker

Keywords: slow activism, slow philosophy, deceleration, complexity, attention

The Slow Movement comprises an eclectic gathering of people devoted to slow activism, the first and most prominent of these being the Slow Food movement. Slow activism calls for a deceleration of the pace of modern technological life, arguing that advanced capitalism is dominated by a logic that equates speed with efficiency. For slow activists, opportunities for a contemplative relation with others and the natural world are decreasing in an ever-accelerating world. Temporally, our very being in the world is challenged by a relentless demand to decide, respond, and act without adequate time to really engage with the complexity of life. A culture of haste infiltrates our twenty-first-century social and political spaces.

In response to this culture, Slow Food was one of the first such movements to emerge in the Western world. In 1989, Carlo Petrini challenged the proliferation of industrialized fast food, championing in its place simple hand-crafted meals that embraced the produce and the traditions of local cuisine. Slow Food has developed from this to celebrate the pleasures of slow cooking, and convivial sharing of food with others in a more leisurely, less commercial context. In addition, the movement raises awareness of the ecological and educational issues associated with the production and consumption of food globally. As such, it provides the basis for a political awareness of issues such as sustainability and cooperative small-scale agriculture as alternatives to fast food and large-scale food production.

The Long Now Foundation, founded in San Francisco in 1996, counters today's accelerating culture by fostering long-term thinking and responsibility. It challenges the nexus between efficiency, productivity and speed, promoting 'slower/better' over 'faster/cheaper'. While the sentiment of 'slower/better' – in the context of food – has at times been criticized as elitist and gourmet-driven, the Slow Food movement actually revives Petrini's early social protest, promoting equitable food policy and justice for those most disadvantaged by global food systems. *Terra Madre*, for example, is an international network promoting sustainable agriculture and biodiversity in order to guarantee good, clean, and fair food. International debates now typically focus on access to local, sustainable and nutritious food for groups in the community who are often overlooked in ethical debates and social policy. The movement in Portland, Oregon, for example, argues that Latino farmworkers must be part of any Slow Food activism, if it is to evolve.

There are now 1,500 Slow Food convivial in 150 countries around the world, in the Global North as well as in Niger, Angola, Bolivia, Sri Lanka, and Indonesia. Further, Slow Food has inspired a series of movements in response to the dehumanizing effects of globalization. The movements in series include Slow Gardening, Slow Cities (Città Slow), Slow Schools, Slow Education, Slow Parenting, Slow Travel, Slow Living, Slow Life, Slow Reading, Slow Goods, Slow Money, Slow Investment, Slow Consulting, Slow Ageing, Slow Cinema, Slow Church, Slow Counselling, Slow Fashion, Slow Media, Slow Communication, Slow Photography, Slow Science, Slow Technology, Slow Design, Slow Architecture, and Slow Art. The last of these, Slow Art, exposes capitalist thinking by acknowledging its complicity with a system that benefits both materially and culturally from exploitation of the non-western world. This self-awareness among people in affluent countries is, more and more, a defining feature of what marks a practice as slow. In the global South, the Slow Movement manifests as a concern with Slow Urbanism and Slow Governance, exploring connections between urban crises, economic downturns, migration, dispossession, expulsion, and exclusion. In these contexts, there is an intimate relation between slow activism and the reclamation of common land.

While there is considerable diversity in the way slowness is embraced by grassroots movements around the world, what unities them is arguably a determination to experience the pleasure of engaging the basic needs of everyday life with a kind of artful slowness. Such movements seek a more substantial and sustained relation with the complexity of the world. Carl Honoré's book *In Praise of Slow*, first published in 2004, explored how industrialized societies could think of slowness in terms of a movement with the potential to challenge the belief that 'faster is always better'. Since this time, the Slow Movement has evolved to more consciously embrace its practice of slow activism throughout the globe. In part, this activism involves challenging our roles as passive consumers in a capitalist system devoted to unchecked economic growth and exchange.

Reclaiming slowness extends also to cultural spaces devoted to 'thought'. The equation of speed and haste with efficiency is embedded in a typical European style of instrumental rational thought, where attention gives way to calculation, and thinking – in general – is reduced to an empty, technical manipulation and application of fact. Slow philosophy is the practice of resisting the kind of thought that is incapable of collecting itself, pausing, considering and contemplating. In this, it is a particularly deep-felt and critically reflective form of slow activism. Just as the Slow Movement draws, in modern and contemporary ways, on non-dominant practices, so too does slow philosophy. Slow philosophy is the practice that challenges an instrumental relation to life; it is, above all, the cultivation of a heightened attentiveness. It provides intense encounters that open us to the beauty and strangeness of the world, and this intensity is what arguably lies at the heart of all slow activism.

Further Resources
Boulous Walker, Michelle (2017), *Slow Philosophy: Reading Against the Institution*. London and New York: Bloomsbury Academic.
Honoré, Carl (2004), *In Praise of Slow: How a Worldwide Movement Is Challenging the Cult of Speed*. London: Orion Books.
Long Now Foundation, http://longnow.org/.
Petrini, Carlo (2007), *Slow Food Nation: Why Our Food Should Be Good, Clean, and Fair*. Clara Furlan and Jonathan Hunt (trans.), New York: Rizzoli International.
—— (2010), *Terra Madre: Forging a New Global Network of Sustainable Food Communities*. John Irving (trans.), White River Junction, VT: Chelsea Green.
Terra Madre Foundation, https://www.terramadre.info/en/.

Michelle Boulous Walker is Head of the European Philosophy Research Group (EPRG) in the School of Historical and Philosophical Inquiry at The University of Queensland, Australia. She is the author of *Philosophy and the Maternal Body: Reading Silence* (1998) and publications in European philosophy, aesthetics, ethics, and feminist philosophy.

———•«•———

SOCIAL ECOLOGY

Brian Tokar

Keywords: direct democracy, ecology, confederation, hierarchy, community, assembly, social movements

Social ecology offers a revolutionary and reconstructive political outlook, challenging conventional views of the relationships between human communities and the natural world, and offering an alternative vision of free, confederated and directly democratic cities, towns and neighbourhoods seeking to re-harmonize those relationships. Social ecology was initially developed by the social theorist Murray Bookchin, working in the United States during the 1960s to early 2000s, and has been further elaborated by his colleagues and many others throughout the world. Social ecology has been influential in various social movements, including the 1970s campaigns against nuclear power, elements of the worldwide alter-globalization and climate justice movements, and the present struggle for democratic autonomy by Kurdish communities in Turkey and Syria.

Social ecology begins with an understanding that environmental problems are fundamentally social and political in nature, and are rooted in the historical legacies of domination and social hierarchy. It is rooted in both anarchist and libertarian socialist currents, questioning capitalism and the nation state and viewing institutions of local democracy as the best antidote to centralized state power. Murray Bookchin was among the first thinkers in the West to identify the growth imperative of capitalism as a fundamental threat to the integrity of living ecosystems and argued that social and ecological concerns are fundamentally inseparable. Through detailed inquiries into history and anthropology, Bookchin challenged the common Western notion that humans inherently seek to dominate the natural world, concluding instead that the domination of nature is a myth rooted in relationships of domination among people that emerged from the breakdown of ancient tribal societies in Europe and the Middle East. Social ecologists are also influenced by elements of Indigenous North American thought and various schools of critical social theory, including the historically rooted approach to ecological feminism pioneered by social ecologists Ynestra King and Chaia Heller.

Following these influences, social ecology highlights various egalitarian social principles that many indigenous cultures – both past and present – have held in common, and elevates these as guideposts for a renewed social order. Such principles have been elevated by critical anthropologists and indigenous thinkers alike, and include concepts of interdependence, reciprocity, unity-in-diversity, and an ethics of complementarity, that is, the balancing of roles among various social sectors, especially by actively compensating for differences among individuals. The inherent conflict between these guiding principles and those of increasingly stratified hierarchical societies has shaped the contending legacies of domination and freedom through much of human history.

Social ecology's philosophical inquiry examines the emergence of human consciousness from within the processes of natural evolution. The perspective of dialectical naturalism examines the dynamic forces of evolutionary history and views cultural evolution as a dialectical development influenced by both natural and social factors. Social ecologists question prevailing views of nature as a 'realm of necessity', suggesting that, as natural evolution has advanced qualities of diversity and complexity, and also seeded the origins of human creativity and freedom, it is imperative for our societies to fully express and elaborate those underlying evolutionary tendencies.

These historical and philosophical explorations provide an underpinning for social ecology's political strategy, which is described as libertarian or confederal municipalism or, more simply, as 'communalism', stemming from the roots of key ideas in the legacy of the Paris Commune of 1871. Social ecology reclaims the ancient Greek roots of the word 'politics' as the democratic self-management of the *polis*, or municipality. Bookchin argued for liberated cities, towns and neighbourhoods, governed by open popular assemblies, freely confederate to challenge parochialism, encourage independence and build a genuine counter-power. He celebrated the lasting Town Meeting traditions in Vermont and throughout the New England region of the United States, describing how the region's Town Meetings assumed an increasingly radical and egalitarian character in the years prior to the American Revolution.

Social ecologists believe that whereas institutions of capitalism and the state heighten social stratification and exploit divisions among people, alternative structures rooted in direct democracy can further the emergence of a general social interest towards social and ecological renewal. People inspired by this view have brought structures of direct democracy and popular assemblies into numerous social movements in the US, Europe and beyond, from popular direct action campaigns against nuclear power in the late 1970s to the more recent global justice/alter-globalization and Occupy Wall Street movements. The prefigurative dimension of these movements – anticipating and enacting the various elements of a liberated society – has encouraged participants to challenge the *status quo* and also advance transformative future visions.

Social ecologists have also sought to renew the utopian tradition in western thought. The Institute for Social Ecology co-founder Dan Chodorkoff argues for a 'practical utopianism', combining social ecology's theoretical insights and political praxis with advanced principles from green building and urban redesign, together with eco-technologies to produce food, energy and other necessities. Ecological design concepts such as permaculture that encourage a more profound understanding of the patterns of the natural world resonate with social ecology's view that human beings can participate in nature in creative, mutually beneficial ways, while seeking to overturn historical legacies of abuse and destruction.

The outlook of social ecology has profoundly influenced international social movement actors, from the early years of Green politics to recent campaigns for local empowerment through popular assemblies in several European and Canadian cities. Social ecologists have influenced efforts towards greener urban design and neighbourhood power in many parts of the world. Perhaps the most striking current influence is on militants in the Kurdish regions of the Middle East, where ethnically diverse populations, long marginalized by colonial and state powers, have created institutions of confederal direct democracy in one of the world's most war-torn regions. Despite persistent sectarian warfare and religious violence, Kurdish towns near the Turkish-Syrian border are working towards gender equity and ecological reconstruction, significantly informed by social ecology and other critical social outlooks rooted in a wide variety of cultural perspectives.

Further Resources
Bookchin, Murray (1982), *The Ecology of Freedom: The Emergence and Dissolution of Hierarchy*. Palo Alto: Cheshire Books (and later editions).

———— (2015), *The Next Revolution: Popular Assemblies and the Promise of Direct Democracy*. New York: Verso Books.

Eiglad, Eirik (ed.) (2015), *Social Ecology and Social Change*. Porsgrunn, Norway: New Compass Press.

Institute for Social Ecology, www.social-ecology.org.

New Compass Press, www.new-compass.net.

Brian Tokar is a lecturer in Environmental Studies at the University of Vermont; a Board member and recent Director of the Institute for Social Ecology in Vermont, USA. His latest book is *Toward Climate Justice: Perspectives on the Climate Crisis and Social Change* (Revised edition; New Compass Press, 2014).

SOCIAL SOLIDARITY ECONOMY

Nadia Johanisova and Markéta Vinkelhoferová

Keywords: social solidarity economy, social enterprise, non-capitalist economies, community ownership, community economies

The social (and) solidarity economy (SSE) is a comprehensive concept referring to a worldwide range of existing economic practices that do not comply with the mainstream economic logic of private businesses competing in abstract markets to maximize profits for self-interested consumers, while reducing nature to a passive resource. Instead, they often involve community ownership, democratic, non-hierarchical, and consensual decision-making, as well as mutual co-operation and embeddedness in a local social and ecological context. The line between consumer and producer may be blurred. Profits and self-interest tend to remain secondary to larger concerns such as equity and solidarity, right to a dignified livelihood and ecological integrity and limits.

Such practices can be formally organized as informal, traditional or new, and may or may not involve money transactions. As opposed to a mainstream understanding of the economy as being part either of the Market or the State, they often operate within other spaces: international solidarity networks, local ecosystems and communities, extended families, villages, and municipalities. Examples of such 'other', 'non-capitalist' economies include reciprocal labour, urban gardening, subsistence farming, some Fair Trade projects, community-supported agriculture projects, collective marketing of handicrafts by local crafts associations, traditional commons regimes, worker-occupied factories, renewable energy cooperatives, some forms of housing cooperatives and self-build associations, community/ social currencies, loan and credit cooperatives, rotating credit and savings associations, non-interest banks, childcare networks, communal waste re-use centres, artistic cooperatives, traditional burial societies, and many more.

While the practice of SSE is widespread and is known by many names, the concept is relatively new. The International Network for the Promotion of the Social Solidarity Economy (RIPESS), made up of continental networks of the SSE, held its first meeting in 1997 in Lima, Peru. The concept and movement gained momentum after the 2001 World Social Forum in Porto Alegre, Brazil. RIPESS explicitly supports economic alternatives to capitalism and to the current flawed development model. The concept (often abbreviated to 'solidarity economy') generates much of its energy from the grassroots, supported by academia in Latin America – see the work of José Luis Coraggio (Ecuador), Louis Razeto (Chile), and Euclides Mance (Brasil) – as well as in France (Jean-Louis Laville), and other Francophone regions, such as Québec and Francophone Africa.

A holistic view of the SSE, recognizing traditional as well as new SSE, challenges the mainstream development discourse. The latter sees economic progress as a march from the non-monetized to the monetized, from the communal to the privatized, from the localized to the global, from the artisanal to the mass-produced. In this perspective, 'traditional' social solidarity economies are seen as backward, unproductive, and they mostly remain invisible. Conversely, some 'new' SSE practices are hailed as 'social innovations', 'social economy', or 'social enterprise'. Such views abound especially in the EU and US. In some ways this attention is welcome and can engender an enabling policy and funding environment for the SSE. However, there are drawbacks: with its emphasis on market behaviour, scaling-up, paid employment, formal structure, and 'innovation', such a discourse can de-radicalize existing SSE organizations, nudging them into the mainstream. Alternatively, it can push both traditional and radical SSE communities into the shadow. In the worst case, the reframing of a SSE as 'social enterprise' can serve as an excuse for business-as-usual. Social enterprises are assumed to provide jobs, mitigate poverty and generally mop up the social costs of the system so can be used to justify and for a roll-back of public social services.

While some SSE initiatives are consciously radical (Conill *et al.* 2012), many SSE entities do not see themselves as such. For instance, urban allotment gardeners in the Czech Republic engage in non-capitalist economies by growing their food and sharing it, but seldom look at this as going against the economic system. Similarly, the Soninké blacksmiths in Kaedi, Mauritania, whose cooperative communities produce metal tools for a local market, use scrap metal as input and grow their own food, but are not consciously radical (Latouche 2007). Yet they might be seen as part of the SSE, having persisted as a non-capitalist enclave in an economic system that destroyed most of the West African blacksmith trade a hundred years ago.

An expanded view of 'the economy' as 'the ways people organise themselves collectively to make a living and the ways a society organises itself to (re)produce its material life and well-being' (Dash 2013) opens up a broad historical and contemporary vista of 'other' economies, hitherto marginalized by the mainstream. An older, 'moral economy' has been documented as prevailing in eighteenth-century Britain by E.P. Thompson and in twentieth-century Southeast Asia by J. Scott. According to this older economic ethos, everybody has a right to life or livelihood, long-term sustainability rather than growth is a priority, and mutual economic support where reciprocity is the norm. Karl Polanyi, David Graeber, and others have argued that such an economic ethos prevailed for most of history in most societies. SSE groups can then be seen not as mere 'innovations', but as returning to an economy with a moral compass after venturing into the dead-end street of mainstream economic practice.

While traditional moral economies can serve as anchor and inspiration, not all have been democratic and equitable. This is one challenge facing a transition to SSE. Another is the issue of ecological integrity and limits: cooperation with non-human nature needs more space in the SEE discourse. The third challenge is the mainstream economy itself, which tends to engulf the SEE via economies of scale, cost externalization and reliance on capital-intensive, fossil-fuel-propelled production. Yet, the SSE as an enduring and expanding grassroots practice remains an important pillar in the transition to an ecologically wise and socially just world.

Further Resources

Conill, Joana, Manuel Castells, Amalia Cardenas and Lisa Sevron (2012), 'Beyond the Crisis: The Emergence of Alternative Economic Practices', in *Manuel Castells, Joao Caraça and Gustavo Cardoso, Aftermath: The Cultures of the Economic Crisis.* Oxford: Oxford University Press.

Dash, Anup (2013), *Towards an Epistemological Foundation for Social and Solidarity Economy*. Draft paper for UNRISD Conference on 'Potential and Limits of Social and Solidarity Economy'. 6–8 May, Geneva, Switzerland.

Latouche, Serge (2007), *La otra Africa: Autogestión y apaño frente al mercado global*. Barcelona: Oozebap.

Red de Educación y Economía Social y Solidaria (Network of Education and Social Solidarity Economy), http://educacionyeconomiasocial.ning.com.

RIPESS (The International Network for the Promotion of the Social Solidarity Economy), www.ripess.org, www.ripess.eu.

Socioeco, www.socioeco.org.

Nadia Johanisova is an ecological economist at the Faculty of Social Studies of Masaryk University, Czech Republic. She is interested in critiques of mainstream economics and in economic alternatives, and authored *Living in the Cracks: A Look at Rural Social Enterprises in Britain and the Czech Republic* (Feasta, NEF, 2005).

Markéta Vinkelhoferová is an activist and practitioner of the social solidarity economy. She works for the Ecumenical Academy, a Prague-based non-profit organization that promotes social justice and environmental sustainability, and is co-founder of the Fair & Bio Roastery, which connects co-operative principles with fair trade and social inclusion.

———

TAO WORLDVIEW

Sutej Hugu

Keywords: inter-species compact, indigenous ontology, eco-calendar, rights of nature, tribal sovereignty

Having been left alone by the colonization of the modern world until the end of the nineteenth century (1896), the Austronesian *Tao* people currently have a population of about 5,000, half of them still living on their home island, *Pongso no Tao* (Lanyu). It is a small volcanic islet of 45 square kilometres off the southeast coast of Taiwan on the western margin of the Pacific Ocean. There are six independent tribal communities speaking the same language, but each with their own origin, myths, and legendary stories.

Nurtured by the richness of the large marine ecosystem of the Kuroshio

Ocean Current, the *Tao* people have lived 'the original affluent society' with their comprehensive traditional ecological knowledge and practices for millennia. There are the non-hierarchical and unspecialized egalitarian tribal communities, without chiefs or ruling elders but functional leaders responsible for guiding various productions, construction and ceremonial activities and events, and a complementary division of labour by gender within the household. Following their unique time-reckoning system called '*ahehep no tao*' (evening of the people), which is an original eco-calendar to keep track of both the monthly lunar cycle and catch up with the annual solar cycle, Taomen alternated their migratory fishing and coral reef–fishing seasonally. *Tao* women grow water taros in private fields with irrigation channels, and work on shifting cultivation – firing and fallowing dry fields, collectively owned by extended families, for taros, sweet potatoes, yams, and millets. They have maintained a community forest through cross-generational care of reserved trees, and ecologically wise 'high alpha-diversity' rainforest timber harvesting for plank boat-building and house building. This adaptive cultural and ecological knowledge and an elaborated '*makaniaw*' (taboo) norm system for the sustainability of the island and its people are still alive.

The kernel of the *Tao* worldview can be delineated as a guide for survival, revival, and sustainability of the coming generations. For example, they use an eco-calendar and interspecies compact as the foundation of governance and rights of nature. Based on a simple but subtle and deep relationship with and close observation of the natural environmental cycles, an eco-calendar is used as the fundamental framework of their governance institution. It counts and gives thirty names for the phasing moon, and twelve names for the lunar cycles around the year that is divided into three seasons. Most amazing is the dynamic intercalation method used to insert a leap month in the right year to calibrate the discrepancy between twelve lunar months and a solar year. This is actually dependent on the biological clock of migratory flying fishes.

Following this eco-calendar along with its ecological and phenomeno-logical knowledge, there are three major ceremonies to initiate each season with critical ethical value:

1. *mivanoa* for *rayon* season, about March to June – all the men, young and old, should gather on the community beach of fishing boats to have a ritual of summoning the flying fish school back, and reconfirming the inter-species compact between the flying fishes and the *Tao* people from the ancient times, to implement the rights of nature and the order

of the living world. In the *Tao* mythology, the ancestor of the noblest black-winged flying fishes had taught the ancestor of the *Tao* people how to harvest sustainably and treat appropriately the flying fishes for the survival of both species. In the same story, there is the first account of the arrangement of works and ceremonies all around the year.

2. *mivaci* for *teyteyka* season, approximately from July to October – all men and women should assemble in the community gathering place to celebrate the harvest of millets from the common fields of consanguineous family and other crops from each conjugal family field, intending to always keep the highest level of agro-biodiversity and food-seed sovereignty of about fifty varieties of different crops through traditional farming system and environmentally sensitive resilient practices.

3. *mipazos* for *amian* season, approximately from November to February – all households should prepare sacrificial offering, go to offering at the community beach first, and then come back to put it on the roof of each main house, intending to restore the harmony with all spirits and reiterate the commitment to the well-being of all beings around us.

In the *Tao* marine governance institution, *rayon* season is for fishing of migratory species only. Coral reef–fishing is absolutely prohibited during this period. Catching flying fish is stopped when it is the peak time of reproduction. While in the other seasons for coral reef–fishing, these fishes are divided into three categories of good *(oyod)*, bad *(rahet)*, and not-for-eating *(jingngana)*, to spread and mitigate the pressure on the marine food chain. 'Good' fishes are for women and children first, and 'bad' fishes are for men and elders only. Obviously, all the above-mentioned practices represent an effective multidimensional zoning system for conservation of natural resources and environment.

A *Tao* **Indigenous ontology names the world and defines a way of connectedness that embodies the trinity of lingual, cultural, and biological diversity.** For the world around their living space, the *Tao* people give names with secular and spiritual meanings. There are about 450 names for the sea and tidal zone creatures, about 350 for the coastal and mountain plants, 120 for land animals, birds, and insects. There are also about 1,200 place names designated all around the island's six tribal community-habitats. The holistic knowledge and value system is connected to each name and interwoven with each other. Besides these, there are dozens of names of winds for navigating orientation, and names of stars as 'eyes on the sky.' Without an existent writing system, *Tao* people live on an oral tradition of collective memory through a continuous lineage of story-telling and ceremonial chanting. This pre-money and pre-market integration of the ecology of knowledge and living way of connectedness could have important lessons for modern societies in crisis. They, therefore, look forward to participating in a paradigm shift and transition of planetary governance, applying creatively indigenous worldviews as a realistic and visionary alternative for our common future.

Further Resources
Arnaud, Véronique (2013), 'Botel Tobago: The Island of Men'. CNRS-CASE & iiAC / CEM, https://archive.org/details/066LIleDesHommesENG.
Benedek, Dezso (1987), 'A Comparative Study of the Bashiic Cultures of Irala, Ivatan, and Itbayat', PhD thesis, Pennsylvania State University.
ICCA Consortium, http://www.iccaconsortium.org.

Sutej Hugu is a tribal sovereignty activist and visionary who organizes alternative collective livelihoods. He was a co-founder and Secretary General of the Taiwan Indigenous Conserved Territories Union (TICTU), and Regional Coordinator for East Asia of the ICCA Consortium. Hugu has also been CEO of the Tao Foundation.

TRANSITION MOVEMENT

Rob Hopkins

Keywords: localization, resilience, REconomy, Transition, climate change

The Transition movement began in 2005 in the UK, initially conceived as a 'detox for the affluent West'. Inspired by the Global Commons Institute's 'Contraction and Convergence' model (Meyer 2000), it aimed to inspire people to conceive the scale of cuts in carbon emissions required by Western nations as a step 'towards' something rather than a move 'away from' something irreplaceable. While it emerged from roots in the permaculture movement, in bioregionalism and the localization movement, it also drew inspiration from the Women's Movement, indigenous cultures, and many others.

Initially framed as being 'a response to climate change and peak oil' (Hopkins 2013), it was driven by the need to see the two challenges as necessitating a deep shift away from fossil fuels and as an historic opportunity to do something new and remarkable. Since its inception, the concept has evolved from its experience as a 'learning network'. There are now over 1,400 initiatives in fifty countries worldwide. Transition Network, the UK-based charity set up to support its evolution, describes Transition as a movement of communities reimagining and rebuilding our world.

The Transition movement has spread based on the Open Source idea. Apart from some key principles and values, communities are invited to take the model, adapt it, shape it, make it their own. There are two elements that stand out in the Transition model. One which has evolved since its inception is that of REconomy. Many Transition groups set up projects, whether to do with food, energy, housing, or whatever, but lack the skills needed to turn them into viable and sustainable enterprises. REconomy develops tools and models for enabling community investment such as Local Entrepreneur Forums and share options, and develops tools like Local Economic Blueprints to allow Transition groups to argue the economic case for what they do.

The other key strand that has evolved within Transition is what is called 'Inner Transition'. This recognizes that 'how' a group does a project matters as much as 'what' they do. Working in a way that prioritizes paying attention to the factors that lead to burnout, develops vital skills around making decisions, running effective meetings, managing conflicts, and so on. Working as activists for a sustainable society but using tools rooted in the very models we are trying to change is ultimately self-defeating. Inner Transition draws its inspiration from a range of psychological and spiritual traditions, as well as from the affinity groups prevalent in the feminist movement of the 1970s.

The Transition movement's approach to development is very different from that being promoted by Western governments. It focuses on:

- **Appropriate localization:** it makes sense to bring production of food, energy generation, and building materials closer to home, nearer the consumer.
- **Resilience:** putting in place infrastructure which enables communities to be better placed to withstand shocks, while taking this as the opportunity to reimagine the local economy and to meet local needs more effectively.
- **Low carbon:** creating projects and enterprises which are inherently low carbon in how they operate and what they create.
- **Community assets:** where possible, it is about bringing assets (land, businesses, energy generation, buildings) into community ownership.

This puts them so much more in control of their ability to shape their future.

- **Natural limits:** recognizing that we no longer live in a world where credit, resources, and energy are infinite.
- **Not purely for personal profit:** a variety of business models are emerging, such as social enterprises, cooperatives, and those focused on maximization of social return.

The Transition approach is spreading (Feola and Nunes 2013). It is being picked up by universities and local governments. It is increasingly being noted by people in public health, adopted by social activists and echoed in some European policy circles. It has been fascinating to see its emergence, alongside the complementary movements of *buen vivir* and Via Campesina, of Transition initiatives across South America and elsewhere. In São Paulo, Transition Brasiliandia is seeing a favela self-organize around social enterprise, ending violence towards women, public health, and urban agriculture. In Greyton, South Africa, a town scarred by the apartheid era, Transition was introduced by Nicola Vernon, who said 'as a driver for social integration it's the best I've encountered in 30 years of working in social welfare. The group has initiated many projects with local schools, including a 'Trash to Treasure' festival, planting thousands of trees, constructing new buildings using 'eco-bricks' – empty plastic bottles filled with non-recyclable trash.

Transition has attracted some criticism. For the Trapese Collective, the focus on individual actions negates the importance of structural change. However, Transitioners argue that their approach of actively building alternatives, seeking common ground rather than political confrontation and seeking a different definition of what constitutes 'political' is an equally valid approach to achieving structural change. Others accuse Transition of only engaging in small-scale actions. This rather condescending view ignores the more ambitious achievements of Transition groups in building social connection and giving people the confidence to take on larger projects. Others have criticized the predominantly white middle class make-up of Transition. This is a challenge noted across many change movements, and one that many Transition groups are working to address, shifting the focus more to local needs and livelihood creation and the potential created for broader engagement.

Transition has been seen from the outset as a movement modelled on mycorrhizal – a fungus – the metaphor does not lose anything by this, that is, one that spreads with its own momentum, self-organizes, creates networks, finds its own way, and fruits sometimes when you expect and sometimes when you don't. It is not a step-by-step process. While there is much Transition can learn from other movements and approaches, it also, after ten years of experimentation, has much to offer them.

Further Resources

Feola, Giuseppe and Richard Nunes (2013), *Failure and Success of Transition Initiatives: A Study of the International Replication of the Transition Movement*, Research Note 4. Walker Institute for Climate System Research Walker Institute, University of Reading.

Hopkins, Rob (2013), *The Power of Just Doing Stuff: How Local Action Can Change the World*. Cambridge: Green Books/UIT.

Meyer, Aubrey (2000), *Contraction and Convergence: The Global Solution to Climate Change*. Totnes, Devon, UK: Green Books (For the Schumacher Society).

REconomy Project (Community-Led Economic Change), http://reconomy.org/.

Transition Network Team (2016), *The Essential Guide to Doing Transition*. Britain: Transition Network, https://transitionnetwork.org/wp-content/uploads/2018/08/The-Essential-Guide-to-Doing-Transition-English-V1.2.pdf.

Transition Network, www.transitionnetwork.org.

Rob Hopkins is the founder of the Transition movement and lives in Totnes, Devon. He holds a PhD from the University of Plymouth, and has authored several books on Transition, the most recent being *21 Stories of Transition*. He writes and speaks extensively on the need for more resilient local economies, and has won several awards for his work. He is also the director of a social enterprise craft brewery, a gardener and tweets as @robintransition.

TRIBUNAL ON THE RIGHTS OF NATURE

Ramiro Ávila-Santamaría

Keywords: Rights of Nature, ethical tribunal, rights enforcement, civilizational crisis.

Poisoned rivers. Violated earth. Dead dolphins. Animals suffering. Mutilated forests. Devastated jungles. Contaminated air. Moribund coral reefs. Birds in extinction. Dying oceans and lakes. Vanishing insects. Non-reproducing mammals. Genetically modified seeds. Surviving *pueblos* (peoples). Persecuted and assassinated persons for defending Life . . .

Millions of living beings have no space in which to express their pain. Conventional tribunals have been designed to address solely some human problems, not even for every human. Those suffering from malnutrition, hunger, poverty, forced migration, shelter, war and loneliness do not have a place to make their agony and needs known either. A small group of humans, those who exploit others by gaining control of nature via property, have tribunals, lawyers, laws, and policies to protect them.

Human beings are the most insensitive and lethal species on the planet. We are going through the sixth extinction and we are not even paying attention to it. The changes effected by humans and their technologies are so rapid that species and nature are incapable of adapting to them. In this extinction, the human is both agent and victim (Kolbert 2015: 267). Nevertheless, the problems caused by the human species have had no overarching institution for assigning responsibility. Given this vacuum, social movements led by enviornmentalists[1] and intellectuals have created an ethical space of civil society where nature is endowed with voice and that voice is listened to. This space for hearing demands and enabling reflection is the International Tribunal on the Rights of Nature (The Tribunal).

The Tribunal met for the first time in January, 2014, in Quito, Ecuador. Those who appeared before the Tribunal included: The Gulf of Mexico, the Ecuadorian Yasuní National Park, the Australian Great Coral Reef, the Ecuadorian Condor Cordillera, the subsoil where fracking is practised in the USA, plus the defenders of nature's rights. Their demands: No to oil spills; no to oil exploitation; no to mining; no to predatory tourism; no to anthropogenic climate change; no to genetic modification; no to the criminalization of activists. Cases were presented by activists and social movements while the Tribunal listened. Other Tribunals followed in Ecuador, the USA, Australia, Peru, and France.[2]

'Mr. President, I ask the Yasuní National Park to appear before this office.'
'Proceed, Ms. Prosecutor.'

And thus Yasuní gains voice through humans who know it and feel with it. Human persons speak on behalf of forests, rivers, dolphins, toads, and insects, and the voices of indigenous peoples are ever present. The humidity of the forest, the chant of the *waorani*, the jaguar's pain, the sorrow of felled trees, the horrors of contamination, the blasts in the oil fields, the misery of the economic exploitation of nature, the impotence of the forest inhabitants, the conflict between those who destroy and those who conserve, and the death of indigenous groups in voluntary isolation – all of these can be intensely felt in the room. Some cry, others scream, others demand, question, inform about the alleged model of development and progress. Voices are heard, scientific reports are discussed, photographs are consulted so that a bit of nature can appear before The Tribunal. The judges deliberate, ponder their votes, and finally state their verdict. Ethically at least, nature is heard and justice is done.

The Tribunal hears cases such as Yasuní's where the Ecuadorian government has proposed oil extraction in the middle of a rich rainforest. It is thought that nature's rights have been violated here and, after considering the proofs, the Tribunal announces the violation, assigns responsibilities, and suggests restorative measures according to the Universal Declaration of the Rights of Mother Earth,[1] the Ecuadorian Constitution (articles 70–3), and other rights derived from nature and on the basis of the knowledge of the Earth's peoples who value her. Mother Earth, and all the living beings that inhabit her, have the right to exist, to be respected, to regenerate without her vital cycles being altered, to maintain their identity and integrity, and her right to an integral restoration.

The tribunals are equipped with a Technical Secretariat, consisting of activists, scientists, politicians, renowned academics, all of whom are aware of the rights of nature and the need to defend her. Among those who have presided over the tribunals are Vandana Shiva, Boaventura de Sousa Santos, Alberto Acosta, Cormac Cullinan, George Caffentzis, Anuradha Mittal, Brendan Mackey, and Tom Goldtooth.

The Tribunal establishes a necessary step for the survival of the planet and the human species. Succinctly, we need an altogether different way of relating to nature. Humans are neither the only nor the finest species on Earth. Our feeling of superiority and dominion over nature has caused the extinction of thousands of species and has placed at risk our very existence as a species. We need to transition from a type of law that sees nature as an object and a physical resource to one in which nature is a subject. Contrary to the conventional civilizational right, this right has been called 'wild' (Cullinan 2003). This new conception of the law involves a new understanding and a new purpose: to create governance systems that simultaneously provide support for humans and to the entire life community. Thus, this 'wild law' recuperates the right to preserve and to retrieve what is wild in our hearts, other forms of being and doing what is right; it protects the wild and the freedom of living communities to self-regulate; and it lends resonance to the creativity of diversity instead of imposing uniformity (Berry 2003).

Wild right, whose holder and legislator are nature itself and human beings as part of it, demands that humans descend from their pedestal in order to find their place on the planet, freeing themselves from what is artificial and superfluous in life, finding a new acceptance of their own animality and finally learning to live again in harmony with the rest of nature. The Tribunal is one space where this transformation is taking place. It is a first step on the path to consolidating an International Tribunal designed, controlled and respected by all of the planet's peoples.

Notes

[1] See http://www.rightsofmotherearth.com/derechos-madre-tierra.

[2] See http://therightsofnature.org/tribunal-internacional-derechos-de-la-naturaleza/.

Further Resources

Berry, Thomas (2003), 'Foreword.; In Cormac Cullinan (2003)', *Wild Law: A Manifesto for Earth Justice*. White River Junction, Vermont: Chelsea Green Publishing.

Cullinan, Cormac (2003), *Wild Law: A Manifesto for Earth Justice*. Vermont: Chelsea Green Publishing.

Kolbert, Elizabeth (2015), *The Sixth Extinction: An Unnatural History*. New York: Picador.

Ramiro Ávila-Santamaría is a judge of Ecuador's Constitutional Court. He has a Masters degree in law from Columbia University; Master and PhD degrees in Juridical Sociology from the University of the Basque Country and from the International Institute of Juridical Sociology, respectively. He also has a law degree from the Pontificia Universidad Católica del Ecuador, and is the author and editor of a number of publications, http://www.uasb.edu.ec/web/area-de-derecho/docente?ramiro-avila-santamaria.

UBUNTU

Lesley Le Grange

Keywords: humanness, interconnectedness, social justice, environmental sustainability

Ubuntu is a southern African concept, which means humanness. Humanness implies both a condition of being and a state of becoming. It concerns the unfolding of the human being in relation to other human beings and the more-than-human world of non-human nature. In other words, the becoming of a human is dependent on other human beings and the cosmos. Moreover, *ubuntu* suggests that a human being is not an atomized individual of the Western tradition, but is embedded in social and biophysical relations. Therefore, *ubuntu* is anti-humanist because it emphasizes the relational existence and becoming of the human being.

Ubuntu is derived from proverbial expressions or aphorisms found in several languages in Africa, south of the Sahara. In the Nguni languages of Zulu, Xhosa, and Ndebele spoken in South Africa, *ubuntu* derives from the expression: *Umuntungumuntungabanye Bantu*, which suggests that a person's humanity is ideally expressed in relationship with others, and, in turn, is a true expression of personhood: 'We are, therefore I am'. *Botho* is its equivalent in Sotho-Tswana languages and is derived from the proverbial expression, *Mothokemothokabathobabang*. *Ubuntu* comprises one of the core elements of a human being. The Zulu word for human being is *umuntu*, who is constituted of the following: *umzimba* (body, form, flesh), *umoya* (breath, air, life), *umphefumela* (shadow, spirit, soul), *amandla* (vitality, strength, energy), *inhliziyo* (heart, centre of emotions), *umqondo* (head, brain, intellect), *ulwimi* (language, speaking), and *ubuntu* (humanness) (Le Roux 2000: 43). *Ubuntu* is, however, not only a linguistic concept, but has a normative connotation embodying how we ought to relate to the other – what our moral obligation is towards the other. *Ubuntu* suggests that our moral obligation is to care for others, because when they are harmed, we are harmed. This obligation extends to all of life, since everything in the cosmos is related: when I harm nature, I am harmed. Like all African cultural values *ubuntu* circulated through orality and tradition – its meaning, interwoven in the cultural practices and lived experiences of African peoples. Such cultural values became eroded or effaced by colonization. However, in post-colonial Africa, *ubuntu* and its equivalents have been re-invoked as a part of a decolonizing project, and also enjoys increasing appeal globally as an alternative to dominant notions of development that threaten the achievement of social justice and environmental sustainability. For example, some Afro-descendent groups in South America are invoking *ubuntu* to gain a more nuanced understanding of *buen vivir*.

Ubuntu conveys the idea that one cannot realize or express one's true self by exploiting, deceiving or acting in unjust ways towards others. Being able to play, to use one's senses, to imagine, to think, to reason, to produce works, to have control over one's environment are not possible without the presence of others. *Ubuntu* therefore depicts solidarity among humans and between humans and the more-than-human world. It can be invoked to build solidarities among humans in the struggle for social justice and environmental sustainability, which are central concerns of social movements across the globe. *Ubuntu* proposes that human creativity and freedom should only be constrained when it harms others. *Ubuntu* is the manifestation of the power within all beings that serves to enhance life, and not thwart it. This is a power that is productive, that connects, and engenders care and compassion – it is the power of the multitude that gives impetus to social movements. This form of power is in contrast to power that imposes, that divides, that colonizes – the power of the sovereign wielded by supranational organizations, governments, the military, and the corporate world. The latter form of power results in the erosion of *ubuntu*.

Ubuntu's transformative potential lies in providing alternative readings to some of the key challenges faced by humanity in the twenty-first century: growing inequality among humans, impending ecological disaster and human's interconnectedness with new technologies to the extent that it is difficult to determine what 'being human' now is. Concerning the latter challenge, the invocation of *ubuntu* foregrounds the importance of affirming humanness, not by defining what it is to be human so as to declare other entities as non-human, but through a process that involves the unfolding of the human in a context of burgeoning new technologies. Addressing inequality in the world suggests a concern about humans only – it is human-centred – whereas addressing the ecological crisis extends the interest to the more-than-human world – it is eco-centric. *Ubuntu* is transformative in that it transcends the human-centred (anthropocentric) and eco-centred (eco-centric) binary (Le Grange 2012). Relationality among human beings should be viewed as a microcosm of relationality within the cosmos. Nurturing the self or caring for other human beings is therefore not antagonistic towards caring for the more-than-human world – *ubuntu* cannot simply be reduced to a category of anthropocentric or eco-centric. The self, community, and nature are inextricably tied up with one another – healing in one domain results in healing in all dimensions and so too is suffering transversally witnessed in all three dimensions. The struggle for individual freedom, social justice and environmental sustainability is one struggle.

Two potential limits of *ubuntu* might be identified. First, a narrow ethnocentric interpretation of the concept could be used politically to exclude others. By this, I mean that certain groups who have gained political power in post-colonial Africa might claim that the concept belongs to them – even though this might contradict the meaning of the term – or hold the view that it cannot be subjected to critical scrutiny. Put differently, *ubuntu* could become reduced to a narrow humanism that has resulted in atrocities such as xenophobia experienced in South Africa in recent times. Second, because of its popular appeal, *ubuntu* could be co-opted by supranational organizations, governments and the corporate world to suit their own agendas, or given the dominance of Western ways of knowing, could become assimilated into a Western cultural archive, thus eroding its 'indigenousness'.

Further Resources

Le Grange, Lesley (2012), 'Ubuntu, Ukama, Environment and Moral Education', *Journal of Moral Education*. 4 (3): 329–40.

Le Roux, Johann (2000), 'The Concept of "Ubuntu": Africa's Most Important Contribution to Multicultural Education?' *Multicultural Teaching*. 18 (2): 43–46.

Novalis Ubuntu Institute, http://novalis.org.za/.

Ubuntu Liberation Movement, http://www.ubuntuparty.org.za/.

Ubuntu Pathways, https://ubuntupathways.org/.

Lesley Le Grange is Distinguished Professor in the Faculty of Education at Stellenbosch University, South Africa. His current research interests include critically 'analysing' sustainability and its relationship to education; developing *ubuntu* as an environmental ethic and its implications for education.

———•···•———

UNDEVELOPING THE NORTH

Aram Ziai

Keywords: undeveloping, post-development, sustainable development, degrowth, internationalism

The concept of 'Undeveloping the North' ('Abwicklung des Nordens') opposes both the discourse of 'development' in general, and its imperative of 'developing the South', as well as the idea of 'sustainable development' as its accompanying ecological modernization. It sees relations of power in global capitalism and its drive for accumulation as the cause of poverty in the South and ecological degradation worldwide. Therefore, it focuses on struggling against these relations of power and this economic system. The concept (Spehr 1996: 209–36; Hüttner 1997; Bernhard *et al.* 1997) arose from a critique of sustainable development, which was seen as an ecological modernization of corporate capitalism reproducing ideas of western superiority, patriarchal faith in science and technology and unjustified trust in planning and development (Hüttner 1997: 141).

The historical roots of the concept lie in debates which took place during the 1990s in the Bundeskongress entwicklungspolitischer Aktionsgruppen (BUKO), the federation of Third World solidarity and internationalist groups in Germany. It built on ecofeminism and world systems theory, but also on post-colonial studies and post-modern internationalism, that is, an international struggle for solidarity, which abandoned traditional concepts of the communist party being the avant-garde, the working class being the revolutionary subject, and state socialism being the solution. The concept was sympathetic towards ecofeminist subsistence approaches, but criticized their proposed solutions as ones that focused too much on agriculture and too little on macro-political alternatives and struggles. Some feared that the valuable ecofeminist critique might end up building non-capitalist or alternative niches while leaving larger structures intact (Bernhard *et al.* 1997: 195f).

Undeveloping the North perceives the North not primarily as a geographical area, but as a model of society and a system of domination, in which some groups are forced to provide their productive, reproductive, and emotional labour for a pittance while others – disproportionally often found in the North, enjoy unjust privileges. It sees the model of society found in the 'developed' world as based on exclusion and thus unsuitable for other parts of the world, as suggested by the discourse of development. The concept is explicitly focused not on the creation of alternative niches, but on general social structures, which are to be tackled from the bottom up, and aims at reducing the amount of work and nature to be exploited within these structures, thus the focus is on strengthening autonomy. Its five principles are the following:

1. Preventing the capacity of the North for military interventions to implement its access to labour and nature ('no blood for oil' was the corresponding slogan against the wars in Iraq);
2. Pushing back the global sector which forces local initiatives into global competition with each other, thus eliminating economic alternatives;
3. Lessening the privilege of formal labour as it excludes major parts of the population from the benefits of the welfare state and should be replaced by the provision of basic social security for all;
4. Direct appropriation of spaces and relationships for the satisfaction of needs ('land and freedom');

5. Measures for securing survival, preventing the use of large areas by the global sector and using it instead for local food security in the South, coupled with rebuilding structures to achieve subsistence also in the North, and thus decolonizing the regions where people have profited from a colonial division of labour until today (Spehr 1996: 214–23).

In contrast to some ideas of sustainable development, Undeveloping the North insists that it is not legitimate for Northern actors to, for instance, prevent deforestation of the Amazon in the name of a global environmental consciousness or 'saving the planet'. In contrast to some post-development approaches, it does not want to prevent Westernization, modernization and industrialization. This is the point where the concept is linked to debates about Zapatista politics and not speaking for others. In contrast to Marxist approaches, Undeveloping the North explicitly avoids statements on how societies should organize and produce, except for the principle that they must not do so on the basis of exploiting other groups' work and resources. However, this principle would severely limit attempts to modernize and industrialize (Spehr 1996: 224). Undeveloping the North does not by itself abolish capitalism, patriarchy and racism, but offers a way of dealing with social and ecological crises that does not reproduce these structures. It merely aims at providing a frame for the future arrangement of society (Spehr 1996: 226).

While the concept has been debated in internationalist and leftist environmentalist circles in Germany since its inception, in recent years it is also increasingly promoted within the degrowth movement as a radical alternative to approaches which seek to overcome growth without confronting capitalism *per se* (Habermann 2012). Instead of suggesting that ecological crises can be solved by technological progress and efficiency revolutions, as suggested by mainstream approaches of sustainable development, the concept tackles their structural causes. Undeveloping the North links the critique of global capitalism and of development discourse with a wider perspective on relations of domination in general. It is an attempt to abolish the 'imperial mode of living' (Brand and Wissen 2013) in the metropolis.

Further Resources

Bernhard, Claudia, Bernhard Fedler, Ulla Peters, Christoph Spehr and Heinz J. Stolz (1997),' Bausteine für Perspektiven', in Schwertfisch (ed.), *Zeitgeist mit Gräten. Politische Perspektiven zwischen Ökologie und Autonomie*. Bremen: Yeti Press.
Brand, Ulrich and Markus Wissen (2013), 'Crisis and Continuity of Capitalist Society-Nature Relationships: The Imperial Mode of Living and the Limits to Environmental Governance', *Review of International Political Economy*. 20 (4): 687–711.

Habermann, Friederike (2012), 'Von Post-Development, Postwachstum und Peer-
    Ecommony: Alternative Lebensweisen als Abwicklung des Nordens', *Journal für
    Entwicklungspoliti.* 28 (4): 69–87.
Hüttner, Bernd (1997), 'Von Schlangen und Fröschen – Abwicklung des Nordens
    statt Öko-Korporatismus', in Schwertfisch (ed.), *Zeitgeist mit Gräten. Politische
    Perspektiven zwischen Ökologie und Autonomie.* Bremen: Yeti Press.
Schwertfisch (ed.) (1997), *Zeitgeist mit Gräten. Politische Perspektiven zwischen
    Ökologie und Autonomie.* Bremen: Yeti Press.
Spehr, Christoph (1996), *Die Ökofalle. Nachhaltigkeit und Krise.* Wien: Promedia.

Aram Ziai is a member of the Bundeskoordination Internationalismus (BUKO)
and Professor of Development and Postcolonial Studies at the University of Kassel,
Germany.

<hr />

WAGES FOR HOUSEWORK

Silvia Federici

Keywords: housework, wages, strategy, feminism

The 1970s Wages for Housework campaign was an important historical
moment in the empowerment of women under capitalism. In underlining
the centrality of reproductive labour, demanding that housework be
recognized as work and remunerated, it pioneered an international political
strategy for women's liberation, put capitalism on trial and exposed the
shallowness of mainstream 'rights based' feminism.

According to Dolores Hayden, the campaign originated in the late
nineteenth century, when after the Civil War some feminists in the United
States called for 'wages for housewives'. In the 1940s, Mary Inman, a
Communist Party member, argued the case in her book *In Woman's Defense*,
but failed to convince the party to adopt it in its programme.

It was in the 1970s that Wages for Housework turned from a demand
that domestic work be monetarily compensated into a political perspective
on the place of women and reproductive work in capitalist accumulation.
This occurred with the formation of an international feminist network
mobilizing in different countries to demand that the state, as collective

capitalist, pay wages to anyone performing this work. The reasoning behind this systemic strategy was provided by Italian political theorist and activist Mariarosa Dalla Costa in 1972. This soon became widely available in translation as the foundational campaign document 'Women and the Subversion of the Community'.

Against the Marxist tradition that pictured domestic work as a personal service, a legacy of the pre-capitalist society to be superseded by full industrialization of the economy, Dalla Costa argued that housework is the pillar of capitalist accumulation, contributing directly to surplus value through the production of labour-power-workers' capacity to work.

As a capitalist construct, housework was invisible and imposed as unpaid labour. This devaluation benefits the capitalist class, which without it, would have to provide social services to enable workers to appear at the workplace. In short, capitalism has built an immense wealth off the backs of women, who have been forced to depend on men for their survival or take on a double-shift inside and outside the home.

The demand of wages for housework unmasked the immense amount of unwaged work that women do for capital, uncovering a whole terrain of exploitation until then naturalized as 'women's labour'. It also revealed the social power that this work potentially confers to those performing it, as domestic work reproduces the worker and is thus the condition of every other form of labour.

By identifying the beneficiaries of reproductive work, the Wages for Housework campaign liberated women from the guilt they experienced when refusing such work. Most importantly, the campaign was a crucial alternative to the dominant liberal feminist demand for equal rights with men and access to traditional male occupations. This mainstream feminism did nothing to destabilize the sexual division of labour and labour hierarchies. Women's free 'gift' to capital remained intact.

Wages for Housework never generated a mass struggle, though the contemporary struggle for women's welfare rights in the United States and defence of the Family Allowance in England demonstrated its importance for proletarian women. Liberal and even socialist feminists opposed the wages campaign, arguing that it would institutionalize women in the home. But three decades of restructuring 'reproductive work' and women's integration as waged workers into the world economy have demonstrated its continuing relevance to the politics of social justice movements.

Access to wage labour has not relieved women from unpaid domestic work, nor has it changed conditions in the 'workplace' enabling female workers to care for their families, and men to share the housework. Women in the United States do not yet have a government mandated maternity leave program, as available in many other countries. Reproductive work, mostly done by women, is still unpaid, though as I argue in *Revolution at Point Zero*, large quotas of it are now performed by low paid, heavily exploited immigrant women.

Several elements of Wages for Housework recommend it as a positive transformative programme. If implemented, it would produce a massive transfer of wealth from the top of society to the bottom. This is needed more than ever given the precarization of work, the dismantling of the welfare state, and the reproductive crisis proletarian communities worldwide are facing. In this, the campaign is akin to the demand for a basic guaranteed income. But the advantage of Wages for Housework is that it puts capital on trial, defining this transfer as a re-appropriation of the wealth produced by women and challenging the expanding capitalist imposition of unpaid labour.

Moreover, as Louise Toup in writes: Wages for Housework opens a new ground of negotiations between women and the state on the question of reproduction, bringing together paid and unpaid houseworkers, to redefine women's relation to work, in and out of the home, marriage, sexuality, procreation, and their identities as women.

Not least, re-centering anti-capitalist struggle on the valorization of activities by which lives are produced is an essential condition for overcoming the logic of capital. A logic which thrives on devaluation – monetary and otherwise – needs to be challenged to recuperate the creativity of such work. Yet this struggle within capitalist wage relations is only a beginning. To be politically transformative, it must be accompanied by a reorganization of housework in a less isolating, more cooperative, socialized way.

If Wages for Housework is embraced by a strong movement, its challenge to the most hidden, and naturalized forms of exploitation, will change power relations – not only between women and capital, but between women and men, and among women themselves – in ways that help to unify the working class.

Further Resources

Dalla Costa, Mariarosa (1972), 'Women and the Subversion of the Community', in Mariarosa Dalla Costa and Selma James (eds), *The Power of Women and the Subversion of the Community*. Bristol: Falling Wall Press.

Federici, Silvia (2012), *Revolution at Point Zero: Housework, Reproduction and Feminist Struggle*. Oakland, CA: PM Press.

Hayden, Dolores (1985), *The Grand Domestic Revolution*. Cambridge, MA: MIT Press.

Inman, Mary (1941), *In Woman's Defense*. Los Angeles: Committee to Organise for the Advancement of Women.
Toupin, Louise (2014), *Le salaire au travail ménager: Chronique d'une lutte féministe international (1972–1977)*. Montréal: Les éditions du remue-ménage.

Silvia Federici is a feminist activist and Professor Emerita at Hofstra University. She is the author of *Caliban and the Witch: Women, the Body and Primitive Accumulation* (2004); *Revolution at Point Zero. Housework, Reproduction, and Feminist Struggle* (2012); editor, *The New York Wages for Housework Committee 1973–76: History, Theory, Documents* (forthcoming).

WORKER-LED PRODUCTION

Theodoros Karyotis

Keywords: self-management, recuperation, cooperatives, workers' control, labour

Worker-led production refers to a diverse set of practices that aim to give protagonism to the subjects of labour: the workers themselves. Throughout the industrial era, with its associated processes of deskilling and mechanization, workers not only have demanded a bigger share of the profits through union struggles but have also strived to participate in decision-making processes at their workplace; they have set up cooperatives based on egalitarian self-management; and ultimately, they have occupied businesses and put them under workers' control.[1]

The cooperative movement, developing alongside the workers' movement in the eighteenth and nineteenth centuries, has been a formidable attempt at calling into question the basic social and economic divisions of industrial modernity. However, in the twentieth century, it was absorbed into the capitalist mode of production, since it largely embraced and naturalized wage-labour relations. However, with the onset of the neoliberal capitalist restructuring in the late twentieth century, a new radical cooperativism emerges in many countries, overlapping to an extent with the nascent movement of social and solidarity economy.

More importantly, around the turn of the twenty-first century, in Latin

American countries such as Argentina, Uruguay, Brazil, and Venezuela, workers respond to the de-industrialization brought about by the restructuring of the economy. They occupy their bankrupt or abandoned companies, resist eviction attempts and restart production relying on their own forces – a practice dubbed 'recuperation'. With the spread to the European periphery of the economic conditions that gave rise to the Latin American movement, a nascent movement of workplace recuperations arises after 2011, with examples in Italy, Greece, Turkey, France, Spain, Croatia, and Bosnia-Herzegovina.

The vision of a future society directed by the 'associated producers' themselves cuts across all historical currents of the left; to this day, democratic self-management at the workplace is for many an effective way to bridge the chasm between this vision of the future and the day-to-day struggle within capitalism, thus becoming an essential component of prefigurative politics, that is, politics which attempts to construct alternative social relations in the present. The replacement of existing hierarchies with horizontal decision-making practices not only helps overcome the alienation inherent in industrial production and liberate the workers' creative powers, but also makes it easier to substitute the myopic profit-seeking motive with humane considerations related to the welfare of the workers and of society at large.

However, elements of worker-led production, divested of their subversive potential, have been gradually introduced in capitalist production. On the one hand, contemporary business management practices pursue increased productivity by allowing – and requiring – some groups of workers to self-direct their activity. On the other hand, as economic restructuring dismantles social welfare provision, commodifies the commons and creates large 'surplus populations' of unemployed and precarious workers, a 'social economy' qua an 'economy of the poor' on the fringes of the mainstream economy is regarded by neoliberal elites as a 'safety net'. This is an inexpensive means of providing a livelihood to the lower social strata, and thus of maintaining social peace. As a kind of social economy, it simply conceals the inability of contemporary capitalism to ensure social and ecological reproduction.

Indeed, in the context of such a social economy, self-managed workers are often victims of self-exploitation: while internal hierarchies may be abolished, competition within the capitalist market determines what is to be produced, as well as the prices, wages, and ultimately the conditions and intensity of labour. The struggle of these endeavours for survival may vitiate their emancipatory character and relegate environmental or social considerations to the background.

Recuperated companies usually face additional obstacles: lack of access to credit; obsolete machinery; a dwindling market share in conditions of recession. More often than not, they are embroiled in long legal battles against the state and the former owners, with very little in the way of legal arguments besides their social legitimacy as ways of preserving livelihoods.

Thus, workers' control over the productive process is a necessary but insufficient condition for social emancipation. However, unlike capitalist businesses, worker-run workplaces do not exist in social isolation but usually form part of wider social movements, which compensate the lack of economic and technological innovation with 'social innovation'. Participation in communities of struggle and networks of worker-run companies helps redirect production towards socially useful products and create alternative avenues of distribution based on solidarity rather than competition. Most newly recuperated companies in Europe have shifted towards environmentally and socially conscious production: Scop-ti and Fabrique du Sud in the south of France towards organic herbal tea and ice-cream respectively; Viome in Greece from chemical building materials to natural cleaning products; Rimaflow and Officine Zero in Italy towards the salvaging and recycling of electronics.

It is precisely the embeddedness of worker-run companies in wider social movements and their attentiveness to the needs and demands of communities that make them important components in a strategy of maximizing social resilience and self-determination. By opening up the company to concerns that are alien to capitalist productivity and profitability, workers call into question the division between the social, the economic and the political spheres, on which capitalist modernity rests. In Latin America and in Europe, occupied factory grounds offer their space to schools, clinics and social centres; they accommodate farmers' markets, bazaars, concerts, and artistic events. In short, 'solidarity ecosystems' are formed around the 'factory commons', helping make the leap from the mere production of commodities to the production of relationships, subjects and collectives, encompassing social life in its entirety and acting as a bulwark against processes of dispossession and enclosure.

Note

[1] Workers' control: an archive of workers' struggle, a multilingual online resource presenting news, debates, analyses and historical accounts. See http://www. workerscontrol.net/.

Further Resouces

Azzelini, Dario and Oliver Ressler (2015), 'Occupy, Resist, Produce', http://www.ressler.at/occupy_resist/.

Azzellini, Dario (2018), 'Labour as a Commons: The Example of Worker-Recuperated Companies'. *Critical Sociology*. 44 (4–5): 763–76.

Barrington-Bush, Liam (2017), 'Work, Place and Community: The Solidarity Ecosystems of Occupied Factories', http://morelikepeople.org/solidarity-ecosystems/.

European Medworkers Economy, http://euromedworkerseconomy.net/.

Karakasis, Apostolos (2015), 'Next Stop: Utopia', http://www.nextstoputopia.com/.

Lewis, Avi and Naomi Klein (2004), 'The Take', http://www.thetake.org/.

Ruggeri, Andrés (2013), 'Worker Self-Management in Argentina: Problems and Potentials of Self-Managed Labor in the Context of the Neoliberal Post-Crisis', in Camila Piñeiro Harnecker (ed.), *Cooperatives and Socialism: A View from Cuba*. London: Palgrave Macmillan.

Theodoros Karyotis is a sociologist, independent researcher, and translator based in Greece. A social activist in grassroots movements related to direct democracy, solidarity economy and the commons, he coordinates the site, workerscontrol.net, a multilingual resource on worker self-management.

ZAPATISTA AUTONOMY

Xochitl Leyva-Solano

Keywords: Zapatismo, autonomous practices, good government, anti-capitalist struggles

Zapatista Autonomy is a central element in the practices of resistance and rebellion of the Zapatista movement. It comprises modes, processes, and networks of struggle, government, and rebel life that together constitute a radical alternative to the established system and its institutions.

Zapatista autonomy emerges from the bottom and the left in times of war. It involves multiple angles.

As resistance. The long resistance of the Zapatista Army of National Liberation (Ejército Zapatista de Liberación Nacional, EZLN) has been mentioned in the First Declaration of the Lacandon Forest on January

1st, 1994: 'We are the product of 500 years of struggle.' Then, the EZLN declared war on the government and called on the Mexican people to join its struggle for labour, land, housing, food, health, education, independence, freedom, democracy, justice, and peace. In 1994, the ELZN also announced the creation of thirty-eight rebel municipalities, and thus broke the military siege and politically confronted the counter-insurgency strategy implemented by the government.

As good government, dignified, and rebel. The EZLN found support for its initial actions in Article 39 of the Mexican constitution, which establishes that 'the people have, always, the inalienable right to alter or to modify its form of government'. The appeal to this article became stronger after the government's unwillingness to fulfil the San Andrés Agreements signed in 1996 with the EZLN. Failing these agreements, the government did not generate a new constitutional framework that would have made the exercise of autonomy and self-determination by indigenous peoples possible in all domains and at all levels.

Confronted with the crescendo of an extended war of attrition, the Zapatista movement mobilized practices and networks of autonomous governments knitted out of people's townships, resulting in The Zapatista Rebel and Autonomous Municipalities (Municipios Autónomos Rebeldes Zapatistas, MAREZ) and Zapatista regions and zones. All of these are organized under the principle of 'governing by obeying', including the following basic premises:

- To serve and not to be served
- To represent and not to supplant
- To build, not to destroy
- To obey, not to command
- To propose and not to impose
- To convince, not to defeat
- To go down, not up.

These foundations in action bring ethics back into the heart of politics and expose the practices of 'bad government' of the Mexican political system including corruption, violence, and impunity.

When 'people command and the government obeys' this involves permanent 'duties' and 'obligations' from both the people and the government. The election of authorities takes place, in general, through assemblies. The authorities of the different levels are the following: the autonomous agents and *commissars*; the members of municipal and regional autonomous councils; the coordinators of the various work areas; and the members of the different commissions and of the Good Government Councils (Juntas de Buen Gobierno, JBG), which operate at the level of each zone and are located in the Caracoles Zapatistas.

The Zapatista autonomous government is organized according to 'work areas' that change over time and from municipality to municipality, but usually include the following: health, education, agroecology, women, agrarian issues, justice, communication, commerce, transportation, administration, and civil registry. In these areas and at other levels of government, the various positions are rotational, collective and unpaid. Each person who participates is connected to others on the basis of his or her own potential and capacity to be, to do, to learn and to unlearn. By doing so, they challenge the dominant forms of social organization and power on the basis of individual ranking and specialized wage labour.

As a radical, comprehensive and life-creating alternative. The Zapatista grassroots support comprises indigenous *campesinos* (*as*) who cultivate the land for their livelihood and reproduction, and thus generate the material conditions for their autonomous struggles. Women occupy a central place, just as the land/territory and the Mother Earth, as creators and givers of life.

The Revolutionary Women's Law incorporated women into the revolutionary struggle by insisting on, and looking after, their political and social rights and their physical and moral integrity. The content of this law would have been meaningless were it not for the women at grassroots support level who – in dialogue with armed EZLN women – came to embody these struggles in every sense: in relation to the army of occupation, by daily cultivating the land with their own hands, by recuperating lost territory, by re-socializing their own sons and daughters, in the organization of cooperatives, as teachers of autonomous education, as promoters of autonomous healing and as radio and videomakers. There is no doubt that the Zapatista struggle grew its roots through the women and men at the grassroots level. Through their support, Zapatista politics gained a strength that many other revolutionary experiences could not achieve, because they did not manage to connect their struggles with the spheres of everyday life and take on board the dimensions of women, the family, the community, ordinary life, collectives and transnationality.

As a central reference to the ongoing globalization from below. Twenty years after the first Zapatista Intergalactic Encounter for Humanity and Against Neoliberalism, Alejandra, a young guardian of the Little Zapatista School (*Escuelita Zapatista*), summarizes the Zapatista glocal planetary consciousness:

[A]s we know, the capitalist system does what it wants, they decide how to govern, how we should live, and that is what we do not want . . .. We are not only struggling for our own sake . . . we want freedom for all. . . As Zapatistas we are not using weapons. . . . we are using our words, our politics. . . . we want to defeat the system, that is our main goal.[1]

Note
[1] Available at *Rebeldía Zapatista 1*, 2014, p. 53.

Further Resources
EZLN, http://enlacezapatista.ezln.org.mx.
———— (2013), *Cuadernos de texto de primer grado del curso*. Mexico: Escuelita Zapatista-EZLN.
———— (2014), *Rebeldía Zapatista:La Palabra del EZLN*,1 and 3, February and September: Mexico, http://enlacezapatista.ezln.org.mx/2014/02/28/editorial-revista-rebeldia/.
———— (2016), *Critical Thought in the Face of the Capitalist Hydra: I', Contributions by the Sixth Commission of the EZLN*. Durham: Duke University Press.
Seminarios CIDECI-UniTierra Chiapas, http://seminarioscideci.org.
ProMedios de Comunicación Comunitaria, http://www.promediosmexico.org.
Radio Zapatista, http://www.radiozapatista.org.

Xochitl Leyva is a co-founder of, and activist in *altermundista* (alter-globalization) collectives and networks. A researcher at CIESAS Sureste, Chiapas, Mexico, she has co-produced multiple videos, multimedia products, and authored several articles and books with women and young indigenous people in resistance. These are used in activist, academic, and community contexts.

POSTSCRIPT
The Global Tapestry of Alternatives

The world is going through an unprecedented crisis engendered by a dominant regime that has resulted in deepening inequalities, increasing in new forms of deprivation, the destruction of ecosystems, climate change, the tearing off of social fabric and the dispossession of all living beings with immense violence.

However, **the past two decades have witnessed the emergence of an immense variety of radical alternatives to this dominant regime and to its roots in capitalist, patriarchal, racist, statist, and anthropocentric forces.** These range from initiatives in specific sectors such as sustainable and holistic agriculture, community-led water/energy/food sovereignty, solidarity and sharing economy, worker control of production facilities, resource/knowledge commons, and inter-ethnic peace and harmony, to more holistic or rounded transformations such as those being attempted by the Zapatista in Chiapas and the Kurds in Rojava, to the revival of ancient traditions or the emergence of new worldviews that re-establish humanity's place within nature and the values of human dignity, equality and the respect of history.

The Global Tapestry of Alternatives is an initiative seeking to create solidarity networks and strategic alliance amongst all these alternatives on local, regional and global levels. It starts in the local interaction among alternatives, to gradually organize forms of agreement on the regional, national and global scale, through diverse and light structures, defined in each space, horizontal, democratic, inclusive and non-centralized, using diverse local languages and other ways of communicating. The initiative has **no central structure or control mechanisms.** It spreads step by step as an ever-expanding, complex set of tapestries, constructed by already existing communal or collective webs, organized as alternatives to the dominant regimes, each of them autonomously weaving itself with other such webs.

It organizes mechanisms of interaction between those regional and national structures and with the societies, in which they exist, in diverse languages and different means, promoting periodically regional, national and global encounters, when the conditions allow for them, as well as close and synergistic linkages with existing organizations, like the World Social Forum.

The Global Tapestry of Alternatives is about **creating spaces of**

collaboration and exchange, in order to learn about and from each other, critically challenge each other, offer active solidarity to each other whenever needed, interweave the initiatives in common actions, give them visibility to inspire other people to create their own initiatives and to go further along existing paths or forge new ones that strengthen alternatives wherever they are, **until the point in which a critical mass of alternative ways can create the conditions for the radical systemic changes we need.**

A small group of activists from several regions of the world started the initiative, which will create its structure as it takes shape in different parts of the world. The initial group will continue supporting the initiative as long as necessary. It has some endorsers who subscribe this document, and will try to weave itself with similar initiatives around the world. Anyone interested in following the evolution of the initiative or participate in it may write a mail to globaltapestryofalternatives@riseup.net.

For more information, browse www.globaltapestryofalternatives.org.